Martin Köhler

Approximate Solution of Non-Symmetric Generalized Eigenvaluem Problems and Linear Matrix Equations on HPC Platforms

Logos Verlag Berlin

λογος

Bibliografische Information der Deutschen Nationalbibliothek

Die Deutsche Nationalbibliothek verzeichnet diese Publikation in der Deutschen Nationalbibliografie; detaillierte bibliografische Daten sind im Internet über http://dnb.d-nb.de abrufbar.

ISBN 978-3-8325-5434-7

Logos Verlag Berlin GmbH
Georg-Knorr-Str. 4, Geb. 10,
D-12681 Berlin
Germany

Tel.:   +49 (0)30 / 42 85 10 90
Fax: +49 (0)30 / 42 85 10 92
http://www.logos-verlag.de

Approximate Solution of Non-Symmetric Generalized Eigenvalue Problems and Linear Matrix Equations on HPC Platforms

Dissertation

zur Erlangung des akademischen Grades

doctor rerum naturalium
(Dr. rer. nat.)

von   **Dipl.-Math. Martin Köhler**

geb. am   **9. August 1986** in Karl-Marx-Stadt

genehmigt durch die Fakultät für Mathematik

der Otto-von-Guericke-Universität Magdeburg

Gutachter: **Prof. Dr. Peter Benner**

Prof. Dr. Hans-Joachim Bungartz

Dr. Jens Saak

eingereicht am: **26. Januar 2021**

Verteidigung am: **22. Juni 2021**

DANKSAGUNG

An erster Stelle möchte ich Herrn Prof. Dr. Peter Benner und Herrn Dr. Jens Saak danken, welche mich nach Magdeburg ans Max-Planck-Institut geholt und mir damit meinen Weg in die Welt der Forschung und des Hochleistungsrechnens ermöglicht haben. Ich bedanke mich bei ihnen für ihre Geduld sowie die Anmerkungen und Hinweise, welche sie mir, gerade in der langwierigen Endphase der Dissertation, gegeben haben. Ich möchte mich auch bei meinen Kollegen und den beiden Sekretärinnen bedanken, die immer ein offenes Ohr hatten, wenn es irgendwo nicht weiter ging. Insbesondere möchte ich Steffen Werner danken, welcher mir immer wieder Denkanstöße rund um das verallgemeinerte Eigenwertproblem gegeben hat. Bedanken möchte ich mich auch bei meinem Bürokollegen Maximilian Behr, der so einige Diskussionen über die komplexen Probleme bei der Optimierung meiner Programmcodes über sich ergehen lassen musste.

Ein besonderer Dank gilt meinen Eltern, deren stetige Ermunterungen und fortwährende Unterstützung mich dorthin gebrachten haben, wo ich heute stehe. Weiterhin gebührt meiner Freundin Kristin Simon größter Dank. Sie hat unzählige Stunden für das Korrekturlesen geopfert und mich immer wieder motiviert weiter zu machen, wenn der Frust mich überkam. Dabei sei auch unser Sohn Konstatin Johann nicht zu vergessen, der mit seiner Spontanität die finale Phase der Arbeit in jeglicher Hinsicht ganz schön auf den Kopf gestellt hat.

Außerdem zu erwähnen sind die vielen Freunde, die mir geholfen haben in Magdeburg anzukommen. Allen voran Markus Reinhold und Bent Arndt, die mit ihrem Unglauben, dass sie diese Dissertation irgendwann sehen werden, immer wieder für Ansporn sorgten, aber auch stets zur Stelle waren, wenn Not am Mann war.

ZUSAMMENFASSUNG

Die Lösung des verallgemeinerten Eigenwertproblems

$$Ax = \lambda Bx$$

und die Berechnung der verallgemeinerten Schur-Zerlegung

$$\left(Q^{H}AZ, Q^{H}BZ\right) = (T, S)$$

gehören zu den zeitaufwendigsten Grundbausteinen im Bereich der numerischen linearen Algebra. Über Jahrzehnte war der QZ-Algorithmus die gängige Technik, um Lösungen für dieses Problem zu berechnen.

In der vorliegenden Arbeit werden Divide-and-Conquer-Verfahren für das Spektrum des verallgemeinerten Eigenwertproblems vorgestellt. Diese setzen, im Gegensatz zum QZ-Algorithmus, auf die Verwendung von schnellen, teilweise mit GPUs beschleunigbaren Grundbausteinen. Auf Basis der Matrix-Sign- und der Matrix-Disc-Funktion mit ihren Projektionseigenschaften bezüglich des Spektrums werden das Divide-Shift-and-Conquer- und das Divide-Scale-and-Conquer-Verfahren entwickelt. Ersteres, basierend auf der verallgemeinerten Matrix-Sign-Funktion, wendet eine rekursive Verschiebungsstrategie auf das Spektrum von (A, B) an. Das zweite Verfahren, welches die verallgemeinerte Matrix-Disc-Funktion benutzt, wendet rekursiv Skalierungen an. Beide Verfahren sind so konstruiert, dass sie problemlos mit Level-3 BLAS- und LAPACK-Routinen, wie zum Beispiel das allgemeine Matrix-Matrix-Produkt, die Lösung von linearen Systemen oder die QR-Zerlegung, implementiert werden können. Für die zeitaufwendigsten Bausteine der neu eingeführten Verfahren, nämlich die Newtonsche Methode für die verallgemeinerte Matrix-Sign-Funktion und die Inverse-Free-Iteration für die verallgemeinerte Matrix-Disc-Funktion, werden verschiedene Implementierungsstrategien vorgestellt. Diese konzentrieren sich auf die Minimierung der Anzahl der benötigten Gleitkommaoperationen sowie die Portabilität auf aktuelle Multi-Core-CPUs oder Multi-GPU-beschleunigte Systeme. Da die verallgemeinerte Matrix-Sign-Funktion und die verallgemeinerte Matrix-Disc-Funktion mit iterativen Verfahren berechnet werden, treten kleinere Ungenauigkeiten während der Berechnung auf. Dennoch zeigen die numerischen Ergebnisse, dass die berechnete Näherung für die meisten Anwendungen ausreichend genau ist. Ein Laufzeitvergleich mit dem QZ-Verfahren zeigt, dass die neu entwickelten Verfahren bei mittleren und großen Problemen um einiges schneller sind. Zum Beispiel kann ein großes reales Problem der Dimension $34\,722$ nun innerhalb von 122 Minuten anstelle von 117 Stunden auf einem aktuellen IBM POWER 8-System mit zwei Nvidia P100-Beschleunigern gelöst werden. Der Rekonstruktionsfehler der verallgemeinerten Schur-Zerlegung, die mit Hilfe des Divide-Shift-and-Conquer-Verfahrens berechnet wurde, beträgt $O(10^{-9})$. Im Falle des QZ-Verfahrens liegt dieser in der Größenordnung von $O(10^{-13})$. Der Wechsel zum Divide-Scale-and-Conquer-Verfahren liefert einen Rekonstruktionsfehler von $O(10^{-14})$, während sich die Laufzeit auf 10,75 Stunden verlängert.

Die verallgemeinerte Schur-Zerlegungen ist eng mit der verallgemeinerten Sylvester-Matrix-Gleichung

$$AXB + CXD = Y$$

verbunden. In den meisten direkten Lösern für diese Gleichung wird die verallgemeinerte Schur-Zerlegung verwendet, um die Anzahl der benötigten Gleitkommaoperationen von $O(n^6)$ auf $O(n^3)$ zu verringern. Der zweite Teil dieser Dissertationsschrift behandelt deshalb die direkte Lösung von Sylvester-ähnlichen Matrix-Gleichungen. Hierfür wird das klassische Bartels-Stewart-Verfahren, welches typischerweise in diesem Fall verwendet wird, zu einem Level-3 BLAS-Verfahren erweitert. Dieses orientiert sich an den Fähigkeiten aktueller CPUs. Außerdem wird gezeigt, dass die meisten der bekannten direkten Lösungsverfahren nur eine spezielle Parametrisierung des vorgestellten Gerüsts sind. Um numerische Probleme, die durch die Ungenauigkeiten bei der Berechnung der verallgemeinerten Schur-Zerlegung mittels der neuen Verfahren auftreten, zu überwinden, wird die iterative Verfeinerung, die vom Lösen des Gleichungssystems $Ax = b$ bekannt ist, auf Sylvester-ähnliche Matrix-Gleichungen angewendet. Im Vergleich zum Bartels-Stewart-Verfahren reduziert sich die Laufzeit für die zum realen Beispielproblem gehörende verallgemeinerte Lyapunov Gleichung, $AXB^\mathsf{T} + BXA^\mathsf{T} = Y$, von 35 Tagen auf 19 Stunden bei gleichzeitiger Senkung des relativen Residuums von $O(10^2)$ auf $O(10^{-10})$.

ABSTRACT

The solution of the generalized eigenvalue problem

$$Ax = \lambda Bx$$

and the computation of the generalized Schur decomposition

$$\left(Q^H AZ, Q^H BZ\right) = (T, S)$$

belong to the computationally most expensive basic building blocks in the field of numerical linear algebra. Over decades, the QZ algorithm was the common technique to handle this problem.

In this thesis spectral divide and conquer algorithms for the generalized eigenvalue problem are presented. These, in contrast to the QZ algorithm, rely on the usage of GPU accelerable high performing building blocks. On top of the matrix sign function and the matrix disc function with their spectral projection properties, the Divide-Shift-and-Conquer and the Divide-Scale-and-Conquer algorithm are developed. The first one, based on the generalized matrix sign function, applies a recursive shifting strategy to the spectrum of (A, B). The second one, belonging to the generalized matrix disc function, uses a recursive scaling strategy. Both algorithms are constructed in such a way that they can be implemented on top of level-3 BLAS and LAPACK routines, such as the general matrix-matrix product, the solution of linear systems, or the QR decomposition. For the most time consuming building blocks of the newly introduced algorithms, namely Newton's method for the generalized matrix sign function and the inverse free iteration for the generalized matrix disc function, different implementation strategies are presented. They focus on minimizing the number of required floating point operations as well as the portability to current multi-core CPUs or multi-GPU accelerated systems. Since the generalized matrix sign function and the generalized matrix disc function are computed with iterative procedures, small inaccuracies occur. Nevertheless, the numerical results show that the computed approximation is sufficient for most applications. A runtime comparison with the QZ algorithm points out that the newly developed algorithms are much faster for medium and large scale problems. For example, a large scale real world problem of dimension 34 722 can now be solved within 122 minutes instead of 117 hours on a current IBM POWER 8 system with two Nvidia P100 accelerators. The reconstruction error of the generalized Schur decomposition computed by the Divide-Shift-and-Conquer algorithm is $O(10^{-9})$, where the QZ algorithm reaches $O(10^{-13})$. Switching to the Divide-Scale-and-Conquer algorithm we shrink the reconstruction error to $O(10^{-14})$ while increasing the runtime to 10.75 hours.

The generalized Schur decomposition is closely connected to the generalized Sylvester matrix equation

$$AXB + CXD = Y.$$

In most direct solvers for this equation, the generalized Schur decomposition is used to reduce the number of required flops from $O(n^6)$ to $O(n^3)$. Hence, the second part of this thesis handles the direct solution of Sylvester-like matrix equations. Therefor, the classical Bartels-Stewart algorithm, typically used in this case, is extended to a level-3 BLAS algorithm, which focuses on the capabilities of current CPUs. Besides, it is shown that most of the known direct solution algorithms are only special parametrizations of the presented framework. In order to overcome numerical problems introduced by the inaccuracies appearing in the computation of the generalized Schur decomposition using our developed algorithms, the iterative refinement procedure known from $Ax = b$, is lifted to Sylvester-like matrix equations. Compared to the Bartels-Stewart algorithm, this approach reduces the runtime for the solution of the generalized Lyapunov equation belonging to the real world problem, $AXB^\mathsf{T} + BXA^\mathsf{T} = Y$, from 35 days to 19 hours while lowering the relative residual from $\mathcal{O}(10^2)$ to $\mathcal{O}(10^{-10})$.

CONTENTS

Contents

LIST OF TABLES

LIST OF ALGORITHMS

Sets and Spaces

$\mathbb{R}, \mathbb{C}$	fields of real and complex numbers
$\overline{\mathbb{R}}, \overline{\mathbb{C}}$	closed set of real and complex numbers
$\mathbb{C}_+, \mathbb{C}_-$	open right/open left complex half plane
$\mathbb{R}_+, \mathbb{R}_-$	strictly positive/negative real line
$\mathbb{R}^n, \mathbb{C}^n$	vector space of real/complex n-tuples
$\mathbb{R}^{m \times n}, \mathbb{C}^{m \times n}$	real/complex $m \times n$ matrices
$\mathcal{A}, \ldots, \mathcal{Z}$	subspaces of $\mathbb{R}^n$ or $\mathbb{C}^n$
$\Gamma, \ldots, \Omega$	sets
$\|\Gamma\|$	number of elements in the set
$C_\alpha(\beta)$	circle with radius α around β,
	$C_\alpha(\beta) = \left\{ \delta \in \mathbb{C} \mid \|\delta - \beta\| = \alpha \right\}$
$C_0(1)$	the unit circle around 0
$\mathcal{X}^\perp$	orthogonal complement of $\mathcal{X}$

Scalars, Vectors, and Matrices

$\alpha, \ldots, \omega$	real and complex scalars
$\|\xi\|$	absolute value of real or complex scalar
$a, \ldots, z$	real and complex vectors or scalar integer values
$\lceil a \rceil$	the smallest integer number which is larger or equal to a
$\lfloor a \rfloor$	the largest integer number which is smaller or equal to a
a_j	j-th entry of the vector a
$A, \ldots, Z$	real and complex matrices or scalar integer values
A_{ij}	the (i, j)-th entry of a matrix
$A_{k:l, i:j}$	the submatrix of A defined by the row k through l and the columns i through j
A^T	the transpose of A
A^H	the conjugate transpose of A

(A, B)	the matrix pair (A, B)
$\Lambda(A)$, $\Lambda(A, B)$	spectrum of the matrix A and the matrix pair (A, B)
$\rho(A)$, $\rho(A, B)$	spectral radius of the matrix A and the matrix pair (A, B)
$\chi_A(x)$, $\chi_{(A,B)}(x)$	characteristic polynomial of A or (A, B)
$\sigma_i(A)$	the i-th singular value of A
λ_{left}	the leftmost eigenvalue of A
λ_{right}	the rightmost eigenvalue of A
$\det(A)$	the determinant of A
$\mathfrak{R}(A)$, $\mathfrak{I}(A)$	real and imaginary part of a complex quantity $A = \mathfrak{R}(A) + \jmath\mathfrak{I}(A) \in \mathbb{C}^{m \times n}$

Norms

$\|x\|_1$, $\|x\|_2$, $\|x\|_\infty$	the one-, two-, and ∞-norm of a vector x
$\|A\|_1$, $\|A\|_2$, $\|A\|_\infty$, $\|A\|_F$	the one-, two-, ∞-norm, and Frobenius-norm of a matrix A
$\|A\|$, $\|x\|$	arbitrary but proper defined norm for A or x
$\kappa_2(A)$	two-norm condition number of A

LIST OF ACRONYMS

CAQR	Communication-Avoiding QR Decomposition
CPU	Central Processing Unit
flop	floating point operation, either an addition or a multiplication
GPU	Graphic Processing Unit
HPC	High Performance Computing
CtoG	CPU to GPU data transfer
ICE	Incremental Condition Estimation
PTSQR	Parallel Tall-Skinny QR Decomposition
TSQR	Tall-Skinny QR Decomposition

Software Packages

ARPACK	The Arnoldi Package
BLAS	Basic Linear Algebra Subroutines
CUDA	Compute Unified Device Architecture
cuBLAS	CUDA implementation of BLAS
cuBLASXT	Automatic Offloading frontend for cuBLAS
EISPACK	Eigenvalue Solver Package, superseded by LAPACK
LAPACK	Linear Algebra Package
MAGMA	Matrix Algebra on GPU and Multicore Architectures
MORLAB	Model Order Reduction LABoratory
OpenCL	Open Compute Language
OpenMP	Open Multi Processing
PThreads	POSIX Threads
RECSY	Recursive Block Sylvester Solver
RRQR	Rank Revealing QR Decomposition
SCASY	Scalable Sylvester Solver
ScaLAPACK	Scalable Linear Algebra Package

SCIPY   Python Library for Scientific Computations
SLICOT   Subroutine Library in Systems and Control Theory

BLAS/LAPACK-like Routines

AXPY   vector scale and add operation
GEBRD   bidiagonal band reduction
GELQF LQ decomposition
GEMM   general matrix-matrix multiply
GEMQRT   multiplication with a compact storage unitary matrix
GEQPX   rank-revealing QR decomposition
GEQP3   level-3 BLAS column pivoted QR decomposition
GEQRF   Householder QR decomposition
GEQRT   compact storage QR decomposition
GER   general rank-1 update
GETRF LU decomposition
GETRI   matrix inversion
GGES QZ algorithm with Hessenberg reduction
GGES3 QZ algorithm with level-3 BLAS Hessenberg reduction
HQRRP   random sampling column pivoted QR decomposition
ILAENV   environment information for LAPACK
KKQZ   Kågström–Kressner QZ algorithm.
LACPY   matrix copy
LARNV   matrix random number generator
LASCL   matrix scaling
LASET   set matrix to a value
LATSQR   tall-skinny QR decomposition
QTRMM3   three operand quasi-triangular matrix-matrix multiplication
SYR2K   symmetric rank-k update
TPQRT   triangular-pentagonal QR decomposition
TRMV   triangular matrix vector product
TRMM   two operand triangular matrix-matrix multiplication
TRMM3   three operand triangular matrix-matrix multiplication
TRSV   triangular solve with a right hand side vector
TRSM   triangular solve with a right hand side matrix

TRTRSM   triangular solve with a triangular right hand side matrix

TRTRI   inversion of a triangular matrix

INTRODUCTION

Contents

1.1. Motivation

This thesis handles two closely connected problems in numerical linear algebra and scientific computing. First, we consider the generalized eigenvalue problem

$$Ax = \lambda Bx, \tag{1.1}$$

which is a common building block in many applications and analysis tasks. Second, we regard the generalized Sylvester equation

$$AXB + CXD = Y, \tag{1.2}$$

which is closely connected to the generalized eigenvalue problem [107], as well as many other applications in numerical linear algebra. For example, its special cases, like the Lyapunov equation, play an important role in system and control theory [12].

The generalized eigenvalue problem consists of finding all pairs of scalar values λ and vectors x fulfilling (1.1) for a given pair of matrices (A, B) of size $m \times m$. One of the most common strategies to solve this problem is based on the generalized Schur decomposition of the matrix pair (A, B)

$$(A, B) = \left(QTZ^\mathsf{H}, QSZ^\mathsf{H} \right). \tag{1.3}$$

The transformation matrices Q and Z in equation (1.3) are unitary and the newly emerged matrix pair (T, S) consists of upper triangular matrices of size $m \times m$ containing the same eigenvalues as (A, B). Since the matrices T and S are upper triangular, the eigenvalues of (T, S) can directly be obtained from the diagonals of T and S.

The determination of the generalized Schur decomposition (1.3) is a computational tough task normally done with the QZ algorithm developed by Moler and Stewart in 1973 [80, 155, 129]. The algorithm is an extension of the Francis QR algorithm [73], which was developed in 1961 to solve the standard eigenvalue problem, i.e. $B = I$ in (1.3). The QZ algorithm became part of the

1

EISPACK library [150] and later it was integrated into LAPACK [10]. Commonly used tools like MATLAB® [127] or GNU Octave [71] and libraries like SLICOT [149], SCIPY [99], or Trillinos [87] rely on this implementation. Since LAPACK wrappers exist for most common programming languages, the QZ algorithm can be interfaced from almost all mathematical software projects. Even newly developed programming languages and environments for scientific computing, like Julia [37], still provide an interface to this implementation. Hence, the QZ algorithm can be easily used from mathematical research projects as well as from industrial applications. Some acceleration techniques developed by Kågström et al. [104] are already included in LAPACK. An algorithmic improvement based on the aggressive early deflation and multi-shift ideas from the QR algorithm [46, 47] is transferred to the QZ algorithm by Kågström, Dackland, and Kressner [57, 103]. Since this algorithm requires a large set of properly selected tuning parameters, it was not yet integrated into standard software like LAPACK.

Although the QZ algorithm is known for decades and many improvements were developed, the algorithm does not benefit from multi-core CPUs, accelerator devices, or massively parallel computer systems very well. These bad utilization of current computers' capabilities leads to runtime issues when using the QZ algorithm. This can be illustrated with a real world example of order 20 209 coming from a finite element model of a heat equation on a steel profile [34]. Using a current high performance computing (HPC) server equipped with two eight core Intel® Xeon® Silver 4110 CPUs, it takes more than 1.4 days to obtain Q and Z with the LAPACK implementation. Even the optimized versions from Kågström, Dackland, and Kressner [57, 103] take more than 13 hours, compare the numerical results in Chapter 5. A possibly available accelerator cannot be used in most parts of the QZ algorithm.

In contrast to the runtime of the QZ algorithm, the computation of an LU decomposition of the same size on the same machine only takes 23.5 seconds without using an additional accelerator. This motivates us to develop a new class of algorithms, which avoids the operations that cause the bottlenecks in the QZ algorithm. Instead of these operations we integrate well performing operations, like the LU decomposition or the QR decomposition, which exist in highly optimized and parallelized variants for shared memory systems [67, 118, 40, 165, 51], GPU accelerated architectures [166, 167, 65], and distributed systems [44, 55]. Furthermore, we expect to easily integrate new computer architectures and accelerator devices due to the fact that optimized versions of these fundamental operations are mostly available soon after a new platform appears.

The first part of this thesis (Chapters 2 to 5) is driven by the derivation of fast algorithms to compute an approximation of the generalized Schur decomposition (1.3) with the help of these operations. Our developed algorithms and the involved optimizations allow to compute a sufficient approximation of the generalized Schur decomposition of the steel profile example [34] within 38 minutes on the same system.

After we handle the fast approximation of the generalized Schur decomposition, we regard the aforementioned generalized Sylvester equation

$$AXB + CXD = Y \tag{1.4}$$

in Chapters 6 and 7. Most of the direct solution strategies for Sylvester-type matrix equations require the computation of two generalized Schur decompositions first [19, 100, 101, 78, 107, 105, 79]. Another connection between the generalized Schur decomposition (1.3) and Sylvester-type matrix equation is given by the coupled generalized Sylvester equation [107, 102]

$$\begin{aligned} AR &- LB = C, \\ DR &- LE = F. \end{aligned} \tag{1.5}$$

It is used to swap single eigenvalues or blocks on the diagonal of the matrix pair (T, S). Thus it can be used to reorder eigenvalues after a generalized Schur decomposition is computed. Sylvester-type matrix equations have a large impact in systems and control theory as well, e.g. [12]. Hence, the Lyapunov equation, a symmetric variant of the generalized Sylvester equation, lies in the field of application of this thesis as well. With $D = A^H$ and $B = C^H$ in the Sylvester equation (1.4) we obtain

$$AXC^H + CXA^H = Y, \tag{1.6}$$

where Y must be a symmetric right-hand side. In contrast to the (coupled) generalized Sylvester equation it only requires the computation of one generalized Schur decomposition in a direct solution process [19, 85, 140].

The computation of the generalized Schur decomposition is one of the most time-consuming operations in the direct solution process of above mentioned equations (1.4) to (1.6). Hence, the improvements made by our algorithms to approximate the generalized Schur decomposition already result in a notable runtime reduction. With the implementation of the Bartels-Stewart algorithm [19] from SLICOT [149], which is the standard solver in MATLAB and GNU Octave, the solution of the Lyapunov equation, arising from the steel profile example, still takes around 11 hours after the generalized Schur decomposition is known. Even newer algorithms, like the recursive blocking approach [100, 101], struggle with the parallel capabilities of current computer architectures. Our goal is the development of an algorithmic framework, which results in fast direct solvers for Sylvester-type matrix equations. Thereby features of current computer architectures and the optimization techniques of the compilers are taken into account. Since our algorithms, derived in the first part of this thesis, only approximate the generalized Schur decomposition (1.3) and the coefficient matrices of the transformed equation get perturbed, we transfer the iterative refinement procedure from standard linear systems to the Sylvester-type matrix equations. The iterative refinement procedure is formulated in a way such that only one generalized Schur decomposition is required over all iteration steps. Using the optimization and parallelization ideas, introduce in Chapter 6, we reduce the overall runtime of solving the Lyapunov equation (1.6) for the steel profile example to 48 minutes instead of 1.86 days with the SLICOT implementation.

Looking into future hardware, the parallel capabilities of the HPC systems will increase more and more, and accelerator devices, like GPUs, will exist in almost all computational hardware. In this way, moving forward to new algorithms and adjust existing ones to support multi-core and accelerator based systems is the aim of the ideas and approaches presented in this thesis.

1.2. Outline of the Thesis

This thesis is organized in two parts. The first part comprises Chapters 2 to 5 and handles the approximation of the generalized Schur decomposition and the optimizations included in the algorithms. The second part, consisting of Chapters 6 and 7, covers the direct solution of the generalized Sylvester equation.

Chapter 2 In Chapter 2, basic properties of the generalized eigenvalue problem and the generalized Schur decomposition are introduced. The Deflation Theorem 2.6 [155, 80], which provides a way to split large eigenvalue problems into smaller ones, is presented as a key building block for our algorithms. Furthermore, we recall the eigenvalue reordering technique develop by Kågström [102]. The reordering strategy is also used later by the solvers for the Sylvester-type matrix equations in Section 6.3 to regularize the memory access patterns. Beside the fact that the spectrum of a generalized eigenvalue problem is invariant under two-sided transformation, as a result of the Deflation Theorem 2.6, we present eigenvalue transformations leaving the deflating subspaces untouched. These transformations are used by our algorithms to move, scale, and modify the spectrum of a matrix pair (A, B).

Chapter 3 The spectral divide-and-conquer approach, which is the main idea behind our algorithmic strategies to compute the generalized Schur decomposition, is derived in Chapter 3. It extends the known spectral division ideas [15, 162, 121] to a recursive scheme. Therefor, we develop a generic spectral division framework, which uses the Deflation Theorem as key building block. In the context of this generic framework we discuss how to extract unitary transformations from a given deflating subspace with the help of two QR decompositions [162]. This extraction method is also used to remove infinite eigenvalues. Since the generic framework requires the computation of a deflating subspace, we suggest two ways to perform this tasks. In Section 3.2, we recall the generalized matrix sign function

$$S = \text{sign}\,(A, B)\,. \tag{1.7}$$

It was introduced by Roberts [144] to solve the algebraic Riccati equation and extended by Gardiner and Laub [77] to its generalized case. The generalized matrix sign function can be used to compute a deflating subspace, which belongs either to the eigenvalues in the open left half-plane or in the open right half-plane. In this way, the generalized matrix sign function is used to split the spectrum at the imaginary axis. Since a recursive application of this generalized matrix sign function, and thus a divide-and-conquer scheme, is only possible if the imaginary axis is enclosed by the spectrum again, we develop a spectral shifting strategy. This results in the *Divide-Shift-and-Conquer* algorithm to compute the generalized Schur decomposition.

The second deflating subspace computation method, shown in Section 3.3, is based on the matrix disc function

$$[U,\,V] = \text{disc}\,(A, B)\,, \tag{1.8}$$

which was introduced by Roberts [144] for a matrix A. Its extension to a matrix pair (A, B) was developed by Malyshev [121], Bai et al. [15], and Benner [21]. The matrix disc function can be used to compute a deflating subspace, which either corresponds to the eigenvalues lying outside the unit circle or inside it. The recursive application of the matrix disc function leads to a divide-and-conquer scheme, which uses eigenvalue scaling to enclose the unit circle again. In this way we obtain the *Divide-Scale-and-Conquer* algorithm. The main advantage of the matrix disc function, over the generalized matrix sign function, is that its iterative computation scheme is free of solving linear systems and computing matrix inverses. A disadvantage is the higher computational cost.

Chapter 4 The implementation and optimization of both algorithms is subject of Chapter 4. Before details of the Divide-Shift-and-Conquer and the Divide-Scale-and-Conquer scheme are discussed, common operations for both algorithms are presented in Section 4.1. The main issue is the rearrangement of the recursive scheme into an iterative one. This enables us to limit the overall memory consumption and provides strategies to manage the available memory. Furthermore, we present a hardware oriented stopping criterion for the iteration [27, 26]. Finally, the section covers common operations like matrix-matrix products and gives an overview about the different QR decompositions used for deflating subspace computations [42, 43, 49, 143, 125].

The efficient computation of the generalized matrix sign function using a Newton method is in the focus of Section 4.2. After recalling how the naive approach [77] is implemented using routines from BLAS and LAPACK, we discuss two strategies to reduce the operation count. First, a QR or QL decomposition based approach by Benner and Quintana-Ortí [31] is evaluated and adjusted to available routines in BLAS and LAPACK. Second, we regard a bi-diagonal reduction approach by Sun and Quintana-Ortí [161], which turns out to have a very reduced operation count, but does not lead to a satisfying performance on current computer architectures, as examples in Section 4.2.3 show. Based on the second idea, we develop a band reduction approach, which leads to a similar low operation count and respects the capabilities of current computer architectures. All regarded algorithms to compute the generalized matrix sign function have in common, that they require the solution of linear systems with many right-hand sides, or the inverse of a matrix. Therefore, we recall the Gauss-Jordan Elimination [142] scheme as an alternative to the LU decomposition. We extend the Gauss-Jordan Elimination to solve linear systems directly and show how multiple GPUs can be used to accelerate this approach. Finally, the Gauss-Jordan Elimination procedure integrates in the different variants of the Newton method to compute the generalized matrix sign function.

The matrix disc function is computed using the inverse free iteration [121, 15, 21], where the QR decomposition is the main operation. In Section 4.3.1, we focus on a naive implementation. Thereby, we regard the influence of different QR decompositions, like the compact storage QR decomposition [40], or the Tile-QR decomposition [51], on the runtime. In order to reduce the number of required operations in the inverse free iteration, we regard the efficiency improving technique by Marqués et al. [124]. Although this approach reduces the operation count and involves level-3 BLAS operations, it breaks with the capabilities of current computer architectures. Using directed acyclic graph scheduling instead, we derive a well-scaling multi-core aware formulation of the Marqués et al. idea in Section 4.3.2. Since the graph scheduling approach is not applicable to GPU accelerators with common compiler features and the OpenMP 4 programming model, it is only used on multi-core CPUs. On GPU accelerated systems, we focus on acceleration of the compact storage QR decomposition instead. Thereby, different efficiency improving strategies, like the communication-avoiding QR (CAQR) decomposition [59] and the Householder vector reconstruction [17], play and important role to overcome the performance issues of the QR decomposition.

Chapter 5   In Chapter 5, the first part of the thesis is closed by numerical results for the Divide-Shift-and-Conquer and the Divide-Scale-and-Conquer scheme. Before the approximation of the generalized Schur decomposition is evaluated, all important building blocks, identified in Chapter 4, are benchmarked and the important runtime parameters are determined. In order to easily adjust these parameters, we integrate the embedded scripting language LUA [95] into the code. It is used to evaluate tuning parameters at runtime. A comparison to the QZ algorithm shows that we obtain a sufficiently good approximation to the generalized Schur decomposition in a fraction of the runtime. Without the usage of GPUs we gain a factor of up to 17 in the runtime with the Divide-Shift-and-Conquer algorithm compared to the QZ algorithm from LAPACK on the already mentioned Intel® Skylake system. Since the Divide-Scale-and-Conquer algorithm is computationally more demanding, we only achieve an acceleration of up to a factor of 10 in the runtime. If GPU accelerators are available we obtain a speed up of up to a factor of 55 in case of the Divide-Shift-and-Conquer algorithm and up to a factor of 28 in case of the Divide-Scale-and-Conquer scheme.

Chapter 6   Chapter 6 opens the second part of the thesis, which deals with the efficient direct solution of Sylvester-type matrix equations. Using the generalized Schur decomposition, the generalized Sylvester equation can be transformed such that all coefficient matrices are of (quasi) upper triangular shape. Therefore, it is sufficient to consider direct solvers for Sylvester equations with (quasi) upper triangular coefficient matrices.

Before we develop our solvers we recall existing algorithms, like the Bartels-Stewart algorithm [19], the Gardiner et al. approach [78], the one- and two-solve schemes by Sorensen and Zhou [152], and the recursive blocking by Jonsson and Kågström [100, 101]. Beside these direct solvers we briefly discuss a generalized matrix sign function based solver [31] due to its close connection to Chapters 3 and 4. High performance implementations for distributed memory systems of this algorithm are already presented by Benner and Quintana-Ortí [32, 33, 30]. On top of the ideas of the existing solvers we develop a level-3 BLAS algorithm. With an experiment-driven approach, we identify the critical points, where the level-3 approach needs special treatment to cover the capabilities of the hardware and the Fortran compilers. The included optimizations are designed to coincide with the capabilities of current computer architectures. Furthermore, the design of the code focuses on building blocks, which can be turned into highly efficient machine code by current Fortran compilers. Beside the level-3 BLAS approach we apply directed acyclic graph scheduling, also used in the QR decomposition for the matrix disc function before, to the Sylvester-type matrix equations as well. This results in an algorithm, whose performance is independent of the parallelization features of the used BLAS implementation. The connection between the developed level-3 approach and the existing solver is given by Theorem 6.8. It shows that many of the solvers that are built on top of the Bartels-Stewart approach [19] are equal to our level-3 approach with a special set of block size parameters.

Due to the fact that both divide-and-conquer algorithms to compute the generalized Schur decomposition only result in a sufficient approximation, the transformation to (quasi) upper triangular coefficient matrices leads to a slightly perturbed equation. This perturbation is minimized or compensated with the help of iterative refinement [174, 128, 156], known from standard linear systems. Thereby, the generalized Schur decomposition of the coefficient matrices is only required once.

Chapter 7   The numerical results for the Sylvester-type matrix equations close the thesis. First, we determine the optimal block size parameters for all covered equations. This includes the generalized Sylvester equation (1.4) and the generalized Lyapunov equation (1.6). This incorporates special cases where some of the coefficient matrices are set to the identity. Using these optimal parameters, we compare the performance of our implementation to RECSY [100, 101], which figured out to be the fastest one of the existing approaches. The parallel capabilities of the used BLAS implementation already results in a runtime gain of a factor between 1.86 and 2.42 on the Intel® Skylake system for generalized Sylvester equations with (quasi) triangular coefficient matrices. The explicitly parallelized algorithm with directed acyclic graph scheduling results in a runtime gain of a factor between 3.06 and 6.38. The iterative refinement procedure to reduce the influence of the approximated generalized Schur decomposition is evaluated with the help of the generalized Lyapunov equation (1.6). Solving the generalized Lyapunov equation corresponding to the aforementioned steel profile example [34] leads to the following results. We only need three steps in the refinement procedure and obtain an overall speed up of 21.25 with a non-GPU version of the Divide-Shift-and-Conquer algorithm compared to the existing implementation in SLICOT [149, 140]. With GPUs enabled, we obtain a speed up of 55.67. In case of the Divide-Scale-and-Conquer method, we only need two steps in the iterative refinement process. This results in a speed up of 9.87 without GPUs and 32.39 with GPUs. For both options we obtain comparable forward errors and relative residuals with respect to the implementation in SLICOT.

1.3. Target Computer Architectures

In the thesis, three different computer systems are used for the evaluations of the algorithms and ideas. The used systems reflect typical compute nodes of high performance clusters from 2014 to 2019 [168]. All systems are equipped with two multi-core CPUs, where both have their own memory controller. Furthermore, the systems are assembled with one or two GPU accelerators, since the portion of accelerator-enabled systems in the Top-500 systems increased from 3.4% in 2010 to 29% in 2019 [168].

In order to cover a wide range of architectures, three different CPUs are considered in this thesis: Intel® Xeon® CPUs with the Haswell and Skylake mirco-architecture and the IBM POWER8 CPUs.

All three architectures are common CPUs for the nodes in the Top-500 cluster systems as well as they are used as standalone compute servers in various kinds of research. Regarding the GPUs we only focus on Nvidia® accelerators, since they are used in 136 of the 145 accelerator-enabled systems in the Top-500 in 2019. Namely, we use Nvidia® K20 and Nvidia® P100 accelerators. The details of the used hardware are given in Section 5.1.1.

Since the typical workflow in the design of high performance algorithms is to have an algorithm running optimal on one node first [84], we only focus on shared memory parallelization techniques in this thesis. An extension of the algorithms to distributed memory systems, such as HPC clusters in the Top-500, should be part of future research.

Although we focus on compute servers and nodes of HPC systems, most of the results are valid for notebooks, desktop computers, and workstations as well. Many of these devices are equipped with CPUs having a similar or the exact micro-architecture as the one in the compute servers. Today, even consumer devices are mostly equipped with accelerator devices, mostly in terms of a real graphic processor. Furthermore, problems, that result from unfavorable memory accesses, affect these systems as well, since the origin of the problem is inherent to the used CPU and memory architectures. Basically, this does not change from small consumer devices to compute servers. The developed algorithms employ many tuning parameters such that they can be easily adjusted to a new architecture by running a set of provide benchmarks and setting the proper parameters. This benchmark driven auto-tuning approach is already known from other mathematical software projects like the ATLAS BLAS implementation [173].

1.4. Related own Publications

Previously developed ideas that lead to this thesis have been published before in the articles and proceedings listed below.

Chapter 3, in particular Section 3.2, extends

> [27]: P. Benner, M. Köhler, and J. Saak, *Fast approximate solution of the non-symmetric generalized eigenvalue problem on multi-core architectures*, in Parallel Computing: Accelerating Computational Science and Engineering (CSE), M. Bader, A. B. H.-J. Bungartz, M. Gerndt, G. R. Joubert, and F. Peters, eds., vol. 25 of Advances in Parallel Computing, IOS Press, 2014, pp. 143–152.

Section 4.1.4 includes the ideas from

> [26]: P. Benner, M. Köhler, and J. Saak, *A cache-aware implementation of the spectral divide-and-conquer approach for the non-symmetric generalized eigenvalue problem*, Proc. Appl. Math. Mech., 14 (2014), pp. 819–820.

Section 4.2.4 summarizes

> [111]: M. Köhler, C. Penke, J. Saak, and P. Ezzatti, *Energy-aware solution of linear systems with many right hand sides*, Comput. Sci. Res. Dev., 31 (2016), pp. 215–223.

> [114]: M. Köhler and J. Saak, *GPU Accelerated Gauss-Jordan Elimination on the Open-POWER platform – A case study*, Proc. Appl. Math. Mech., 17 (2017), pp. 845–846.

> [115]: M. Köhler and J. Saak, *Frequency scaling and energy efficiency regarding the gauss-jordan elimination scheme with application to the matrix-sign-function on openpower 8*, Concurrency and Comput.: Pract. Exper., (2018).

Chapter 6 has its origin in

[112]: M. Köhler and J. Saak, *On GPU acceleration of common solvers for (quasi-) triangular generalized Lyapunov equations*, Parallel Comput., 57 (2016), pp. 212–221.

[113]: M. Köhler and J. Saak, *On BLAS level-3 implementations of common solvers for (quasi-) triangular generalized Lyapunov equations*, ACM Trans. Math. Software, 43 (2016), pp. 3:1–3:23.

MATHEMATICAL BASICS OF THE GENERALIZED EIGENVALUE PROBLEM

Contents

The introduction has already mentioned that the solution of a generalized eigenvalue problem is necessary for the direct solution of dense Sylvester-type matrix equations. Therefore, this chapter is intended to give a brief overview of the mathematical basics behind the generalized eigenvalue problem, which are necessary for the derivation of our algorithms in Chapter 3. For additional information, like error estimation or perturbation theory, we refer to the literature [88, 155, 129, 80].

The generalized eigenvalue problem used in this work is given by the following definition:

Definition 2.1 (Generalized Eigenvalue Problem):
*We call $(A, B) \in \mathbb{C}^{m \times m} \times \mathbb{C}^{m \times m}$ a **matrix pair of order** m. The **characteristic polynomial** $\chi_{(A,B)}$ of the matrix pair (A, B) for $\gamma \in \overline{\mathbb{C}}$ is given by*

$$\chi_{(A,B)}(\gamma) := \det(A - \gamma B) \tag{2.1}$$

*and the **spectrum**, i.e., the **set of eigenvalues**, $\Lambda(A, B)$ of the matrix pair (A, B) is defined as*

$$\Lambda(A, B) := \left\{ \lambda \in \mathbb{C} \mid \chi_{(A,B)}(\lambda) = 0 \right\}. \tag{2.2}$$

If $\lambda \in \Lambda(A, B)$ and $x \in \mathbb{C}^m \setminus \{0\}$ is a nontrivial solution of the equation

$$Ax = \lambda B x, \tag{2.3}$$

*then x is called a **(right) eigenvector** of (A, B) corresponding to the eigenvalue λ. The pair (λ, x) is referred to as **(right) eigenpair** of (A, B). A vector $y \in \mathbb{C}^m \setminus \{0\}$, which satisfies*

$$y^H A = \lambda y^H B \tag{2.4}$$

is said to be a **left eigenvector** of (A, B) corresponding to the eigenvalue λ and (λ, y) is called a **left eigenpair** of (A, B).

The determination of $\Lambda(A, B)$ by computing the roots of $\chi_{(A,B)}$ is not possible in general because direct solution formulas can only be derived for polynomials of degree less or equal to four [36]. Even if we have obtained $\Lambda(A, B)$ from $\chi_{(A,B)}$, the computation of the corresponding eigenvector will require to find a nontrivial solution of

$$(A - \lambda B)x = 0 \tag{2.5}$$

for each eigenvalue $\lambda \in \Lambda(A, B)$. Thereby, $A - \lambda B$ is singular by the definition of λ. If (A, B) is not well-structured, for example A is upper Hessenberg and B is upper triangular, this becomes a numerically tough task as well. For example, the algorithms implemented in LAPACK [10] rely on this property to compute the generalized eigenvectors.

Depending on the properties of the characteristic polynomial $\chi_{(A,B)}(\gamma)$ we divide the matrix pairs (A, B) into two classes using the following definition:

Definition 2.2 (e.g. [14]):
Let (A, B) be a matrix pair of order m. If there exists at least one $\lambda \in \mathbb{C}$ such that

$$\chi_{(A,B)}(\lambda) = \det(A - \lambda B) \neq 0, \tag{2.6}$$

the matrix pair (A, B) is said to be **regular**. If the characteristic polynomial instead fulfills

$$\chi_{(A,B)}(\lambda) = 0 \quad \forall \lambda \in \mathbb{C}, \tag{2.7}$$

the matrix pair (A, B) is said to be **singular**.

If the matrix pair is singular, Van Dooren [169] and Wilkinson [175] showed that the problem can be handled by splitting the matrix pair into a singular and a regular one and treat both of them separately. Furthermore, having a singular matrix pairs means that A and B have a common nontrivial null space, they are nonsingular as well and the spectrum of (A, B) contains the whole complex plane. This violates the solution conditions for the Sylvester type matrix equations covered in Chapter 6. For these reasons, we focus only on regular matrix pairs in this work. The solution of the generalized eigenvalue problem for singular matrix pairs is a separate area of research [61, 60, 175, 169].

Remark 2.3:
If B is invertible, then the generalized eigenvalue problem

$$Ax = \lambda Bx$$

is equivalent to the standard eigenvalue problem

$$B^{-1}Ax = \lambda x, \tag{2.8}$$

and $\Lambda(A, B) = \Lambda(B^{-1}A)$.

2.1. Invariant Subspaces and Deflation

The basic definition of the generalized eigenvalue problem shows that if we find an eigenpair (λ, x), x spans a distinguished one dimensional subspace in $\mathbb{C}^m$. From the Equation (2.3) it follows:

$$\text{span}(Ax) = \text{span}(Bx), \tag{2.9}$$

which means that the images of x under A and B lie in the same one dimensional subspace. Especially, we can conclude that

$$\text{span}(Ax + Bx) = \text{span}(Ax).\tag{2.10}$$

The generalization of this special subspace to higher dimensional subspaces of $\mathbb{C}^m$ is defined as follows:

Definition 2.4:
*Let (A, B) be a matrix pair of order m and let $\mathcal{Z}$ and $\mathcal{Q}$ be p-dimensional subspaces of $\mathbb{C}^m$. Then $\mathcal{Z}$ is called **right deflating subspace** of (A, B) if*

$$\dim\left(A\mathcal{Z} + B\mathcal{Z}\right) \leq p\tag{2.11}$$

*and $\mathcal{Q}$ is called **left deflating subspace** if*

$$\dim\left(A^H\mathcal{Q} + B^H\mathcal{Q}\right) \leq p\tag{2.12}$$

holds.

Remark 2.5:
If the matrix pair (A, B) is regular, equality holds in (2.11) and (2.12) [155, 129].

Based on this definition, Stewart [155] proved the following theorem concerning the connection between a given right deflating subspace and a block decomposition of the corresponding eigenvalue problem.

Theorem 2.6 (Deflation-Theorem [155, Theorem 2.1]):
Let (A, B) be a matrix pair of order m and $\mathcal{Z}$ be a p-dimensional right deflating subspace of (A, B). Then there exist unitary matrices Q and Z such that the first p columns of Z span the subspace $\mathcal{Z}$ and

$$Q^H A Z = \begin{bmatrix} A_{11} & A_{12} \\ 0 & A_{22} \end{bmatrix} \quad \text{and} \quad Q^H B Z = \begin{bmatrix} B_{11} & B_{12} \\ 0 & B_{22} \end{bmatrix}\tag{2.13}$$

holds, where A_{11} and B_{11} are of order p. The right eigenvectors $x \in \mathbb{C}^m \setminus \{0\}$ belonging to $\lambda \in \Lambda\left(A_{11}, B_{11}\right)$ fulfill

$$x \in \mathcal{Z}.\tag{2.14}$$

Corollary 2.7 ([162, Theorem 2.3, Corollary 2.4, Corollary 2.5]):
Let (A, B), Q, Z, and $\mathcal{Z}$ be given as in Theorem 2.6. Then we conclude:

1. *The first p columns of Q span a p-dimensional left deflating subspace $\mathcal{Q}$ corresponding to $\Lambda\left(A_{11}, B_{11}\right)$.*

2. *The subspace spanned by the last $(m - p)$ columns of Q is the orthogonal complement of $A\mathcal{Z} + B\mathcal{Z}$.*

3. *The set of eigenvalues $\Lambda\left(A, B\right)$ deflates into*

$$\Lambda\left(A, B\right) = \Lambda\left(A_{11}, B_{11}\right) \cup \Lambda\left(A_{22}, B_{22}\right).\tag{2.15}$$

4. *$\Lambda\left(A, B\right)$ is independent of (A_{12}, B_{12}).*

From the previous corollary together with Theorem 2.6 we conclude that the solution of the generalized eigenvalue problem (2.2) can be replaced by two smaller problems (2.15) if a deflating subspace $\mathcal{Z}$ is known. This introduces additional costs of only four matrix-matrix products (2.13). This fundamental result is used in Chapter 3 to derive a family of divide-and-conquer algorithms.

2.2. Generalized Schur Decomposition

The Deflation Theorem shows that there exist matrices such that the matrix pair (A, B) can be transformed to upper block triangular structure using two-sided unitary transformations. Thereby, the eigenvalues stay untouched and the problem decomposes into two independent smaller subproblems. However, we still have to solve two eigenvalue problems. Before we obtain a more general variant of the Deflation Theorem, we use the following lemma to proof that the set of eigenvalues $\Lambda(A, B)$ stays invariant under the transformations used for deflation and more general ones.

Lemma 2.8:
Let (A, B) be a matrix pair of order m. Furthermore, the matrices $U \in \mathbb{C}^{m \times m}$ and $V \in \mathbb{C}^{m \times m}$ are assumed to be nonsingular. Then it follows that

$$\Lambda(A, B) = \Lambda(UAV, UBV),\qquad(2.16)$$

where the right and left eigenvectors of (A, B) for $\lambda \in \Lambda(A, B)$ are given by

$$x = V\hat{x} \quad and \quad y^H = \hat{y}^H U.\qquad(2.17)$$

The vectors $\hat{x}$ or $\hat{y}$ are the corresponding right and left eigenvectors to the eigenvalue $\lambda \in \Lambda(UAV, UBV)$.

The lemma shows that we can transform the matrix pair (A, B) without losing eigenvalue information. Especially, if U and V are unitary matrices, this allows numerically stable transformations [155, 80].

Since the determinant of a triangular matrix is computed by

$$\det(T) = \prod_{k=1}^{m} t_{kk},\qquad(2.18)$$

we try to find nonsingular matrices U and V such that $UTV = A$ and $USV = B$ and both T and S are upper (or lower) triangular. Then we can compute the roots of $\chi_{(A,B)}(\lambda)$ from

$$\chi_{(A,B)}(\lambda) = \det(U)\det(T - \lambda S)\det(V) = \underbrace{\det(UV)}_{\neq 0}\prod_{k=1}^{m}(T_{kk} - \lambda S_{kk}).\qquad(2.19)$$

Hence, the set of eigenvalues $\Lambda(A, B) = \Lambda(T, S)$ is given by

$$\Lambda(A, B) = \left\{\lambda := \frac{T_{kk}}{S_{kk}} \text{ if } S_{kk} \neq 0, \quad \lambda := \infty \text{ if } S_{kk} = 0\right\}.\qquad(2.20)$$

The existence of these transformation matrices (U, V) is guaranteed by the following theorem. Furthermore, it proves that beside being nonsingular, these transformation matrices are unitary:

Theorem 2.9 (Generalized Schur Decomposition [155, 80, 162]):
Let (A, B) be a matrix pair of order m. Then, there exist unitary matrices $Q \in \mathbb{C}^{m \times m}$ and $Z \in \mathbb{C}^{m \times m}$ such that

$$T = Q^H A Z \quad and \quad S = Q^H B Z\qquad(2.21)$$

are upper triangular and the eigenvalues of (A, B) are $\lambda_k = T_{kk}/S_{kk}$. If $T_{kk} \neq 0$ and $S_{kk} = 0$ then $\lambda_k = \infty$. If $T_{kk} = 0$ and $S_{kk} \neq 0$ then $\lambda_k = 0$.

Remark 2.10:

Theorem 2.9 does not give any information about the order of the eigenvalues λ_k on the diagonals of T and S. For this reason, Q and Z are not uniquely defined. However, for each renumbering of the eigenvalues λ_k, one can find matrices Q and Z such that they appear in the desired order on the diagonals of T and S [102].

The generalized Schur decomposition is typically computed using the QZ algorithm based on the ideas of Moler and Stewart [129]. Due to the complexity of the algorithm and the work which was done to improve it, especially by Kågström et al. [103, 104], only a sketch of these ideas is given in this work. The principal idea is to port the ideas of the Francis-QR algorithm for the standard eigenvalue problem [73], and its improvements, to the generalized eigenvalue problem. This results in a two-step scheme. At first, the matrix pair (A, B) is reduced using two unitary transformations $\tilde{Q}$ and $\tilde{Z}$ such that $\tilde{A} = \tilde{Q}^H A \tilde{Z}$ is upper Hessenberg and $\tilde{B} = \tilde{Q}^H B \tilde{Z}$ is upper triangular. On this, now structured problem, an iterative process based on unitary transformations [80, 129, 155] is used to annihilate the subdiagonal elements in $\tilde{A}$ and $\tilde{B}$. This yields the generalized Schur decomposition (2.21) in Theorem 2.9. A variant of the algorithm, but without aggressive early deflation and multi-shift strategies [103], is implemented in LAPACK [10] since version 3.0 as GGES. An updated variant with accelerated Hessenberg-triangular reduction appeared as GGES3 in LAPACK 3.6.0. A distributed memory parallel version of the algorithm was presented by Adlerborn, Kågström, and Kressner in [8].

These theoretical results rely on complex arithmetic to solve the generalized eigenvalue problem. Even with a real matrix pair $(A, B) \in \mathbb{R}^{m \times m} \times \mathbb{R}^{m \times m}$, the previously presented results, especially Theorem 2.9, may engender complex eigenvalues and complex eigenvectors and thus require the use of complex arithmetic. In many practical applications we want to avoid switching between real and complex arithmetic if the input data are real. On the one hand, this reduces the necessary memory for storing the input data, the output data, and the intermediate results. On the other hand a multiply-add operation $a + b \cdot c$ costs two flops in real arithmetic but eight flops if we use complex numbers, which slows down the computations if we switch to the complex case.

Because complex roots of a polynomial with real coefficients only appear as complex conjugate pairs, we can obtain them as roots of a real polynomial of degree two. As a consequence of Theorem 2.6 we can separate these eigenvalue pairs in 2×2 blocks on the diagonal of T and S. The resulting real version of Theorem 2.9 then reads as follows:

Theorem 2.11 (Real Generalized Schur Decomposition [80, 155]):

Let (A, B) be a real matrix pair of order m. Then, there exists orthogonal matrices $Q \in \mathbb{R}^{m \times m}$ and $Z \in \mathbb{R}^{m \times m}$, such that

$$T := Q^T A Z \tag{2.22a}$$

is quasi upper triangular, i.e. 2×2 blocks can appear on the diagonal, and

$$S := Q^T A Z \tag{2.22b}$$

is upper triangular.

Remark 2.12:

Theorem 2.6 and Corollary 2.7 are further valid for real matrix pairs (A, B) and orthogonal matrices $Q \in \mathbb{R}^{m \times m}$ and $Z \in \mathbb{R}^{m \times m}$.

2.3. Reordering of Eigenvalues

The generalized Schur decomposition from Theorem 2.9 provides a way to obtain all eigenvalues of (A, B) but only the first column of Z represents a right eigenvector for the eigenvalue $\lambda = T_{11}/S_{11}$ and the last column of Q gives us the left eigenvector for T_{mm}/S_{mm}. In the real case, see Theorem 2.11, the (T_{11}, S_{11}) element may belong to 2×2 block representing a complex conjugate eigenvalue pair. Then, the first two columns of Z form a deflating subspace containing the eigenvector for this pair. These can be extracted easily, for example by switching to complex arithmetic for the remaining 2×2 problem. The left eigenvectors behave similarly in the real case.

We can think of two different strategies to obtain all eigenvectors from the generalized Schur decomposition. The first one, which is implemented as TGEVC in LAPACK [10], solves Equation (2.5) for each eigenvalue λ of (T, S) and then uses Lemma 2.8 together with Equation (2.17) to obtain the eigenvectors of (A, B) from its generalized Schur decomposition. An alternative to this approach can be realized if we can reorder the diagonal elements in T and S since Theorem 2.9 (and Theorem 2.11) does not prescribe any order of the eigenvalues on the diagonals. Only the arrangement of the deflating subspaces in Q and Z decides about the positions if the eigenvalues on the diagonals of T and S.

This motivates the idea of computing unitary or orthogonal transformations such that a selected eigenvalue is moved to the top or the bottom of the diagonals of T and S. Obviously, it is enough to study the swapping of the diagonal blocks in a 2×2 block upper triangular matrix with a block size of at most 2×2. This includes the cases of swapping two real or complex eigenvalues or a complex conjugate eigenvalue pair with a real one. An algorithm which performs this operation was developed by Kågström in 1993 [102]. This algorithm is briefly recalled here since it shows again the connection between the generalized Schur decomposition and the generalized Sylvester equation. We consider the following block partitioning of our already computed generalized Schur decomposition (T, S):

$$T = \begin{bmatrix} T_{11} & T_{12} \\ & T_{22} \end{bmatrix} \quad \text{and} \quad S = \begin{bmatrix} S_{11} & S_{12} \\ & S_{22} \end{bmatrix}, \tag{2.23}$$

where $(T_{11}, S_{11}) \in \mathbb{C}^{p \times p} \times \mathbb{C}^{p \times p}$ and $(T_{22}, S_{22}) \in \mathbb{C}^{q \times q} \times \mathbb{C}^{q \times q}$ with $p, q \in \{1, 2\}$. Furthermore, we assume (T_{11}, S_{11}) and (T_{22}, S_{22}) to have no eigenvalues in common, otherwise swapping is not necessary.

In order to swap the diagonal blocks in (T, S) in (2.23), we first have to block-diagonalize the matrices using

$$\left(\begin{bmatrix} T_{11} & T_{12} \\ & T_{22} \end{bmatrix}, \begin{bmatrix} S_{11} & S_{12} \\ & S_{22} \end{bmatrix} \right) = \underbrace{\begin{bmatrix} I_p & L \\ & I_q \end{bmatrix}}_{\hat{L}} \left(\begin{bmatrix} T_{11} & \\ & T_{22} \end{bmatrix}, \begin{bmatrix} S_{11} & \\ & S_{22} \end{bmatrix} \right) \underbrace{\begin{bmatrix} I_p & -R \\ & I_q \end{bmatrix}}_{\hat{R}}, \tag{2.24}$$

where L and R solve the generalized coupled Sylvester equation [107, 105]

$$\begin{aligned} T_{11}R - LT_{22} &= -T_{12} \\ S_{11}R - LS_{22} &= -S_{12}, \end{aligned} \tag{2.25}$$

which is uniquely solvable if (T_{11}, S_{11}) and (T_{22}, S_{22}) do not have any common eigenvalue [107]. By introducing two permutations

$$P_L = \begin{bmatrix} & I_q \\ I_p & \end{bmatrix} \quad \text{and} \quad P_R = \begin{bmatrix} & I_p \\ I_q & \end{bmatrix}, \tag{2.26}$$

from Equation (2.24) we obtain

$$\left(\begin{bmatrix} T_{11} & T_{12} \\ & T_{22} \end{bmatrix}, \begin{bmatrix} S_{11} & S_{12} \\ & S_{22} \end{bmatrix}\right) = \hat{L}P_L\left(\begin{bmatrix} T_{22} & \\ & T_{11} \end{bmatrix}, \begin{bmatrix} S_{22} & \\ & S_{11} \end{bmatrix}\right)P_R\hat{R}, \tag{2.27}$$

where the positions of the eigenvalues on the diagonal have been swapped. Unfortunately, this transformation is neither unitary nor orthogonal and thus vulnerable to round-off errors [88]. This issue is solved by block-wise orthogonalization of X and Y. Setting

$$X = \hat{L}P_L = \begin{bmatrix} L & I_p \\ I_q & \end{bmatrix} \quad \text{and} \quad Y = P_R\hat{R}\begin{bmatrix} & I_q \\ I_p & -R \end{bmatrix}, \tag{2.28}$$

the first block column of X and the last block row of Y have full column rank. This enables the computation of two unitary matrices Q_X and Z_Y using a QR or RQ decomposition [80, 11] to obtain

$$Q_X^H\begin{bmatrix} L \\ I_q \end{bmatrix} = \begin{bmatrix} R_L \\ 0 \end{bmatrix} \quad \text{and} \quad \begin{bmatrix} I_p & -R \end{bmatrix}Z_Y = \begin{bmatrix} 0 & R_R \end{bmatrix}. \tag{2.29}$$

With the block partitioning

$$Q_X = \begin{bmatrix} Q_{11} & Q_{12} \\ Q_{21} & Q_{22} \end{bmatrix} \quad \text{and} \quad Z_Y = \begin{bmatrix} Z_{11} & Z_{12} \\ Z_{21} & Z_{22} \end{bmatrix},$$

this yields

$$Q_X^H\left(\begin{bmatrix} T_{11} & T_{12} \\ & T_{22} \end{bmatrix}, \begin{bmatrix} S_{11} & S_{12} \\ & S_{22} \end{bmatrix}\right)Z_Y = Q_X^H X\left(\begin{bmatrix} T_{22} & \\ & T_{11} \end{bmatrix}, \begin{bmatrix} S_{22} & \\ & S_{11} \end{bmatrix}\right)Y Z_Y$$

$$= \left(\begin{bmatrix} R_L T_{22}Z_{21} & R_L T_{22}Z_{22} + Q_{11}^H T_{11}R_R \\ & Q_{12}^H T_{11}R_R \end{bmatrix}, \begin{bmatrix} R_L S_{22}Z_{21} & R_L S_{22}Z_{22} + Q_{11}^H S_{11}R_R \\ & Q_{12}^H S_{11}R_R \end{bmatrix}\right)$$

$$= \left(\begin{bmatrix} \tilde{T}_{22} & \tilde{T}_{12} \\ & \tilde{T}_{11} \end{bmatrix}, \begin{bmatrix} \tilde{S}_{22} & \tilde{S}_{12} \\ & \tilde{S}_{11} \end{bmatrix}\right), \tag{2.30}$$

where the transformed matrix pairs $\left(\tilde{T}_{11}, \tilde{S}_{11}\right)$ and $\left(\tilde{T}_{22}, \tilde{S}_{22}\right)$ have the same eigenvalues as (T_{11}, S_{11}) and (T_{22}, S_{22}) through Lemma 2.8. The error analysis for this scheme was done by Kågström in [102]. The scheme is implemented as TGEX2 in LAPACK [10] to swap two neighboring eigenvalues and as TGEXC for swapping two arbitrary ones. In this case one has to create a sequence of subsequent neighboring swaps until the desired eigenvalues are exchanged.

Remark 2.13:
The scheme can be used for $p, q \geq 2$ as well. In this case, the transformed matrix pairs $\left(\tilde{T}_{11}, \tilde{S}_{11}\right)$ and $\left(\tilde{T}_{22}, \tilde{S}_{22}\right)$ have to be transformed to generalized Schur form again after swapping the diagonal blocks.

Remark 2.14:
If an eigenvalue has an algebraic multiplicity larger than one, the approach to compute all eigenvectors by swapping them to the first or last diagonal element is not possible because Equation (2.25) is not longer uniquely solvable. In this case we have to find the nontrivial solutions of Equation (2.5) for the affected eigenvalues as done by LAPACK.

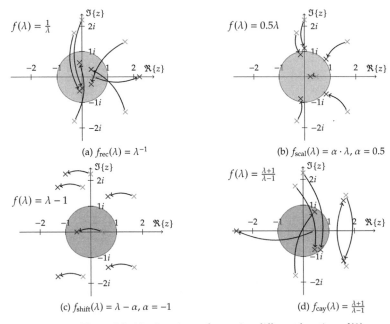

Figure 2.1.: Moving eigenvalues using different functions $f(\lambda)$.

2.4. Transformation of Eigenvalues

In the previous section we have seen that the eigenvalues of a matrix pair (A, B) are invariant under two-sided transformations as proved in Lemma 2.8. These transformations only affect the eigenvectors and the deflating subspaces but leave the eigenvalues untouched. Here, we present transformations which retain the deflating subspaces but move the eigenvalues within the complex plane. Therefore, we recall a theorem concerning matrix functions [89, 39]:

Theorem 2.15 ([39, Lemma 1.7]):
  Let Γ be an open subset of the complex plane $\mathbb{C}$ and $f(z) : \Gamma \mapsto \mathbb{C}$ a holomorphic function. Let $A \in \mathbb{C}^{m \times m}$ be an arbitrary matrix with $\Lambda(A) = \{\lambda_k\}_{k=1}^{m} \subset \Gamma$. Then the eigenvalues of the function f applied to the matrix A are equal to $f(\lambda_k)$, $k = 1, \ldots, m$. Furthermore, if A decomposes into

$$AV = V\Lambda,$$

where $\Lambda = \text{diag}(\lambda_1, \ldots, \lambda_k)$ and $V \in \mathbb{C}^{m \times k}$, $1 \leq k \leq m$, then it holds

$$f(A)V = Vf(\Lambda). \tag{2.31}$$

Due to this theorem, the spectra of standard eigenvalue problems can be modified using a suitable function without altering invariant subspaces. Because the previous theorem only covers the standard eigenvalue problem we discuss the functions needed later on and their usage with the generalized eigenvalue problem in the next subsections.

2.4.1. Reciprocal Function

First, we regard the reciprocal function $f_{rec} : \overline{\mathbb{C}} \mapsto \overline{\mathbb{C}}$:

$$f_{rec}(\lambda) = \lambda^{-1} = \frac{\mathcal{R}(\lambda)}{|\lambda|^2} - \frac{\mathcal{I}(\lambda)}{|\lambda|^2}i. \tag{2.32}$$

It inverts the eigenvalues at the unit circle but with conjugating their imaginary parts as seen in Figure 2.1a. Using Lemma 2.8 we can extend this function to $\overline{\mathbb{C}}^{m \times m} \mapsto \overline{\mathbb{C}}^{m \times m}$. For this purpose we assume that neither A nor B are singular matrices. This allows to replace the generalized eigenvalue problem from (2.3) by

$$B^{-1}Ax = \lambda x. \tag{2.33}$$

Now, we use Theorem 2.15 to plug in the reciprocal function (2.32) and end up with:

$$\left(B^{-1}A\right)^{-1}x = f_{rec}\left(B^{-1}A\right)x = f_{rec}(\lambda)x = \frac{1}{\lambda}x$$

$$\Longleftrightarrow \qquad A^{-1}Bx = \frac{1}{\lambda}x$$

$$\Longleftrightarrow \qquad Bx = \frac{1}{\lambda}Ax. \tag{2.34}$$

In this way the reciprocal function (2.32) transforms the generalized eigenvalue problem for the matrix pair (A, B) into the generalized eigenvalue problem for the matrix pair (B, A) with $\Lambda(B, A) = \left\{\frac{1}{\lambda} \mid \lambda \in \Lambda(A, B)\right\}$. If B is singular we use A^{-1} in (2.33).

Remark 2.16:

The reciprocal function (2.32) on $\overline{\mathbb{C}}$ applied to the spectrum of the matrix pair (A, B) has the following properties:

1. *eigenvalues on the real axis are mapped onto the real axis,*

2. *eigenvalues on the imaginary axis are mapped onto the imaginary axis,*

3. *the infinity eigenvalues $\lambda = \pm\infty$ are mapped into the origin, and*

4. *eigenvalues outside the unit circle are mapped into it and vice versa.*

2.4.2. Scale Function

The second regarded function is the scale function $f_{scal} : \mathbb{C} \mapsto \mathbb{C}$ with a real scaling factor:

$$f_{scal}(\lambda) = \alpha\lambda, \quad \alpha \in \mathbb{R} \setminus \{0\}. \tag{2.35}$$

This function moves eigenvalues towards or away from zero as shown in Figure 2.1b. Under the same assumptions as in Subsection 2.4.1 we obtain:

$$\alpha B^{-1}Ax = f_{scal}\left(B^{-1}A\right)x = f_{scal}(\lambda)x = \alpha\lambda x$$

$$\Longleftrightarrow \qquad \alpha Ax = \alpha\lambda Bx, \tag{2.36}$$

which results in the generalized problem for the matrix pair $(\alpha A, B)$ or $(A, 1/\alpha B)$ with the spectrum $\Lambda(\alpha A, B) = \Lambda(A, 1/\alpha B) = \{\alpha\lambda \mid \lambda \in \Lambda(A, B)\}$.

Remark 2.17:

The scale function (2.35) applied to the spectrum of the matrix pair (A, B) has the following properties:

1. *zero and infinity eigenvalues stay zero or infinity, and*

2. *all other eigenvalues are moved on a line passing through them and the origin.*

2.4.3. Shift Function

The third function we are using to modify the spectrum is the shift function $f_{\text{shift}} : \mathbb{C} \mapsto \mathbb{C}$ with a real shift coefficient:

$$f_{\text{shift}}(\lambda) = \lambda + \alpha, \quad \alpha \in \mathbb{R}. \tag{2.37}$$

The function moves the eigenvalues to the left, if $\alpha < 0$, or to the right, if $\alpha > 0$, as shown in Figure 2.1c. Applied to the generalized eigenvalue problem (2.3), under the assumptions of Subsection 2.4.1 we obtain:

$$\left(B^{-1}A + \alpha I_m\right) x = f_{\text{shift}}\left(B^{-1}A\right) x = f_{\text{shift}}(\lambda)x = (\lambda + \alpha)x$$
$$\Longleftrightarrow \qquad (A + \alpha B) x = (\lambda + \alpha)Bx. \tag{2.38}$$

This results in the generalized eigenvalue problem for the matrix pair $(A + \alpha B, B)$ with the spectrum $\Lambda (A + \alpha B, B) = \{\lambda + \alpha \mid \lambda \in \Lambda (A, B)\}$.

Remark 2.18:

The shift function (2.37) applied to the spectrum of the matrix pair (A, B) has the following properties:

1. *zero eigenvalues are moved to $\lambda = \alpha$,*

2. *purely imaginary eigenvalues get the real part α,*

3. *if $-\alpha \in \Lambda (A, B)$ then $0 \in \Lambda (A + \alpha B, B)$, and*

4. *the infinity eigenvalues stay untouched.*

2.4.4. Cayley Transformation

The last function we regard is the Cayley transformation [39] $f_{\text{cay}} : \mathbb{C} \setminus \{1\} \mapsto \mathbb{C} \setminus \{1\}$:

$$f_{\text{cay}}(\lambda) = \frac{\lambda + 1}{\lambda - 1}. \tag{2.39}$$

It maps the left half-plane onto the inner of the unit circle and the right half-plane to the outer, as we see for the example spectrum in Figure 2.1d. The Cayley transformation is a special case of the Möbius transformation [45], which can be used for advanced eigenvalue transformations as well. The advantage of this parametrization of the Cayley transformation (2.39) used here is that it coincides with its inverse. The generalized eigenvalue problem under the assumptions of Subsection 2.4.1 is transformed into:

$$\left(B^{-1}A - I_M\right)\left(B^{-1}A + I_M\right) x = f_{\text{cay}}\left(B^{-1}A\right) x = f_{\text{cay}}(\lambda)x = \frac{\lambda + 1}{\lambda - 1}x$$
$$\Longleftrightarrow \qquad \left(B^{-1}A + I_M\right) x = \frac{\lambda + 1}{\lambda - 1}\left(B^{-1}A - I_M\right) x$$
$$\Longleftrightarrow \qquad (A + B) x = \frac{\lambda + 1}{\lambda - 1}(A - B) x. \tag{2.40}$$

The resulting generalized eigenvalue problem for the matrix pair $(A + B, A - B)$ has the spectrum $\Lambda (A + B, A - B) = \left\{\frac{\lambda + 1}{\lambda - 1} \mid \lambda \in \Lambda (A, B)\right\}$.

Remark 2.19 ([39]):

The Cayley transformation (2.37) applied to the spectrum of the matrix pair (A, B) has the following properties:

1. $f_{\text{cay}}(0) = -1$ and $f_{\text{cay}}(\infty) = 1$,

2. *eigenvalues on the real axis are mapped onto the real axis,*

3. *eigenvalues on the imaginary axis are mapped onto the unit circle,*

4. *the left half-plane is mapped onto the unit disc, and*

5. *the right half-plane is mapped to the outer of the closed unit disc.*

SPECTRAL DIVIDE AND CONQUER ALGORITHMS

Contents

In this chapter we derive a set of generalized eigenvalue solvers, which focus on a rich usage of level-3 BLAS operations. Especially, we will increase the usage of matrix-matrix products (GEMM), linear equation solves and unitary decompositions. The reason for this special focus is that the GEMM operation is mostly the first routine available in a highly optimized version for scientific computing on new computer architectures and accelerator devices. In most cases, the optimized GEMM implementations reach up to 90% of the theoretical floating-point peak performance. Optimized versions are available in open-source software [81, 137, 171, 172, 173] as well as parts of proprietary libraries [96, 97, 171] for all major CPU architectures. On accelerator devices specialized libraries with a BLAS-like interface exist [4, 6, 132]. Regarding distributed memory systems, the parallel BLAS implementations inside ScaLAPACK [44] or libELEMENTAL [141] make scalable parallel versions of these routines available.

This motivates to build a generalized eigenvalue solver which focuses on the usage of the GEMM operation and closely related operations to benefit easily from the improvements on new hardware architectures without time-consuming adjustments of the algorithms.

The widely used QZ algorithm [80, 103, 104, 129] determines eigenvalues and deflating subspaces of the matrix pair (A, B) within the following three steps:

1. The matrix B is transformed to upper triangular form $\tilde{B}$ using the QR decomposition. The resulting matrix Q is used to compute the updated matrix pair $\left(\tilde{A}, \tilde{B} \right) = (Q^H A, Q^H B)$.

2. The matrix pair $\left(\tilde{A}, \tilde{B}\right)$ is transformed via unitary transformations Q_H and Z_H to Hessenberg-triangular form, i.e. the matrix $Q_H^H \tilde{A} Z_H$ is upper Hessenberg and the matrix $Q_H^H \tilde{B} Z_H$ is upper triangular.

3. Apply QZ steps [129] to $\left(\hat{A}, \hat{B}\right) = \left(Q_H^H \tilde{A} Z_H, Q_H^H \tilde{B} Z_H\right)$ such that the non-zero subdiagonal of $\hat{A}$ vanishes and $\hat{B}$ is kept upper triangular at the same time.

The Hessenberg-triangular form, obtained from the first two steps, accelerates the third step since this structure is invariant under the QZ steps.

The steps differ in the way they utilize the capabilities of current computer architectures. The main aspects are the following:

1. The QR decomposition in the initial step can be performed using a level-3 BLAS enabled algorithm [40] as well as a multi-core parallel algorithm [50], or with the help of a GPU accelerated one [177]. The resulting matrix pair has the same eigenvalues as (A, B), compare Lemma 2.8.

2. The Hessenberg-triangular reduction in the original algorithm [129] was performed with the help of Givens rotations [80]. These have to be applied sequentially, and they involve vector-vector operations only. These two facts make it hard to create a parallel implementation. Furthermore, processing vector after vector with only performing few operations on each dataset leads to a bandwidth bound algorithm. Overall this causes the algorithm to perform badly on current computer architectures. A level-3 BLAS enabled variant of the Hessenberg-triangular reduction was presented by Kågström et. al. [104], where the Givens rotations are grouped to increase the data locality and reduces the cache misses. Another variant based on block-wise applied QR decompositions was presented by Dackland and Kågström [57].

3. The QZ steps can be regarded as the implicit application of the Francis-QR algorithm [73] to $M = \hat{A}\hat{B}^{-1}$ without forming M. The algorithm originally proposed by Moler and Stewart [129] implements this using Givens rotations and small Householder transformations. There, the data locality and the cache reuse is very limited. This issue is partly fixed by the aggressive early deflation and multi-shift implementation by Kågström and Kressner [103], which is – due to many tuning parameters – not part of LAPACK, yet. A bad selection of the involved tuning parameters even can cause the algorithm to fail.

Even though the mentioned improvements result in a partial level-3 enabled algorithm, it does not provide a sufficient performance increase in most cases since the sizes of the level-3 operations stay small. Most operations, except of the QR decomposition of B in the first step, will not benefit from the properties of current computer hardware, due to their sequential nature. For this reason we develop an alternative strategy.

In this chapter we will derive a family of generalized eigenvalue solvers. Theorem 2.6 is used to split the full generalized eigenvalue problem into a set of smaller generalized eigenvalue problems by only employing multi-core and multi-GPU aware level-3 enabled algorithms. Thereby, we restrict ourselves to matrix-matrix products, the solution of linear systems, and variants of the QR decomposition as long as possible. Only for small-scale problems, sophisticated algorithms like the QZ algorithm or direct solution techniques are utilized. From a mathematical point of view, the problem is small enough and the roots of the characteristic polynomial can be computed directly. The Abel-Ruffini Theorem [36] states that this is only possible for arbitrary polynomials of degree four and less. A hardware related characterization of small and trivial-to-solve generalized eigenvalue problems, is presented in Section 4.1.4 later.

3.1. Basic Concept

The key building block of our scheme is Theorem 2.6. Once we have found a deflating subspace $\mathcal{Z}_1$, we decompose the original generalized eigenvalue problem

$$Ax = \lambda Bx \tag{3.1}$$

for the matrix pair (A, B) into

$$\left(\underbrace{[Q_1 \ Q_2]}_{Q^{(1)}}^{\mathsf{H}} A \underbrace{[Z_1 \ Z_2]}_{Z^{(1)}} \right) x = \lambda \left(\underbrace{[Q_1 \ Q_2]}_{Q^{(1)}}^{\mathsf{H}} B \underbrace{[Z_1 \ Z_2]}_{Z^{(1)}} \right) x$$

$$\implies \begin{bmatrix} A_{11} & A_{12} \\ 0 & A_{22} \end{bmatrix} x = \lambda \begin{bmatrix} B_{11} & B_{12} \\ 0 & B_{22} \end{bmatrix} x, \tag{3.2}$$

where $\operatorname{span}(Z_1) = \mathcal{Z}_1$ and $\Lambda(A, B) = \Lambda(A_{11}, B_{11}) \cup \Lambda(A_{22}, B_{22})$. Now, the problem can be handled for (A_{11}, B_{11}) and (A_{22}, B_{22}) independently. If we find a deflating subspace $\mathcal{Z}_2$ for (A_{11}, B_{11}) (or (A_{22}, B_{22}) respectively) the Deflation Theorem 2.6 can be used again to split the matrix pair (A_{11}, B_{11}) into

$$\left(\underbrace{[\hat{Q}_1 \ \hat{Q}_2]}_{Q^{(2)}}^{\mathsf{H}} A_{11} \underbrace{[\hat{Z}_1 \ \hat{Z}_2]}_{Z^{(2)}}, \underbrace{[\hat{Q}_1 \ \hat{Q}_2]}_{Q^{(2)}}^{\mathsf{H}} B_{11} \underbrace{[\hat{Z}_1 \ \hat{Z}_2]}_{Z^{(2)}} \right)$$

$$= \left(\begin{bmatrix} \hat{A}_{11} & \hat{A}_{12} \\ 0 & \hat{A}_{22} \end{bmatrix}, \begin{bmatrix} \hat{B}_{11} & \hat{B}_{12} \\ 0 & \hat{B}_{22} \end{bmatrix} \right), \tag{3.3}$$

where $\operatorname{span}(\hat{Z}_1) = \mathcal{Z}_2$ and $\Lambda(A_{11}, B_{11}) = \Lambda\left(\hat{A}_{11}, \hat{B}_{11}\right) \cup \Lambda\left(\hat{A}_{22}, \hat{B}_{22}\right)$. Thereby, the application of $Q^{(2)}$ and $Z^{(2)}$ to the matrix pair (A_{11}, B_{11}) can be expressed in terms of the original matrix pair (A, B) using

$$\begin{bmatrix} Q^{(2)} & \\ & I_x \end{bmatrix}^{\mathsf{H}} Q^{(1)\mathsf{H}} A Z^{(1)} \begin{bmatrix} Z^{(2)} & \\ & I_x \end{bmatrix} = \begin{bmatrix} \hat{A}_{11} & \hat{A}_{12} & \\ & \hat{A}_{22} & * \\ & & A_{22} \end{bmatrix} \tag{3.4}$$

and

$$\begin{bmatrix} Q^{(2)} & \\ & I_x \end{bmatrix}^{\mathsf{H}} Q^{(1)\mathsf{H}} B Z^{(1)} \begin{bmatrix} Z^{(2)} & \\ & I_x \end{bmatrix} = \begin{bmatrix} \hat{B}_{11} & \hat{B}_{12} & \\ & \hat{B}_{22} & * \\ & & B_{22} \end{bmatrix}. \tag{3.5}$$

In this way, all applications of Theorem 2.6 on a subproblem (A_{11}, B_{11}) can be expressed in terms of higher-up levels. Together with Lemma 2.8 this enables the solution of the generalized eigenvalue problem (3.1) and the computation of the generalized Schur decomposition (2.21) by searching for a sequence of deflating subspaces $\mathcal{Z}_i$ and the recursive application of Theorem 2.6. The recursion stops for small problems (A_*, B_*), which are trivial-to-solve. Here, trivial means either the roots of the characteristic polynomial $\chi_{(A_*, B_*)}$ can be computed directly, compare the Abel-Ruffini Theorem [36], or the usage of the QZ algorithm is "cheap" in a hardware related sense. Details are discussed in Section 4.1.4.

Algorithm 1 Spectral Divide and Conquer Framework for the Generalized Schur Decomposition

---

Input: $(A, B) \in \mathbb{C}^{m \times m} \times \mathbb{C}^{m \times m}$
Output: Unitary $Q \in \mathbb{C}^{m \times m}$ and $Z \in \mathbb{C}^{m \times m}$ such that $(Q^H A Z, Q^H B Z)$ is generalized Schur form of (A, B)

1: **if** m is small enough **then**
2:     Compute Q and Z directly using the QZ algorithm [129]
3: **else**
4:     Determine a deflating subspace $\mathcal{Z}$ for (A, B).
5:     Compute Q_D and Z_D from $\mathcal{Z}$ to obtain

$$(Q_D^H A Z_D, Q_D^H B Z_D) = \left(\begin{bmatrix} \hat{A}_{11} & \hat{A}_{12} \\ 0 & \hat{A}_{22} \end{bmatrix}, \begin{bmatrix} \hat{B}_{11} & \hat{B}_{12} \\ 0 & \hat{B}_{22} \end{bmatrix} \right).$$

6:     Compute Q_L and Z_L from $\left(\hat{A}_{11}, \hat{B}_{11} \right)$ using Algorithm 1 again.
7:     Compute Q_R and Z_R from $\left(\hat{A}_{22}, \hat{B}_{22} \right)$ using Algorithm 1 again.
8:     Setup Q and Z via

$$Q = Q_D \begin{bmatrix} Q_L & \\ & Q_R \end{bmatrix} \quad \text{and} \quad Z = Z_D \begin{bmatrix} Z_L & \\ & Z_R \end{bmatrix}.$$

9: **end if**

The above procedure yields the divide-and-conquer approach shown in Algorithm 1. It generalizes the formulation of the algorithm used in our proof-of-concept for the fast approximation of the generalized Schur decomposition [27]. This general framework consists of two phases. First, a deflating subspace $\mathcal{Z}$ for the matrix pair (A, B) is required. We present two different ways to obtain a deflating subspace $\mathcal{Z}$ for Step 4 using the generalized matrix sign function or the matrix disc function in Sections 3.2 and 3.3. Second, the deflating transformations Q and Z have to be computed from the subspace $\mathcal{Z}$. A robust way to achieve this is presented in the next subsection. Since the spectrum of (A, B) only depends on the diagonal blocks, the parts $\hat{A}_{12}$ and $\hat{B}_{12}$ are not computed in Step 5 and not included in future considerations. If the eigenvectors are required, the $\hat{A}_{12}$ and $\hat{B}_{12}$ need to be updated, as well, or the initial matrix pair (A, B) needs to be multiplied with the final matrices Q and Z again. The later option results in large matrix-matrix products, which again utilize current computing hardware very well.

3.1.1. Construction of Q and Z from a Deflating Subspace

Theorem 2.6 does not describe how to construct the matrices Q and Z from a given deflating subspace $\mathcal{Z}$. Even the proof shown in [155] is not constructive. Following the ideas of Sun and Quintana-Ortí [162], we show an algorithm to obtain these matrices. This algorithm only involves operations that are mentioned at the beginning of this chapter. The main advantage of this approach is that only a right deflating subspace $\mathcal{Z}$ is required. Other approaches like the one by Bai, Demmel, and Gu [15] require the additional knowledge of the left deflating subspace Q to compute Q. In addition to the need for both, a right and a left deflating subspace, the separate computation neglects the interaction between Q and Z described in Corollary 2.7. This may cause numerical instabilities, e.g. the lower left block in $Q^H A Z$ or $Q^H B Z$ does not vanish exactly. Furthermore, it increases the computational cost drastically, since two subspaces are required.

Our goal is to find $Q = \begin{bmatrix} Q_1 & Q_2 \end{bmatrix}$ and $Z = \begin{bmatrix} Z_1 & Z_2 \end{bmatrix}$ from a given deflating subspace $\mathcal{Z}$, such that the spectrum is split as in Equation (3.2) (or in Theorem 2.6). The p-dimensional deflating subspace $\mathcal{Z}$ is given as kernel of a matrix $X \in \mathbb{C}^{m \times n}$, with $\ker X = \mathcal{Z}$ and $p \leq n$. We assume that p is not a priori known, here. The matrix X is provided by an external source, like the matrix sign function, which is discussed later in this chapter.

Now, the first p columns of Z, namely Z_1, must span the kernel of X. That means Z_1 fulfills:

$$X Z_1 = 0 \tag{3.6}$$

and the remaining Z_2 satisfies

$$X Z_2 \neq 0. \tag{3.7}$$

Combined, this yields that Z has to fulfill:

$$P_Z X \begin{bmatrix} Z_1 & Z_2 \end{bmatrix} = \begin{bmatrix} 0 & R_Z^{\mathsf{H}} \end{bmatrix}, \tag{3.8}$$

where P_Z is a permutation matrix and R_Z is an arbitrary matrix with rank $m-p$. The expression (3.8) is rearranged into

$$X^{\mathsf{H}} P_Z^{\mathsf{H}} = \begin{bmatrix} Z_1 & Z_2 \end{bmatrix} \begin{bmatrix} 0 \\ R_Z \end{bmatrix}$$

and by introducing an identity to rearrange the terms we obtain finally:

$$\begin{aligned} X^{\mathsf{H}} P_Z^{\mathsf{H}} &= \begin{bmatrix} Z_1 & Z_2 \end{bmatrix} \begin{bmatrix} & I \\ I & \end{bmatrix} \begin{bmatrix} & I \\ I & \end{bmatrix} \begin{bmatrix} 0 \\ R_Z \end{bmatrix} \\ &= \begin{bmatrix} Z_2 & Z_1 \end{bmatrix} \begin{bmatrix} R_Z \\ 0 \end{bmatrix}. \end{aligned} \tag{3.9}$$

Since Z needs to be unitary, we compute $[Z_2, Z_1]$, R_Z, and P_Z from X^{H} using the rank revealing QR decomposition [42, 43, 49, 54, 143, 162].

The matrix Q is partitioned into $[Q_1, Q_2]$ such that Q_2 spans the orthogonal complement of $A\mathcal{Z} + B\mathcal{Z}$, where $\mathcal{Z} = \operatorname{span} Z_1$ (see Corollary 2.7, [162]). Together with the previously computed matrix $Z = \begin{bmatrix} Z_1 & Z_2 \end{bmatrix}$, the matrix Q_2 has to fulfill

$$Q_2^{\mathsf{H}} \begin{bmatrix} A Z_1 & B Z_1 \end{bmatrix} = 0. \tag{3.10}$$

Since Q is required to have full rank the matrix Q_1 has to satisfy

$$Q_1^{\mathsf{H}} \begin{bmatrix} A Z_1 & B Z_1 \end{bmatrix} = R_Q, \tag{3.11}$$

where $\operatorname{rank}(R_Q) = \operatorname{rank}(Q_1) = m - \operatorname{rank}(Q_2) = p$.

Bringing (3.10) and (3.11) together and requiring Q to be unitary, this leads to a rank-revealing QR decomposition for Q:

$$Q^{\mathsf{H}} \begin{bmatrix} A Z_1 & B Z_1 \end{bmatrix} P_Q^{\mathsf{T}} = \begin{bmatrix} R_Q \\ 0 \end{bmatrix}, \tag{3.12}$$

where P_Q is a permutation matrix. Because the dimension p of the deflating subspace is not known a priori, we have to determine it from the rank of R_Z in (3.9) or use an externally provided information.

The involved rank-revealing QR decomposition can be realized using different schemes. We focus on three of them. First, the sophisticated RRQR decomposition of Bischof and Quintana-Ortí [42, 43] is used. Second, the column pivoted QR decomposition by Quintana-Ortí, Bischof, and Sun [143] is taken into account since it is a level-3 reformulation of the original Businger and Golub approach [49]. The third variant is the column pivoted QR decomposition with random sampling by Martinsson et. al. [125], which is a level-3 enabled algorithm and achieves a high performance on current hardware architectures. For a given matrix $X \in \mathbb{C}^{m \times n}$, $m \geq n$, the first two variants compute $XP^\mathsf{T} = QR$, where P is a permutation matrix, such that the absolute values of the diagonal elements decrease:

$$|R_{11}| \geq |R_{22}| \geq \ldots \geq |R_{nn}|. \tag{3.13}$$

Since the column pivoted variant with random sampling does not fulfill this condition, it is only used if the rank of X (and R) is known a priori. Both variants to determine a matrix X containing the deflating subspace $\mathcal{Z}$, namely the generalized matrix sign function presented in Section 3.2 and matrix disc function presented in Section 3.2, provide this feature. Alternatively, more expensive strategies, like singular value decomposition based approaches [80], are required. These will exceed the range of the desirable level-3 enabled operations and thus, are not covered in this work. Beside this approach, cheaper strategies to estimate p exist.

A simple criterion to determine the numerical rank p of R (and in this way the rank of X) is taking the smallest value k, $0 \leq k \leq n$, such that

$$|R_{kk}| < \tau_{QR}|R_{11}|. \tag{3.14}$$

Thereby, τ_{QR} is a tolerance factor chosen in the magnitude of the machine precision. A more reliable technique is based on the Incremental Condition Estimation (ICE) by Bischof [41], which approximates the largest and the smallest singular value of a triangular matrix in $O(m)$ operations. It is used in the RRQR decompositions [42] and advanced least squares solvers in LAPACK [10, 41], as well.

The order of the diagonal values in R as shown in (3.13) enables us to determine the rank by regarding the sequence

$$R^{(k)} = R_{1:k,1:k}, \quad 1 \leq k \leq n, \tag{3.15}$$

of upper left blocks of R. The ICE procedure estimates singular values $\sigma_{max}^{(k+1)}$ and $\sigma_{min}^{(k+1)}$ of $R^{(k+1)}$ from the singular values $\sigma_{max}^{(k)}$ and $\sigma_{min}^{(k)}$ of $R^{(k)}$ only accessing the newly added column $R_{:,k+1}$. The numerical rank of R is then set to the smallest value k such that the condition number of $R^{(k)}$ fulfills

$$\kappa_2\left(R^{(k)}\right) = \frac{\sigma_{max}^{(k)}}{\sigma_{min}^{(k)}} < \frac{1}{\tau_{ICE}}, \tag{3.16}$$

where τ_{ICE} is a tolerance factor again properly chosen according to the machine precision.

Remark 3.1:

Both tolerances τ_{QR} and τ_{ICE} may depend on the dimension of the matrix R as well. A commonly used value for a matrix $A \in \mathbb{C}^{m \times n}$ decomposed by one of the above QR decomposition schemes is

$$\tau_{QR} = \tau_{ICE} = \max(m, n) \cdot u, \tag{3.17}$$

where u is the machine precision.

A common problem of both rank decision strategies appears if the computed deflating subspace $\mathcal{Z}$ contains the whole space $\mathbb{C}^m$ or $\mathbb{R}^m$. In this case we have $\ker X = \mathbb{C}^m$ or $\ker X = \mathbb{R}^m$. This requires

---

Algorithm 2 Compute Q and Z from a given Deflating Subspace $\mathcal{Z}$

Input: $X \in \mathbb{C}^{m \times n}$ with $\ker X = \mathcal{Z}$, matrix pair (A, B) of order m.
Output: Unitary $Q \in \mathbb{C}^{m \times m}$ and $Z \in \mathbb{C}^{m \times m}$ such that (A, B) deflates as in Equation (3.2) with $\Lambda(A_{11}, B_{11})$ belongs to $\mathcal{Z}$.

1: Compute a rank-revealing QR decomposition of X^T:
$$X^\mathsf{T} P_Z^\mathsf{T} = \begin{bmatrix} Z_2 & Z_1 \end{bmatrix} R_Z$$

2: Determine the rank p of X^T using Incremental Condition Estimation on R_Z:
$$p = \operatorname{rank} R_Z$$

3: Compute Q from a rank-revealing QR decomposition:
$$\begin{bmatrix} AZ_1 & BZ_1 \end{bmatrix} P_Q = QR_Q$$

4: Build Z:
$$Z = \begin{bmatrix} Z_1 & Z_2 \end{bmatrix}$$

5: **return** Q and Z

X to be zero. Due to the finite precision computation of a computer we only obtain $X \approx 0$ and thus we only have $R \approx 0$. The diagonal elements of R are in magnitude of the machine precision and both rank detection criteria will probably fail. In this case, external rank sources are required as in the case of the column pivoted QR decomposition with random sampling. If R has either full rank or rank zero, we set the matrices Q and Z to the identity, but our algorithm stagnates, and we have to search for a new deflating subspace.

The overall procedure to compute deflating transformations Q and Z out of a matrix X spanning a deflating subspace $\mathcal{Z}$ is shown in Algorithm 2. Except of the rank determination, whose complexity depends on the way it is computed, all operations are available as level-3 enabled algorithms. For performance reasons, we use the column pivoted QR [143] decomposition in practical implementations although there are artificial examples, where this procedure fails. This problem is minimized if the used QR decomposition is aware of underflows in the downdate of the column norms [68]. LAPACK implements this since version 3.1 and thus it is included in all newer versions. The column pivoted QR decomposition with random sampling [125] can be used as accelerated replacement, if a computationally cheap external rank decision is available.

3.1.2. Decoupling Residual

The determination of the numerical rank of R_Z is a procedure based on a heuristic, which may fail. Especially, this happens if there is no large gap in the absolute value of the diagonal elements of R_Z. Another reason to fail is a wrong selection of the tolerances τ_{QR} or τ_{ICE}. If they are too small or too large, the rank of the matrix R_Z is over or underestimated. Because the selection of both quantities and the behavior of the absolute values on the diagonal of R_Z depends strongly on the input data, there is no proper algorithmic way to guarantee a suitable determination of the rank here. If we rely on an external rank determination, this can be inaccurate as well. In order to detect the error resulting from a wrong rank decision, we check the defining property of the deflation, i.e. that Q_2 spans the orthogonal complement of $A\mathcal{Z} + B\mathcal{Z}$. Regarding the matrix pair (A, B), this means that for

$$Q^H(A, B)Z = \left(\begin{bmatrix} A_{11} & A_{12} \\ E & A_{22} \end{bmatrix}, \begin{bmatrix} B_{11} & B_{12} \\ F & B_{22} \end{bmatrix} \right) \qquad (3.18)$$

the blocks E and F need to be zero. Since we are not able to use infinite precision arithmetic, these blocks are only close to zero. This motivates the introduction of the *relative decoupling residual* [162]:

$$\text{rdr} = \left\| \begin{bmatrix} A & B \end{bmatrix} \right\|_F^{-1} \left\| \begin{bmatrix} E & F \end{bmatrix} \right\|_F. \qquad (3.19)$$

This allows an a posteriori check if the computation of Q and Z results in a proper deflation of the matrix pair (A, B) with respect to a given deflating subspace and its determined dimension. If the relative decoupling residual is larger than a threshold τ_{rdr}, the computation of Q and Z have not been successful due to numerical inaccuracies. A typical value for τ_{rdr} is

$$\tau_{\text{rdr}} = 100 \cdot m \cdot u \qquad (3.20)$$

similar to τ_{ICE} and τ_{QR} in Remark 3.1. The additional security factor 100 is added to capture small numerical instabilities. If the tolerance is exceeded, the spectral division failed and the eigenvalue problem need to be handled with a different algorithm, like the QZ algorithm.

If the matrix R_Z has rank zero or full rank, the relative decoupling residual cannot be used since the blocks E and F in Equation (3.18) do not exist and (A_{11}, B_{11}) or (A_{22}, B_{22}) is of order m. In this case, the successful computation of the deflating transformation cannot be checked. Then, we have to rely on the fact that our external rank decision obtained from the deflating subspace computation in Sections 3.2 and 3.3 provides this information correctly.

3.1.3. Infinite Eigenvalues

The theory about generalized eigenvalue problems in Chapter 2 shows that, in contrast to the standard eigenvalue problem, infinite eigenvalues can appear. From a theoretical point of view these eigenvalues do not affect the ideas behind Algorithm 1 or the computation of Q and Z. Nevertheless, the determination of the deflating subspace $\mathcal{Z}$ and the transformation of the spectra (see Section 2.4) might cause problems. In order to remove the infinite eigenvalues, we have to compute deflating transformations Q_∞ and Z_∞ such that

$$(Q_\infty^H A Z_\infty, Q_\infty^H B Z_\infty) = \left(\begin{bmatrix} A_\infty & * \\ & A_{\text{fin}} \end{bmatrix}, \begin{bmatrix} B_\infty & * \\ & B_{\text{fin}} \end{bmatrix} \right), \qquad (3.21)$$

where $\Lambda(A_\infty, B_\infty) = \{\pm\infty\}$ and $\pm\infty \notin \Lambda(A_{\text{fin}}, B_{\text{fin}})$. The following theorem shows that such decompositions are constructible although Theorem 2.9 does not provide any information about the order of the eigenvalues.

Theorem 3.2:

Let (A, B) be a matrix pair of order m, with A nonsingular and B singular. Then $\infty \in \Lambda(A, B)$ is an eigenvalue of (A, B) and unitary matrices $Q \in \mathbb{C}^{m \times m}$ and $Z \in \mathbb{C}^{m \times m}$ exists, such that $B_{11} = 0$ in Equation (2.13) holds. Furthermore, if B has rank p then B_{11} is a $m - p \times m - p$ matrix and the matrix pair (A_{22}, B_{22}) is of order p.

Proof. From Theorem 2.9, it follows that there exist Q and Z such that

$$(Q^H A Z, Q^H B Z) = (T, S),$$

where T and S are upper triangular. $\infty \in \Lambda(A, B)$ implies that there is at least one k, $1 \le k \le m$, such that $S_{kk} = 0$. This yields $p = \text{rank}(S) = \text{rank}(B) < m$.

Algorithm 3 Detect and Remove Infinite Eigenvalues from (A, B)

Input: Matrix pair (A, B) of order m.
Output: Unitary $Q_\infty \in \mathbb{C}^{m \times m}$ and $Z_\infty \in \mathbb{C}^{m \times m}$ such that (A, B) deflates as in Equation (3.21) and p
    is the number of infinite eigenvalues.
1: $p = 0$, $Q_\infty = I$, $Z_\infty = I$, $\hat{A} = A$, and $\hat{B} = B$.
2: **repeat**
3:     Compute Q_∞ and Z_∞ such that

$$\left(\hat{Q}_\infty^H \hat{A} \hat{Z}_\infty, \hat{Q}_\infty^H \hat{B} \hat{Z}_\infty \right) = \left(\begin{bmatrix} A_\infty & * \\ & A_{\text{fin}} \end{bmatrix}, \begin{bmatrix} B_\infty & * \\ & B_{\text{fin}} \end{bmatrix} \right)$$

    from $\mathcal{Z} = \ker \hat{B}$ using Algorithm 2, where (A_∞, B_∞) is of order k.
4:     Set $\hat{A} = A_{\text{fin}}$ and $\hat{B} = B_{\text{fin}}$.
5:     Update $Q_\infty \leftarrow Q_\infty \begin{bmatrix} I_p & \\ & \hat{Q}_\infty \end{bmatrix}$ and $Z_\infty \leftarrow Z_\infty \begin{bmatrix} I_p & \\ & \hat{Z}_\infty \end{bmatrix}$
6:     $p \leftarrow p + k$.
7: **until** $k = 0$

Let span $Z_1 = \ker B$ and span $Z_2 = (\text{span } Z_1)^\perp$. Then $Z = \begin{bmatrix} Z_1 & Z_2 \end{bmatrix}$ has rank m and without loss of generality Z can be chosen to be unitary. Furthermore, we can find a matrix $Q_2 \in \mathbb{C}^{m \times p}$ with

$$Q_2^H A Z_1 = 0 \quad \text{and} \quad Q_2^H B Z_1 = 0,$$

where the second condition is trivially fulfilled. Setting span $Q_1 = (\text{span } Q_2)^\perp$ provides a full rank matrix $Q = \begin{bmatrix} Q_1 & Q_2 \end{bmatrix}$. The matrices Q_1 and Q_2 can be chosen to be unitary. By this construction we have

$$Q^H A Z_1 \neq 0 \quad \text{and} \quad Q^H B Z_1 = 0,$$

since A is nonsingular. $\qquad\qquad\qquad\qquad\qquad\qquad\qquad\qquad\qquad\qquad\qquad\qquad\quad\square$

Remark 3.3:
Theorem 3.2 and its proof present a constructive way to remove infinite eigenvalues. Unfortunately, the theorem does not guarantee that the remaining matrix pair (A_{22}, B_{22}) does not have infinite eigenvalues anymore. If infinity is still contained in the $\Lambda(A_{22}, B_{22})$, the theorem needs to be applied again.

The previous proof shows that the right deflating subspace corresponding to the infinite eigenvalue is spanned by $\ker B$. This fits with the input data for Algorithm 2 which uses the kernel of the input matrix as deflating subspace. Since Theorem 3.2 does not guarantee to get all infinite eigenvalues at once we have to apply it iteratively as shown in Algorithm 3. This is a consequence of Remark 3.3. A different approach to remove the infinite eigenvalues is presented by Kågström and Van Dooren [106] and is implemented in the Model Order Reduction Laboratory (MORLAB) [35]. This approach derives from a system theoretic application and uses a set of numerically expensive functions, like the QZ algorithm and the singular value decomposition, which are avoided here to obtain a fast algorithm. If $\hat{B}$ has full rank, Algorithm 2 only performs one QR decomposition to detect this and sets $Q = I$ and $Z = I$. Then, a single step of Algorithm 3 only costs $4/3(m - p)^3$ flops. In the worst case, where $\hat{B}$ has a very small rank, the number of flops is bounded from above by $80/3(m - p)^3$. Compared to the QZ algorithm which takes $66m^3$ operations [80, 129], all operations in Algorithm 2 and 3 are level-3 BLAS enabled. Thus, our approach is well suited for multi-core and GPU accelerated systems.

After we have shown how to use Theorem 2.6 and ker B as deflating subspace to remove infinite eigenvalues, we will discuss more advanced strategies. The different ways to determine a deflating subspace of the matrix pair (A, B) with respect to properties of its spectrum are presented in the next sections.

3.2. Deflating Subspaces via the Matrix Sign Function

The first way to compute a deflating subspace of (A, B) splits the spectrum of (A, B) at the imaginary axis. To this end, we will use the matrix sign function and its properties since the involved computations heavily rely on level-3 operations.

The matrix sign function was originally introduced by Roberts to solve algebraic Riccati equations [144]. There, its properties concerning the distribution of the eigenvalues play an import role.

Definition 3.4 (Matrix Sign Function, e.g. [20, 63]):
Let $A \in \mathbb{C}^{m \times m}$ be a matrix without eigenvalues on the imaginary axis and

$$A = S \begin{bmatrix} J_- & \\ & J_+ \end{bmatrix} S^{-1} \tag{3.22}$$

its Jordan Canonical form, where the Jordan block J_- and J_+ correspond to the eigenvalues in the left and in the right half plane. Then the **matrix sign function** of A is given by

$$\text{sign}(A) = S \begin{bmatrix} -I_- & \\ & I_+ \end{bmatrix} S^{-1}, \tag{3.23}$$

where I_- and I_+ are identities of the same size as J_- and J_+.

The matrix sign function provides a set of important properties that make it useful in the context of eigenvalue computations. The ones, which are of interest in our application, are given in the next theorem.

Theorem 3.5 (e.g. [109]):
Let $A \in \mathbb{C}^{m \times m}$ be a matrix without purely imaginary eigenvalues and $\text{sign}(A)$ its sign. Then the following conditions are fulfilled:

1. $\text{sign}(A)$ is diagonalizable with eigenvalues ± 1.

2. $\text{sign}(A)^2 = I$.

3. For a nonsingular $V \in \mathbb{C}^{m \times m}$ it holds $\text{sign}(VAV^{-1}) = V \text{sign}(A) V^{-1}$.

4. $P_- = \frac{1}{2}(I - \text{sign}(A))$ is a projection onto the invariant subspace for eigenvalues with negative real part.

5. $P_+ = \frac{1}{2}(I + \text{sign}(A))$ is a projection onto the invariant subspace for eigenvalues with positive real part.

Proof. See [109, 89, 144]. $\qquad\square$

The projection properties presented in the last theorem provides a strategy to obtain a subspace as input for our algorithmic framework if we can transfer them from the standard eigenvalue problem to the generalized one. Furthermore, the projection properties allow us to determine the dimension of the subspaces for P_+ and P_-. This can be used to determine whether the computed deflating subspace is empty or spans the full space. Furthermore, it is used as external rank source, compare the heuristic in Section 3.1.1. In this way, we also enable the column pivoted QR decomposition with random sampling to be used in the subspace extraction.

Lemma 3.6 (e.g. [13]):

Let $A \in \mathbb{C}^{m \times m}$ be a matrix without purely imaginary eigenvalues and $\mathrm{sign}\,(A)$ its sign. Then it holds:

1. *$p_+ = \frac{1}{2}\left(m + \mathrm{tr}(\mathrm{sign}\,(A))\right)$ is the number of eigenvalues with positive real part*

2. *and $p_- = \frac{1}{2}\left(m - \mathrm{tr}(\mathrm{sign}\,(A))\right) = m - p_+$ is the number of eigenvalues with negative real part.*

P_+ and P_-, defined in Theorem 3.5, are of rank p_+ and p_-, respectively.

Proof. The projections P_+ and P_- defined in Theorem 3.5 are idempotent by definition. For each idempotent matrix P it holds [16]
$$\mathrm{tr}(P) = \mathrm{rank}(P)\,.$$

It follows:

$$\mathrm{rank}(P_+) = \mathrm{tr}(P_+) = \mathrm{tr}\left(\frac{1}{2}\left(I + \mathrm{sign}\,(A)\right)\right) = \frac{1}{2}\left(\mathrm{tr}(I) + \mathrm{tr}(\mathrm{sign}\,(A))\right)$$

$$= \frac{1}{2}\left(m + \mathrm{tr}(\mathrm{sign}\,(A))\right) = p_+,$$

which proves the lemma for p_+. The second case works straight forward on P_-. $\qquad\square$

Remark 3.7:

Even if the evaluation of p_+ or p_- is not an integer due to round off errors or numerical instabilities its nearest integer value can be used as educated guess for the rank decision in Formulas (3.14) and (3.16).

Computation   Typically, the matrix sign function is computed by applying Newton's method to the property
$$\mathrm{sign}\,(A)^2 = I. \tag{3.24}$$

This leads to an easy to implement iterative scheme [89, 109]:

$$X_0 = A$$
$$X_{k+1} = \frac{1}{2}\left(X_k + X_k^{-1}\right) \quad \text{for } k = 0, 1, \ldots, \tag{3.25}$$

which converges to $\mathrm{sign}\,(A) = \lim_{k \to \infty} X_k$ if A has no purely imaginary eigenvalues [144]. The convergence is globally quadratic [89]. In this iterative scheme, the inverse of a matrix has to be computed explicitly [88]. The convergence can be accelerated by several scaling strategies on X_k, which changes the iteration to:

$$X_0 = A$$
$$X_{k+1} = \frac{1}{2}\left(\gamma_k X_k + \gamma_k^{-1} X_k^{-1}\right) \quad k = 0, 1, \ldots, \tag{3.26}$$

where $\gamma_k \in \mathbb{R}$ is a scaling factor. The literature [13, 89, 108, 109] provides the following values for γ_k:

- *Determinantal Scaling:* $\gamma_k = |\det X_k|^{-\frac{1}{m}}$,

- *Spectral Scaling:* $\gamma_k = \sqrt{\dfrac{\rho\left(X_k^{-1}\right)}{\rho\left(X_k\right)}}$,

- or *Norm Scaling:* $\gamma_k = \sqrt{\dfrac{\left\|X_k^{-1}\right\|}{\left\|X_k\right\|}}$.

Thereby, only the determinantal scaling and the norm scaling are used in practice since the determinant is obtained from the inversion procedure and the norm can be selected freely to avoid the expensive 2-norm.

3.2.1. Generalized Matrix Sign Function

The generalization of the matrix sign function to matrix pairs (A, B) was shown by Gardiner and Laub in [77] in order to solve a generalized algebraic Riccati equation. For a given matrix pair (A, B) they assume B to be factorized as $B = B_L B_R$. Now, the Newton iteration (3.26) is performed for $X_0 = B_L^{-1} A B_R - 1$. By defining $Y_k = B_L X_k B_R$, the iteration (3.25) becomes:

$$Y_0 = B_L X_0 B_R = A$$

$$
\begin{aligned}
Y_{k+1} = B_L X_k B_R &= \frac{1}{2} B_L \left(\gamma_k X_k + \gamma_k^{-1} X_k^{-1}\right) B_R \\
&= \frac{1}{2}\left(\gamma_k B_L X_{k+1} B_R + \gamma_k^{-1} B_L \left(B_L^{-1} Y_k B_R^{-1}\right)^{-1} B_R\right) \\
&= \frac{1}{2}\left(\gamma_k Y_k + \gamma_k^{-1} B_L B_R Y_k^{-1} B_L B_R\right) \\
&= \frac{1}{2}\left(\gamma_k Y_k + \gamma_k^{-1} B Y_k^{-1} B\right), \quad k = 0, 1, \dots
\end{aligned}
\tag{3.27}
$$

This scheme converges globally quadratic to $\lim\limits_{k\to\infty} Y_k = B \operatorname{sign}\left(B^{-1}A\right)$ [77, 162]. Again, the scaling factor γ_k is used to accelerate the convergence. In contrast to the scaling factors presented for the computation of $\operatorname{sign}(A)$, only a modified determinantal scaling is known for matrix pairs [77]. In this case the scaling factor γ_k is set to:

$$\gamma_k = \left(\frac{|\det B|}{|\det Y_k|}\right)^{\frac{1}{m}}.\tag{3.28}$$

Recalling that $\Lambda(A, B) = \Lambda\left(B^{-1}A\right)$ (as long as B is invertible) and using Newton's method for $B_L^{-1} A B_R^{-1}$ from (3.27) leads to a generalized definition of the sign of a matrix pair. Thereby the factorization $B_L = B$ and $B_R = I$ is chosen.

Definition 3.8 ([77]):

*Let (A, B) be a regular matrix pair of order m without infinite eigenvalues. Assuming that (A, B) has no eigenvalues on the imaginary axis. Then the **generalized matrix sign function** $\operatorname{sign}(A, B)$ is defined as*

$$\operatorname{sign}(A, B) = B \operatorname{sign}\left(B^{-1}A\right).\tag{3.29}$$

Remark 3.9:

An alternative but compatible variant of the definition of $\operatorname{sign}(A, B)$ is given by Sun and Quintana-Ortí [162], which extends Definition 3.4: For a given matrix pair (A, B) there exist nonsingular matrices $U \in \mathbb{C}^{m\times m}$ and $V \in \mathbb{C}^{m\times m}$ such that

$$V^{-1}AU = \begin{bmatrix} A_{11} & \\ & A_{22} \end{bmatrix} \quad and \quad V^{-1}BU = \begin{bmatrix} B_{11} & \\ & B_{22} \end{bmatrix},$$

where $\Lambda(A_{11}, B_{11})$ lies in the open left half plane and $\Lambda(A_{22}, B_{22})$ lies in the open right half plane. Then the generalized matrix sign function is given by

$$\text{sign}(A, B) = V \begin{bmatrix} -B_{11} & \\ & B_{22} \end{bmatrix} U^{-1}. \tag{3.30}$$

In order obtain a deflating subspace $\mathcal{Z}$ from the generalized matrix sign function as input for Algorithm 1 we have to extend the properties collected in Theorem 3.5 and Lemma 3.6 to the case of the matrix pair (A, B).

Corollary 3.10:

Let (A, B) be a matrix pair of order m without purely imaginary eigenvalues. Furthermore, let B be nonsingular. Then the generalized matrix sign function S=sign(A, B) fulfills the following conditions:

1. $\Lambda(S, B) \subseteq \{-1, 1\}$,

2. *the matrix pair (S, B) has the same deflating subspaces as (A, B),*

3. $P_+ = \frac{1}{2}B^{-1}(B + S)$ *is a projector onto the right deflating subspace corresponding to the eigenvalues with positive real part,*

4. $P_- = \frac{1}{2}B^{-1}(B - S)$ *is a projector onto the right deflating subspace corresponding to the eigenvalues with negative real part,*

5. $p_+ = \frac{1}{2}\left(m + \text{tr}(B^{-1}S)\right)$ *is the number of eigenvalues in the right half plane,*

6. *and $p_- = \frac{1}{2}\left(m - \text{tr}(B^{-1}S)\right)$ is the number of eigenvalues in the left half plane.*

Proof. For properties 1 and 2 see Gardiner and Laub [77] and for properties 3 and 4 see Sun and Quintana-Ortí in [162]. The properties 5 and 6 directly follow from 3 and 4 together with Lemma 3.6. □

Remark 3.11:

The assumption that B is nonsingular is no restriction for the computation of the deflating subspace. Although it is shown that the Newton scheme (3.27) loses its quadratic convergence if B is singular, we know that a singular matrix B means to have infinite eigenvalues in $\Lambda(A, B)$. Theorem 3.2 together with Algorithm 3 shows how to remove them from the spectrum of a matrix pair. This yields a matrix pair $\left(A_{fin}, B_{fin}\right)$, where B_{fin} is nonsingular and Corollary 3.10 holds again.

3.2.2. Spectral Splitting

Following Corollary 3.10 we can obtain a right deflating subspace for the matrix pair (A, B) from P_+ or P_-. Unfortunately, the computation of these projectors requires the solution of a linear system with B, which might be ill conditioned. In order to avoid this additional solution step, several rank revealing strategies were developed [42, 43, 143]. These ideas coincide with the computation of a deflating transformation Q and Z from a given deflating subspace $\mathcal{Z}$ in Subsection 3.1.1.

Following the ideas from Sun and Quintana-Ortí [161], we see that the projector kernel ker P_+ is obviously equivalent to the range of P_- and vice versa. It follows

$$\ker P_+ = \text{range}(P_-) \quad \text{and} \quad \ker P_- = \text{range}(P_+) \tag{3.31}$$

and finally we have

$$\ker\left(B + \text{sign}(A, B)\right) = \text{range}\left(B - \text{sign}(A, B)\right) \tag{3.32}$$

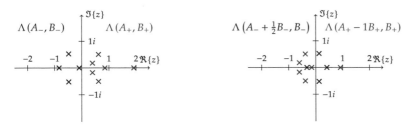

Figure 3.1.: Split spectrum of a matrix pair (A, B).

Figure 3.2.: Shifted split spectrum of a matrix pair (A, B).

together with

$$\ker\left(B - \text{sign}\,(A, B)\right) = \text{range}\left(B + \text{sign}\,(A, B)\right). \tag{3.33}$$

If we now set the input matrix X of Algorithm 2 to

$$X = B + \text{sign}\,(A, B), \tag{3.34}$$

we can split the spectrum of the matrix pair (A, B) at the imaginary axis. The resulting matrix pairs (A_{11}, B_{11}) and (A_{22}, B_{22}) have spectra in the open left or open right half plane, respectively. Their positions can be swapped if $X = B - \text{sign}\,(A, B)$ is used in the computation of Q and Z.

By embedding this idea into the general spectral divide-and-conquer framework from Section 3.1 we end up with the following situation: After computing $\text{sign}\,(A, B)$ and the transformation matrices Q and Z from $X = B + \text{sign}\,(A, B)$ we obtain

$$(Q^H A Z, Q^H B Z) = \left(\begin{bmatrix} A_- & * \\ & A_+ \end{bmatrix}, \begin{bmatrix} B_- & * \\ & B_+ \end{bmatrix}\right) \tag{3.35}$$

with $\Lambda\,(A_-, B_-) \subset \mathbb{C}_-$ and $\Lambda\,(A_+, B_+) \subset \mathbb{C}_+$. If the generalized matrix sign function is applied to (A_-, B_-) (and (A_+, B_+)) again, we obtain the whole space as deflating subspace, i.e. we cannot use the generalized matrix sign function in a recursive manor. Figure 3.1 shows this for an example matrix.

In order to obtain a recursive scheme, as it was the idea behind Algorithm 1, we have to modify the matrix pairs (A_+, B_+) and (A_-, B_-) such that we again obtain nontrivial deflating subspaces from $\text{sign}\,(A_+, B_+)$ and $\text{sign}\,(A_-, B_-)$. Without loss of generality we only regard (A_+, B_+) here.

After splitting the spectrum into the open left and open right half plane using the deflating subspace $\mathcal{Z} = \ker\left(B + \text{sign}\,(A, B)\right)$ we have

$$\Lambda\,(A_+, B_+) \subset \mathbb{C}_+. \tag{3.36}$$

In order to apply the spectral splitting at the imaginary axis again we need

$$\Lambda\,(A_+, B_+) \cap \mathbb{C}_- \neq \emptyset. \tag{3.37}$$

Therefore, we can use the eigenvalue transformations presented in Section 2.4. The shift function from Subsection 2.4.3 is suitable for the spectral splitting since it moves the eigenvalues along the real axis. In contrast to the scaling function it allows moving eigenvalues across the imaginary axis. An example is shown in Figure 3.2. The best split with respect to the depth of the recursion is achieved if the numbers of eigenvalues in the left and in the right half-plane are equal. Because eigenvalues with the same real part are moved the same distance along the real axis, the equality cannot be reached exactly in some cases. Hence, the optimal shift parameter $\theta \in \mathbb{R}$ should fulfill

$$|\Lambda (A_+ + \theta B_+) \cap \mathbb{C}_-| \approx |\Lambda (A_+ + \theta B_+) \cap \mathbb{C}_+|. \tag{3.38}$$

This requires the knowledge of $\Lambda (A_+, B_+)$ to select θ and thus the solution of a large generalized eigenvalue problem, which we have to avoid here. The ideas in [27] suggest computing the spectral abscissa, i.e. the real part of the rightmost eigenvalue, of (A_+, B_+)

$$\lambda_{\text{right}} (A_+, B_+) = \max_{\lambda \in \Lambda(A_+, B_+)} \mathfrak{R} (\lambda) \tag{3.39}$$

and use

$$\theta = -\frac{1}{2}\lambda_{\text{right}} (A_+, B_+) \tag{3.40}$$

as shift parameter. This yields a fast recursion if all eigenvalues of (A_+, B_+) are distributed equally along the real axis between 0 and λ_{right}. Analogously, the second problem (A_-, B_-) is handled using the real part of the leftmost eigenvalue

$$\lambda_{\text{left}} (A_-, B_-) = \min_{\lambda \in \Lambda(A_-, B_-)} \mathfrak{R} (\lambda) \tag{3.41}$$

and the shift parameter

$$\theta = -\frac{1}{2}\lambda_{\text{left}} (A_-, B_-). \tag{3.42}$$

Since, both values $\lambda_{\text{left}} (A_-, B_-)$ and $\lambda_{\text{right}} (A_+, B_+)$ are not available in general, the next section will show an easy approximation strategy.

3.2.3. The Divide-Shift-and-Conquer Algorithm

The previous section shows how to use the generalized matrix sign function to split spectra at the imaginary axis and how to modify the emerging smaller subproblems such that a recursive application of this scheme is possible. Thereby, the knowledge of the leftmost or the rightmost eigenvalue is required to obtain the next subproblems. Since the leftmost and rightmost eigenvalues are difficult to compute, we consider (A_+, B_+) and provide a bound to the rightmost eigenvalue by

$$\lambda_{\text{right}} \leq \rho (A_+, B_+) = \max \{|\lambda| \mid \lambda \in \Lambda (A_+, B_+)\}, \tag{3.43}$$

where $\rho (A_+, B_+)$ is the spectral radius of (A_+, B_+). The spectral radius is bounded from above by common matrix norms

$$\rho (A_+, B_+) \leq \rho_{\text{appx}} (A_+, B_+) = \begin{cases} \left\| B_+^{-1} A_+ \right\|_1 \\ \left\| B_+^{-1} A_+ \right\|_2 \\ \left\| B_+^{-1} A_+ \right\|_\infty \\ \left\| B_+^{-1} A_+ \right\|_F \end{cases} . \tag{3.44}$$

The two-norm is easy to estimate using iterative eigensolvers [145] and software packages like ARPACK [120]. Although ARPACK is made for large and sparse problems, it allows a fast and reliable estimation for dense matrices as well since it only requires a sophisticated matrix-vector product. This avoids to setup $B_+^{-1} A_+$ in this case. Furthermore, ARPACK allows to estimate the eigenvalue with the largest real part as well. If this procedure converges we use this result instead of one of the above approximations. However, if the shift parameter is computed using any of the mentioned approximations in every recursive step, the following problem can appear. Let λ^* be an eigenvalue of (A_+, B_+) fulfilling

$$|\lambda^*| = \rho(A_+, B_+) \quad \text{and} \quad \Re(\lambda^*) < \frac{1}{2}\lambda_{\text{right}}.$$

Furthermore, the rightmost eigenvalue fulfills

$$2\lambda_{\text{right}} < \rho(A_+, B_+).$$

If the shift parameter θ is set to $\theta = -\frac{1}{2}\rho(A_+, B_+)$, the spectrum of $(A_+ + \theta B_+, B_+)$ lies completely in the left half plane and it holds

$$\rho(A_+, B_+) < \rho(A_+ + \theta B_+, B_+).$$

Thus, in the next recursion step the whole problem is moved from the left half plane to the right once again. In this way the spectrum is no longer split using the generalized matrix sign function and the recursion stagnates. We avoid this issue and the multiple approximation of the spectral radius by a scheme, which guarantees that the sequence of shift parameters decreases.

We regard (A_+, B_+) (or (A_-, B_-)) after the first spectral division and assume

$$\theta \leq -\frac{1}{2}\lambda_{\text{right}}(A_+, B_+) \quad \left(\text{or } \theta \geq -\frac{1}{2}\lambda_{\text{left}}(A_-, B_-)\right). \tag{3.45}$$

Then the spectrum $\Lambda(S)$ of $S = (A_+ + \theta B_+, B_+)$ (or $S = (A_- + \theta B_-, B_-)$) fulfills

$$\Re(\Lambda(S)) \subset [-|\theta|, |\theta|]. \tag{3.46}$$

This gives $-|\theta| \leq \lambda_{\text{left}}(S)$ and $\lambda_{\text{right}}(S) \leq |\theta|$. Hence, setting $\theta = -\frac{1}{2}\rho(A_+, B_+) \leq -\frac{1}{2}\lambda_{\text{right}}(A_+, B_+)$ (or $\theta = \frac{1}{2}\rho(A_-, B_-) \geq -\lambda_{\text{left}}$) for the first spectral division let us bound the left and right most eigenvalue of the shifted problem with the help of Equation (3.46). Thus, we do not need to compute a new approximation of bounds in the newly emerged subproblems. On the one hand, we only have to compute the approximation to the left and right most eigenvalues after the first spectral division step. On the other hand, this scheme avoids the over-estimation described before. If an initial approximation of the overall spectral radius for (A, B) is known, this can be used applying the same strategy to avoid the approximation of $\rho(A_+, B_+)$ and $\rho(A_-, B_-)$ in the first spectral division step.

Multiple Eigenvalues. A general problem of the spectral splitting idea is that eigenvalues with a multiplicity larger than one or eigenvalues with the same real part lead to an infinite recursion, since the splitting using this ansatz does not work. Especially if the input matrices are real this appears in terms of complex conjugate eigenvalue pairs. A similar situation occurs if clusters of eigenvalues appear in the spectrum. In this case, many recursion steps are required to separate them. Therefore, we have to specify a set of situations where we switch back to the QZ algorithm [129] to avoid a long-running or infinite recursion.

The complex conjugate eigenvalue pairs are covered by our definition of a trivial-to-solve generalized eigenvalue problem in Section 3.1. This covers all problems of order $m \leq 4$. For eigenvalues with a larger multiplicity than four or the same real part, the recursion is stopped if the shift parameter θ gets too small, i.e. all remaining eigenvalues lie in a small bar $[-|\theta|, |\theta|]$ surrounding the imaginary axis. This includes clusters of eigenvalues as well. Again we use the QZ algorithm in this case. A criterion whether θ is small enough uses the following theorem:

Theorem 3.12:
Let $m = 2^M$, $M \in \mathbb{N}$, (A, B) be a matrix pair of order m. Assume the recursion splits it in two equally sized problems in each step. Furthermore, let $m_t = 2^{M_t}$, $M_t \in \mathbb{N}$ be the trivial problem size to use a direct solver. Then the recursion stops after

$$d(m, m_t) = \log_2 \frac{m}{m_t}, \quad m \leq m_t \tag{3.47}$$

steps.

Proof. If $M = M_t$ then we solve the problem directly and

$$d(m_t, m_t) = \log_2 \frac{m_t}{m_t} = \log_2 1 = 0$$

recursive steps are done. Since we require an equal distribution of the eigenvalues along the spectra $d(\cdot, \cdot)$ for both subproblems. Now, let $M = M_t + 1$ and after the first spectral division we have two equal sized trivial problems and it took

$$d(m, m_t) = 1 + d(m_t, m_t) = 1$$

steps in the recursion. The induction $M \to M + 1$ requires one additional deflation step to split the problem into two problems. This yields

$$d(2m, m_t) = 1 + d(m, m_t) = 1 + \log_2 \frac{m}{m_t} = \log_2 2\frac{m}{m_t}$$

which completes the proof. □

Using the maximum depth of the optimal recursion tree from Theorem 3.12, we formulate a stopping criterion on top of this. If the depth of the recursion tree gets too large, that means there are too many non equally sized spectral divisions, we stop the recursion and solve the remaining problems with the help of the QZ algorithm. Using the optimal recursion depth from Theorem 3.12 as lower bound, we use

$$d \geq \tau_{\text{rec}} \log_2 \frac{m}{m_t} \tag{3.48}$$

as stopping criterion based on the current recursion depth d. Thereby, $\tau_{\text{rec}} \geq 1$ is a tolerance factor to allow more than the optimal number of recursions steps.

Other stopping criteria. Depending on the involved operations, we define other stopping criteria for the recursion. These are:

- The computation of $\text{sign}(A, B)$ failed, i.e. the Newton iteration (3.27) does not converge within a maximum number of iteration steps. Furthermore, the Newton iteration fails if the matrices A and B are too ill-conditioned [161, 162].

- The spectral division based on Algorithm 2 does not work, i.e. the dimension of the deflating subspace could not be determined properly.

- The relative decoupling residual (3.19) detects a failed deflation.

In all cases, we solve the remaining subproblem with the help of the QZ algorithm.

The trivial-to-solve property, which is until now defined in terms of the direct computation of the roots of the characteristic polynomial, is a stopping criteria as well. The hardware related

---

Algorithm 4 Divide-Shift-and-Conquer Algorithm

Input: $A \in \mathbb{C}^{m \times m}$ and $B \in \mathbb{C}^{m \times m}$ nonsingular, $\Lambda(A,B) \cap \imath\mathbb{R} = \{\}$,
  $\theta \in \mathbb{R}$ spectral radius approximation, and recursion depth d (both 0 at start).
Output: $Q \in \mathbb{C}^{m \times m}$ and $Z \in \mathbb{C}^{m \times m}$ unitary, such that $(Q^H AZ, Q^H BZ)$ as in (2.21).

1: **if** m small enough or $d \geq \tau_{\text{rek}} \log_2 \frac{m}{m_l}$ **then**
2:   Compute Q, Z directly using QZ algorithm.
3:   **return** $[Q, Z]$
4: **end if**
5: Compute $S = \text{sign}(A, B)$ using the iteration (3.27).
6: Compute Q_D and Z_D from $X = B + S$ using Algorithm 2.
7: Transform A and B:

$$(Q_D^H AZ, Q_D^H BZ) = \left(\begin{bmatrix} A_- & * \\ & A_+ \end{bmatrix}, \begin{bmatrix} B_- & * \\ & B_+ \end{bmatrix} \right)$$

  and check the decoupling residual (3.19). If the spectral split fails, use the QZ algorithm to compute Q and Z and return.

8: **if** $d = 0$ **then**
9:   Approximate $\lambda_{\text{left}}(A_-, B_-)$, $\lambda_{\text{right}}(A_+, B_+)$, and set $\theta_- = \frac{1}{2}|\lambda_{\text{left}}(A_-, B_-)|$ and $\theta_+ = -\frac{1}{2}|\lambda_{\text{right}}(A_+, B_+)|$.
10: **else**
11:   Set $\theta_- = \frac{1}{2}|\theta|$ and $\theta_+ = -\frac{1}{2}|\theta|$.
12: **end if**
13: Compute Q_- and Z_- using Algorithm 4 for $(A_- + \theta_- B_-, B_-)$ with $\theta \leftarrow \theta_-$ and $d \leftarrow d+1$.
14: Compute Q_+ and Z_+ using Algorithm 4 for $(A_+ + \theta_+ B_+, B_+)$ with $\theta \leftarrow \theta_+$ and $d \leftarrow d+1$.
15: **return** $Q = Q_D \begin{bmatrix} Q_- & \\ & Q_+ \end{bmatrix}$ and $Z = Z_D \begin{bmatrix} Z_- & \\ & Z_+ \end{bmatrix}$.

---

aspects of this property are shown in Chapter 4. There, we will see that is not necessary to perform the recursion until we reach $m \leq 4$.

Algorithm 4 summarizes the previous ideas and combines them with the general framework from Algorithm 1. Due to the fact that the divide part operates with the shift function from Section 2.4.3, we call the algorithm *Divide-Shift-and-Conquer* algorithm for the generalized eigenvalue problem. Since most time-consuming operations involve matrix-matrix product like structures, the algorithm can be implemented using scalable level-3 BLAS operations [66], which fulfills our requirements for an algorithm on current computers. The implementation details are discussed in Chapter 4.

Operation Count Although Algorithm 4 is an irregular divide and conquer scheme, we show the complexity for the best case situation with a naive evaluation of the generalized matrix sign function. Since almost all involved level-3 operations are of cubic complexity, we only count them. For the best case it is assumed, as in Theorem 3.12, that the recursion splits into two equally sized subproblems in each step. Furthermore, we assume that the generalized matrix sign function is computed using 15 iterations of scheme (3.27). Then, one spectral division step in Algorithm 4 costs

$$\underbrace{70m^3}_{\text{Computation of } S} + \underbrace{\left(\frac{16}{3}m^3 + 2m^3 + 4m^3 \right)}_{\text{Compute } Q \text{ and } Z.} + \underbrace{4m^3 + 2m^3}_{\text{Apply } Q \text{ and } Z} = \left(87 + \frac{1}{3} \right)m^3 \tag{3.49}$$

flops. Together with the recursive call to Algorithm 4 and the final updates in Q and Z we obtain

$$\text{flops}(m) = \left(87 + \frac{1}{3}\right) m^3 + m^3 + 2 \, \text{flops}\left(\frac{m}{2}\right). \tag{3.50}$$

Because a problem of order $m = 1$ is solved by definition, we set flops $(1) = 0$. Setting $m = 2^M$, $M \in \mathbb{N}$, and neglecting the lower order terms leads us to the best case flop count:

$$\text{flops}(m) = \frac{1060}{9} m^3 = \left(117 + \frac{7}{9}\right) m^3 \approx 118 m^3. \tag{3.51}$$

Compared to the QZ algorithm [80, 129], which needs $66m^3$ flops, our approach needs nearly twice the number of operations. On the other hand, all operations are done using level-3 BLAS operations which scale well on multi-core CPU, as well as on accelerator devices. During Chapter 4, we will see how this is done and how to improve the required number of flops to compute sign (A, B).

Finally, the next example demonstrates how Algorithm 4 transforms a small matrix pair of order 5 to obtain its eigenvalues. Thereby, we assume that only problems of at most order two can be directly solved.

Example 3.13:

Let (A, B) be the matrix pair of order $m = 5$ with

$$A = \begin{bmatrix} 22.50 & 9.50 & -33.00 & 25.00 & -6.50 \\ 1.75 & -2.75 & -25.50 & -4.50 & -20.75 \\ 32.00 & 20.00 & -18.00 & 45.00 & 14.00 \\ 50.25 & 22.75 & -22.50 & 33.50 & -18.25 \\ 6.00 & 0.00 & -20.00 & 1.00 & -15.00 \end{bmatrix}$$

and

$$B = \begin{bmatrix} 65.00 & 59.00 & 66.000 & 124.000 & 129.000 \\ 41.50 & 30.50 & 47.000 & 60.000 & 69.500 \\ 73.00 & 63.00 & 62.000 & 138.000 & 149.000 \\ 51.00 & 45.00 & 58.000 & 86.000 & 85.000 \\ 28.50 & 23.50 & 33.000 & 46.000 & 50.500 \end{bmatrix}.$$

The trivial-to-solve problem size is set to two. The first call to Algorithm 4 computes Q and Z such that we obtain:

$$(A_-, B_-) = \left(\begin{bmatrix} -3.5569e+01 & 2.9536e+01 & 6.1369e+01 \\ -1.7123e+00 & -5.5024e-02 & -1.7208e-15 \\ -1.0803e+00 & 9.3087e-01 & -1.3323e-15 \end{bmatrix}, \right.$$
$$\left. \begin{bmatrix} 3.7173e+01 & -2.4818e+01 & -5.3485e+01 \\ 1.1115e+01 & -3.3025e+00 & -3.2258e+00 \\ 0.0000e+00 & -1.6991e-01 & 3.0084e+00 \end{bmatrix} \right) \tag{3.52}$$

and

$$(A_+, B_+) = \left(\begin{bmatrix} 2.0800e+01 & 2.1584e+01 \\ 8.1899e+00 & -6.4270e+00 \end{bmatrix}, \right.$$
$$\left. \begin{bmatrix} -7.1063e+00 & 3.3913e+00 \\ 2.9920e+01 & -1.7965e+00 \end{bmatrix} \right), \tag{3.53}$$

where (A_+, B_+) is solved directly. This gives

$$\Lambda(A_+, B_+) = \{0.5, 7\}.$$

For the left part of the spectrum we approximate $\left\|B_-^{-1} A_-\right\|_2 = 5.6142$ and we set $\theta_- = 2.8071$. In this way we obtain

$$\Lambda(A_- + \theta_- B_-, B_-) \subset \mathbb{C}_+, \quad with \quad \rho(A_- + \theta_- B_-, B_-) \le \theta_-.$$

Since the 2-norm over-estimates the spectral radius, we have to shift twice to the left and $\theta_+ = \frac{1}{2} \cdot 2.8071 = 1.4036$ gets the new shift parameter. From this, Algorithm 4 computes the spectral splitting

$$(A_-, B_-) = \left([-8.5057e - 01], [8.8192e + 00]\right)$$

and

$$(A_+, B_+) = \left(\begin{bmatrix} 6.8904e - 01 & -4.5581e + 00 \\ 1.5020e + 01 & -6.5391e + 00 \end{bmatrix}, \begin{bmatrix} 8.1006e - 02 & -4.3773e + 00 \\ 1.4106e + 01 & -4.7730e + 00 \end{bmatrix}, \right)$$

which directly gives

$$\Lambda(A_-, B_-) = \{-9.6445e - 02\} \quad and \quad \Lambda(A_+, B_+) = \{9.0355e - 01, 1.1536e + 00\}.$$

By adding the overall shift parameter $\tilde{\theta} = (2.8071 - 1.4036)$ to both spectra we obtain

$$\{-1.5\} \quad and \quad \{-0.5, -0.25\}.$$

Combined with the previously computed eigenvalues we obtain the spectrum

$$\Lambda(A, B) = \{7, 0.5, -0.25, -0.5, -1.5\}$$

for the generalized eigenvalue problem of the matrix pair (A, B).

3.3. Deflating Subspaces via the Matrix Disc Function

A second variant to split the spectrum of a matrix pair (A, B), in our generic framework from Section 3.1, is along the unit circle $C_1(0)$. To this end we apply the matrix disc function originally introduced by Roberts in his fundamental paper [144]. It has its origins in a derivation of the generalized matrix sign function given in Definition 3.4.

Definition 3.14 ([144]):

*Let A be a matrix of order m and $\mathrm{sign}^+(A) = \frac{1}{2}(I + \mathrm{sign}(A))$ be the matrix sign function (3.23) restricted to the eigenvalues with positive real part. Then the **matrix disc function** with center $c \in \mathbb{C}$ and radius $r \in \mathbb{R}^+$ is defined as:*

$$\mathrm{disc}(A, c, r) = \mathrm{sign}^+\left((A - cI_m - rI_m)^{-1}(A - cI_m + rI_m)\right), \tag{3.54}$$

or equivalently for scalars $z \in \mathbb{C}$ as

$$\mathrm{disc}(z, c, r) = \begin{cases} 1 & if \: |z - c| < r \\ 0 & if \: |z - c| > r \end{cases}. \tag{3.55}$$

We see that the matrix disc function is only a shifted and Cayley transformed variant of the matrix sign function (see Subsections 2.4.3 and 2.4.4). The common case, where we regard it with respect to the unit circle $C_1(0)$ ($c = 0$ and $r = 1$), leads to

$$\mathrm{disc}(A) := \mathrm{disc}(A, 0, 1) = \mathrm{sign}^+\left((A - I_m)^{-1}(A + I_m)\right). \tag{3.56}$$

An alternative representation as used for the matrix sign function in (3.23) was shown by Benner [21]:

Theorem 3.15:

Let $A \in \mathbb{C}^{m \times m}$ be a matrix without eigenvalues on the unit circle $C_1(0)$ and the Jordan canonical form

$$A = T \begin{bmatrix} J_0 & 0 \\ 0 & J_\infty \end{bmatrix} T^{-1}, \tag{3.57}$$

where $J_0 \in \mathbb{C}^{p \times p}$ corresponds to the eigenvalues inside the unit circle $C_1(0)$ and $J_\infty \in \mathbb{C}^{m-p \times m-p}$ corresponds to the eigenvalues outside the unit circle $C_1(0)$. Then the matrix disc function $\operatorname{disc}(A)$ is given by

$$\operatorname{disc}(A) = T \begin{bmatrix} I_p & 0 \\ 0 & 0 \end{bmatrix} T^{-1} \tag{3.58}$$

and its relation to the matrix sign function by

$$\operatorname{disc}(A) = \frac{1}{2} \left(I - \operatorname{sign} \left((A - I)^{-1} (A + I) \right) \right). \tag{3.59}$$

Proof. See [21]. □

The matrix disc function has properties similar to those of the matrix sign function given in Theorem 3.5.

Theorem 3.16:

Let $A \in \mathbb{C}^{m \times m}$ be a matrix without eigenvalues on the unit circle $C_1(0)$ and $\operatorname{disc}(A)$ its matrix disc function, then the following conditions are fulfilled:

1. *$\operatorname{disc}(A)$ is diagonalizable with eigenvalues 1 and 0.*

2. *$\operatorname{disc}(A)^2 = \operatorname{disc}(A)$.*

3. *For an arbitrary nonsingular matrix $V \in \mathbb{C}^{m \times m}$ it holds $\operatorname{disc}(VAV^{-1}) = V \operatorname{disc}(A) V^{-1}$.*

4. *$P_0 = \operatorname{disc}(A)$ is a projector onto the invariant subspace for the eigenvalues inside the unit circle $C_1(0)$ and $p_0 = \operatorname{tr}(\operatorname{disc}(A))$ is the corresponding number of eigenvalues.*

5. *$P_\infty = I - \operatorname{disc}(A)$ is a projector onto the invariant subspace for the eigenvalues outside the unit circle $C_1(0)$ and $p_\infty = m - \operatorname{tr}(\operatorname{disc}(A))$ is the corresponding number of eigenvalues.*

Proof. 1. Follows directly from (3.58) in Theorem 3.15.

2. From (3.59) we have

$$\begin{aligned} \operatorname{disc}(A)^2 &= \frac{1}{4} \left(I - \operatorname{sign} \left((A - I)^{-1}(A + I) \right) \right)^2 \\ &= \frac{1}{4} \left(2I - 2 \operatorname{sign} \left((A - I)^{-1}(A + I) \right) \right) \\ &= \frac{1}{2} \left(I - \operatorname{sign} \left((A - I)^{-1}(A + I) \right) \right) = \operatorname{disc}(A). \end{aligned}$$

3. Using the matrix sign function based definition of the matrix disc function, again we obtain

$$\begin{aligned} \operatorname{disc}(VAV^{-1}) &= \frac{1}{2} \left(I - \operatorname{sign} \left((VAV^{-1} - I)^{-1} (VAV^{-1} + I) \right) \right) \\ &= \frac{1}{2} \left(I - \operatorname{sign} \left(V(A - I)^{-1} V^{-1} V (A + I) V^{-1} \right) \right) \\ &= \frac{1}{2} \left(VV^{-1} - V \operatorname{sign} \left((A - I)^{-1}(A + I) \right) V^{-1} \right) \\ &= V \operatorname{disc}(A) V^{-1}. \end{aligned}$$

4. Let $A = T \begin{bmatrix} J_0 & 0 \\ 0 & J_\infty \end{bmatrix} T^{-1}$ be the Jordan canonical form of A with J_0 and J_∞ as in (3.57), then we get

$$P_0 A = \text{disc}\,(A)\,A = T \begin{bmatrix} I_p & 0 \\ 0 & 0 \end{bmatrix} T^{-1} T \begin{bmatrix} J_0 & 0 \\ 0 & J_\infty \end{bmatrix} T^{-1} = T \begin{bmatrix} J_0 & 0 \\ 0 & 0 \end{bmatrix} T^{-1} = A_0$$

where $\Lambda\,(A_0)$ lies inside the unit circle $C_1(0)$. From 2 it follows that $\text{disc}\,(A)$ is a projector. Furthermore, we have

$$p_0 = \text{tr}(\text{disc}\,(A)) = \text{tr}\!\left(T \begin{bmatrix} I_p & 0 \\ 0 & 0 \end{bmatrix} T^{-1} \right) = \text{tr}\!\left(T^{-1} T \begin{bmatrix} I_p & 0 \\ 0 & 0 \end{bmatrix} \right) = p.$$

5. Analogous to 4. $\qquad\qquad\qquad\qquad\qquad\qquad\qquad\qquad\qquad\qquad\qquad\qquad\qquad\square$

Similar to the generalization of the matrix sign function in Subsection 3.2.1, the concept of the matrix disc function can be generalized to matrix pairs by the following definition:

Definition 3.17 ([21]):
Let (A, B) be a matrix pair of order m that has the following Weierstrass/Kronecker canonical form [75, 76]:

$$(A, B) = \left(V \begin{bmatrix} J_0 & 0 \\ 0 & J_\infty \end{bmatrix} U^{-1}, V \begin{bmatrix} I & 0 \\ 0 & N \end{bmatrix} U^{-1} \right), \tag{3.60}$$

*where the matrix pair (J_0, I) of order p contains the Kronecker blocks corresponding to the eigenvalues inside the unit circle $C_1(0)$ and the matrix pair (J_∞, N) of order $m - p$ contains the Kronecker blocks corresponding to the eigenvalues outside the unit circle $C_1(0)$. Then we define the **right matrix disc function** for the matrix pair (A, B) as*

$$\text{disc}_R\,(A, B) = \left(U \begin{bmatrix} I_p & 0 \\ 0 & 0 \end{bmatrix} U^{-1}, U \begin{bmatrix} 0 & 0 \\ 0 & I_{m-p} \end{bmatrix} U^{-1} \right) \tag{3.61}$$

*and the **left matrix disc function** as*

$$\text{disc}_L\,(A, B) = \left(V \begin{bmatrix} I_p & 0 \\ 0 & 0 \end{bmatrix} V^{-1}, V \begin{bmatrix} 0 & 0 \\ 0 & I_{m-p} \end{bmatrix} V^{-1} \right). \tag{3.62}$$

Since Subsection 3.1.1 shows how to construct the left deflating subspace from a given right deflating subspace, we regard only $\text{disc}_R\,(A, B)$, here. In order to connect the right matrix disc function to the properties shown in Theorem 3.16 we set

$$P_0\,(A, B) := U \begin{bmatrix} I_p & 0 \\ 0 & 0 \end{bmatrix} U^{-1} \tag{3.63}$$

and

$$P_\infty\,(A, B) := I - U \begin{bmatrix} I_p & 0 \\ 0 & 0 \end{bmatrix} U^{-1} := U \begin{bmatrix} 0 & 0 \\ 0 & I_{m-k} \end{bmatrix} U^{-1}. \tag{3.64}$$

Obviously, the matrices $P_0\,(A, B)$ and $P_\infty\,(A, B)$ are first and second component of $\text{disc}_R\,(A, B)$. In this way, we can use all properties from Theorem 3.16 for matrix pairs in a straight forward manner.

An alternative definition of the right matrix disc function, which shows the connection to the matrix sign function, was given by Benner and Byers [22] and is used in the next theorem:

Theorem 3.18 ([22]):

Let (A, B) be a matrix pair of order m with p eigenvalues inside the unit circle $C_1(0)$ and no eigenvalue on the unit circle $C_1(0)$. Furthermore, the right matrix disc function of (A, B) is given by

$$\mathrm{disc}_R\,(A, B) = \left(U \begin{bmatrix} I_p & 0 \\ 0 & 0 \end{bmatrix} U^{-1}, U \begin{bmatrix} 0 & 0 \\ 0 & I_{m-p} \end{bmatrix} U^{-1} \right) = (W, Y), \tag{3.65}$$

as in Definition 3.17. Then it holds:

$$W + Y = I, \tag{3.66}$$

$$W - Y = \mathrm{sign}\left((A - B)^{-1}(A + B)\right), \tag{3.67}$$

and W and Y are given by

$$W = \frac{1}{2}\left(I - \mathrm{sign}\left((A - B)^{-1}(A + B)\right)\right) \tag{3.68}$$

and

$$Y = \frac{1}{2}\left(I + \mathrm{sign}\left((A - B)^{-1}(A + B)\right)\right). \tag{3.69}$$

Proof. Equation (3.66) is trivially satisfied:

$$\underbrace{U \begin{bmatrix} I_k & 0 \\ 0 & 0 \end{bmatrix} U^{-1}}_{W} + \underbrace{U \begin{bmatrix} 0 & 0 \\ 0 & I_{m-k} \end{bmatrix} U^{-1}}_{Y} = U \left(\begin{bmatrix} I_k & 0 \\ 0 & 0 \end{bmatrix} + \begin{bmatrix} 0 & 0 \\ 0 & I_{m-k} \end{bmatrix} \right) U^{-1} = I.$$

Next, we consider

$$\mathrm{sign}\left((A - B)^{-1}(A + B)\right) = S \begin{bmatrix} -I_k & 0 \\ 0 & I_{m-l} \end{bmatrix} S^{-1},$$

where k is the number of eigenvalues in the left half-plane of $(A - B)^{-1}(A + B)$. The Cayley-Transform (2.40) $B^{-1}A \mapsto (A - B)^{-1}(A + B)$ maps the inside of the unit circle $C_1(0)$ onto the open left half-plane and vice versa. In this way, the p eigenvalues inside the unit circle $C_1(0)$ correspond to the k eigenvalues in the open left half plane. Thus, $k = p$. Furthermore,

$$P_- = \frac{1}{2}\left(I - \mathrm{sign}\left((A - B)^{-1}(A + B)\right)\right)$$

defines a projection onto the subspace corresponding to the eigenvalues in the open left half-plane of $(A - B)^{-1}(A + B)$, i.e., the subspace corresponding to the eigenvalues inside the unit circle $C_1(0)$ of $B^{-1}A$. This yields

$$\frac{1}{2}\left(I - S \begin{bmatrix} -I_p & 0 \\ 0 & I_{m-p} \end{bmatrix} S^{-1}\right) = \underbrace{S}_{U} \begin{bmatrix} I_p & 0 \\ 0 & 0 \end{bmatrix} \underbrace{S^{-1}}_{U^{-1}} = W.$$

Analogously, we show

$$\frac{1}{2}\left(I + S \begin{bmatrix} -I_p & 0 \\ 0 & I_{m-p} \end{bmatrix} S^{-1}\right) = \underbrace{S}_{U} \begin{bmatrix} 0 & 0 \\ 0 & I_{m-p} \end{bmatrix} \underbrace{S^{-1}}_{U^{-1}} = Y,$$

which defines a projection onto the subspace belonging to the eigenvalues in the right half-plane of $(A - B)^{-1}(A + B)$, i.e., the eigenvalues outside the unit circle $C_1(0)$ of $B^{-1}A$. Finally,

$$W - Y = \text{sign}\left((A - B)^{-1}(A + B)\right)$$

is fulfilled by construction. □

Theorem 3.18 shows a way to compute the right matrix disc function of (A, B) using the matrix sign function.

3.3.1. Inverse Free Iteration

The previous section showed that the matrix disc function can be computed using the matrix sign function either for a matrix A using Equation (3.54) or for a matrix pair (A, B) using Theorem 3.18. Especially for the matrix pair (A, B) this requires to compute

$$Z = (A - B)^{-1}(A + B) \tag{3.70}$$

and find sign (Z) using the Newton scheme (3.25). If $(A - B)$ is ill-conditioned this way introduces additional errors to the computation of $\text{disc}_R(A, B)$. A modification of the Newton scheme for the computation of the matrix disc function was shown by Benner in [21]. This is, however, affected by ill-conditioned matrices and is therefore unfortunately not feasible for a practical implementation.

An improved algorithm, without the need for inverting a matrix, was developed by Malyshev [121, 122] to solve some spectral division problems for a matrix pair, without mentioning the matrix disc function. This approach was extended by Bai, Demmel, and Gu in [15]. The main result is presented in the following theorem.

Theorem 3.19 (Inverse Free Iteration, [15]):
Let (A, B) be a matrix pair of order m without eigenvalues on the unit circle $C_1(0)$ and $\text{disc}_R(A, B) = (W, Y)$ its right matrix disc function. Then the sequence (W_k, Y_k), given by

$$W_0 = A, \quad Y_0 = B$$
$$W_{k+1} = Q_{k,12}^H W_k, \quad Y_{k+1} = Q_{k,22}^H Y_k, \tag{3.71}$$

with $Q_{k,12}$ and $Q_{k,22}$ the matrices of order m in the QR factorization

$$\begin{bmatrix} Y_k \\ -W_k \end{bmatrix} = \begin{bmatrix} Q_{k,11} & Q_{k,12} \\ Q_{k,21} & Q_{k,22} \end{bmatrix} \begin{bmatrix} R_k \\ 0 \end{bmatrix}, \tag{3.72}$$

converges quadratically and (W, Y) are given by

$$W = \lim_{k \to \infty} (W_k + Y_k)^{-1} Y_k = (W_\infty + Y_\infty)^{-1} Y_\infty,$$
$$Y = \lim_{k \to \infty} (W_k + Y_k)^{-1} W_k = (W_\infty + Y_\infty)^{-1} W_\infty. \tag{3.73}$$

The iteration shown in Theorem 3.19 provides a way to compute W_k and Y_k, $k \to \infty$, without directly inverting a matrix. Still, the solution of two linear systems, to obtain W and Y, is necessary. On the one hand, sophisticated replacements for

$$W = (W_\infty + Y_\infty)^{-1} Y_\infty \quad \text{and} \quad Y = (W_\infty + Y_\infty)^{-1} W_\infty$$

based on the variants of the QR decomposition exist [15, 21]. The condition number of $(W_\infty + Y_\infty)$ still plays an important role in these cases. On the other hand, our divide-and-conquer idea in Section 3.1 only requires the knowledge of deflating subspaces. In order to obtain them directly from W_∞ and Y_∞, we extend an idea, proposed by Sun and Quintana-Ortí [162], to the following theorem:

Theorem 3.20:

Let (A, B) be a matrix pair of order m without eigenvalues on the unit circle $C_1(0)$. Furthermore, let W_∞ and Y_∞ be the converged matrices from Theorem 3.19. Then the deflating subspaces $\mathcal{D}_0$ and $\mathcal{D}_\infty$ corresponding to the eigenvalues of (A, B) inside and outside the unit circle $C_1(0)$ are given by

$$\mathcal{D}_0 = \ker W_\infty \tag{3.74}$$

and

$$\mathcal{D}_\infty = \ker Y_\infty. \tag{3.75}$$

Proof. We recall the definition of the inverse free iteration (3.71)

$$W_{k+1} = Q_{k,12}^H W_k \quad \text{and} \quad Y_{k+1} = Q_{k,22}^H Y_k,$$

and from the involved QR decomposition (3.72) it follows

$$Q_{k,12}^H Y_k = Q_{k,22}^H W_k. \tag{3.76}$$

Furthermore, let $\mathcal{D}_0 = \operatorname{span} D_0$ and L be a matrix with $\Lambda(L) = \Lambda(A, B) \cap \mathcal{B}_1(0)$, with $\mathcal{B}_1(0) = \{z \in \mathbb{C} \,|\, |z| < 1\}$. Then we have

$$AD_0 = BD_0 L. \tag{3.77}$$

Together with the definition of W_k and Y_k in (3.71) for $k = 1$ we get:

$$W_1 D_0 = Q_{0,12}^H A D_0 \overset{(3.77)}{=} Q_{0,12}^H B D_0 L \overset{(3.76)}{=} Q_{0,22}^H A D_0 L$$
$$\overset{(3.77)}{=} Q_{0,22}^H B D_0 L^2 \overset{(3.71)}{=} Y_1 D_0 L^2. \tag{3.78}$$

Extending this to any $k > 1$ by induction yields

$$W_{k+1} D_0 \overset{(3.71)}{=} Q_{k,12}^H W_k D_0 \overset{(3.78)}{=} Q_{k,12}^H Y_k D_0 L^{2^k}$$
$$\overset{(3.76)}{=} Q_{k,22}^H W_k D_0 L^{2^k} \overset{(3.78)}{=} Q_{k,22}^H Y_k D_0 L^{2^{k+1}} \overset{(3.71)}{=} Y_{k+1} D_0 L^{2^{k+1}}. \tag{3.79}$$

Since the absolute values of eigenvalues of L are smaller than one, $k \to \infty$ gives us

$$Y_{k+1} D_0 L^{2^{k+1}} \overset{k \to \infty}{\longrightarrow} 0$$

and this yields

$$W_{k+1} D_0 \overset{k \to \infty}{\longrightarrow} 0,$$

which means

$$\mathcal{D}_0 \subseteq \ker W_\infty.$$

Analogously, we obtain from the eigenvalue problem of the matrix pair (B, A) (see Subsection 2.4.1)

$$\mathcal{D}_\infty \subseteq \ker Y_\infty.$$

Theorem 3.18 shows that under our assumptions the matrix pair (W, Y) is regular and hence the matrix pairs $\left((W_\infty + Y_\infty)^{-1} Y_\infty, (W_\infty + Y_\infty)^{-1} W_\infty\right)$ and (Y_∞, W_∞) are regular as well. This yields

$$\mathcal{D}_0 \cap \mathcal{D}_\infty = \ker W_\infty \cap \ker Y_\infty = \{0\},$$

otherwise $\Lambda(W_\infty, Y_\infty)$ contains zero. Finally, we know that $\mathcal{D}_0 \cup \mathcal{D}_\infty$ must be the whole m dimensional space and therefore

$$\mathcal{D}_0 = \ker W_\infty \quad \text{and} \quad \mathcal{D}_\infty = \ker Y_\infty. \qquad \square$$

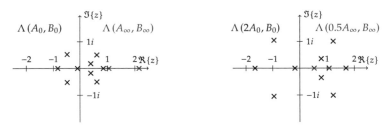

Figure 3.3.: Split spectrum of a matrix pair   Figure 3.4.: Scaled split spectrum of a matrix pair
(A, B). (A, B).

Theorem 3.20 shows that the inverse free iteration provides two different deflating subspaces. Their characterization in terms of W_∞ and Y_∞ perfectly match the requirements to compute the matrices Q and Z. By setting either $X = W_\infty$ or $X = Y_\infty$ in Algorithm 2, the output of the inverse free iteration is directly used to compute the deflating transformation without solving the additional linear systems (3.73). This step is only required to compute the number of eigenvalues inside and outside the unit circle, which are used as external rank determination in the subspace extraction.

Remark 3.21:

Quintana-Ortí and Sun suggested using a preconditioner-like scheme to make the computation of $\ker W_\infty$ *and* $\ker Y_\infty$ *more stable [162]. Therefore, they compute the RQ decomposition*

$$\begin{bmatrix} W_\infty & Y_\infty \end{bmatrix} = R \begin{bmatrix} \hat{W}_\infty & \hat{Y}_\infty \end{bmatrix}. \tag{3.80}$$

Thereby, the matrix R is nonsingular under the assumptions of Theorem 3.20. This yields

$$\ker W_\infty = \ker \hat{W}_\infty \quad and \quad \ker Y_\infty = \ker \hat{Y}_\infty \tag{3.81}$$

and the singular values of $\hat{W}_\infty$ *and* $\hat{Y}_\infty$ *are clustered at 0 and 1.*

As the generalized matrix sign function is not defined if there are eigenvalues on the imaginary axis, the matrix disc function does not work with its eigenvalues on the Cayley transformed counterpart, the unit circle, although it converges [162].

3.3.2. The Divide-Scale-and-Conquer Algorithm

Like for the generalized matrix sign function, the computation of the matrix disc function and the subsequent spectral splitting will only work once for a given matrix pair (A, B). If we want to obtain a recursive scheme, which splits the spectra in every step again, we have to modify the spectra similar to the Divide-Shift-Conquer Algorithm in Section 3.2.3. Here, it is necessary to move eigenvalues from inside the unit circle to its outside and vice versa. Regarding the basic eigenvalue transformations presented in Section 2.4, the scaling function f_{scal} is sufficient to reach this goal.

First, we assume that we computed a deflating transformation using the matrix disc function such that

$$(Q^H AZ, Q^H BZ) = \left(\begin{bmatrix} A_0 & A_* \\ 0 & A_\infty \end{bmatrix}, \begin{bmatrix} B_0 & B_* \\ 0 & B_\infty \end{bmatrix} \right), \tag{3.82}$$

and $\Lambda(A_0, B_0)$ lies inside and on the unit circle and $\Lambda(A_\infty, B_\infty)$ lies outside the unit circle. Figure 3.3 shows this for an example matrix. Now, we have to move the spectra of (A_0, B_0) and (A_∞, B_∞)

such that both enclose the unit circle again. That means we have to find a θ_0 and θ_∞ such that $\Lambda\left(\theta_0 A_0, B_0\right)$ and $\Lambda\left(\theta_\infty A_\infty, B_\infty\right)$ fulfill this property. Like for the matrix sign function based Divide-Shift-and-Conquer scheme, we use the spectral radius here. Having the spectral radius $\rho\left(A_\infty, B_\infty\right) \geq 1$, we set the scaling parameter θ_∞ to

$$\theta_\infty = \frac{2}{1 + \rho\left(A_\infty, B_\infty\right)}. \tag{3.83}$$

and introduce θ_∞^+ and θ_∞^- by

$$\theta_\infty^+ = 1 + \rho\left(A_\infty, B_\infty\right) \quad \text{and} \quad \theta_\infty^- = 1 + \rho\left(A_\infty, B_\infty\right)^{-1}.$$

Since θ_∞^+ is bounded from below by two and $\theta_\infty^{(-)}$ is bounded from above by two, we obtain for the scaled matrix pair $\left(\theta_\infty A_\infty, B_\infty\right)$

$$\left\{|\lambda| \,\middle|\, \lambda \in \Lambda\left(\theta_\infty A_\infty, B_\infty\right)\right\} \subset \left[\frac{2}{\theta_\infty^+}, \frac{2}{\theta_\infty^-}\right] \subset [0,\, 2]. \tag{3.84}$$

In this way, the scaled matrix pair possibly encloses the unit circle again. The second subproblem lying inside the unit circle needs a slightly different treatment. There, we need the eigenvalue with the smallest magnitude. This is equivalent to the reciprocal of the spectral radius of the inverted matrix pair $\left(B_0, A_0\right)$, see Subsection 2.4.1). Then, we set θ_0 to

$$\theta_0 = \frac{2}{1 + \rho\left(B_0, A_0\right)^{-1}} \tag{3.85}$$

and introduce θ_0^+ and θ_0^- by

$$\theta_0^+ = 1 + \rho\left(B_0, A_0\right) \quad \text{and} \quad \theta_0^- = 1 + \rho\left(B_0, A_0\right)^{-1}.$$

Now, θ_0^+ is bounded from below by two and θ_0^- is limited from above by two, the spectrum of the scaled matrix pair $\left(\theta_0 A_0, B_0\right)$ fulfills

$$\left\{|\lambda| \,\middle|\, \lambda \in \Lambda\left(\theta_0 A_0, B_0\right)\right\} \subset \left[\frac{2}{\theta_0^+}, \frac{2}{\theta_0^-}\right] \subset [0,\, 2]. \tag{3.86}$$

This possibly encloses the unit circle again. Using the scaling parameters θ_0 and θ_∞, we can apply the matrix disc function and the spectral splitting to $\left(\theta_0 A_0, B_0\right)$ and $\left(\theta_\infty A_\infty, B_\infty\right)$ again. The quantities $\rho\left(A_\infty, B_\infty\right)$ and $\rho\left(B_0, A_0\right)^{-1}$ are estimated using the same techniques presented for the matrix sign function based splitting in Section 3.2.2.

Similar to the Divide-Shift-Conquer algorithm we can use Equations (3.83) and (3.86) to obtain estimates for $\rho\left(B_0, A_0\right)^{-1}$ and $\rho\left(A_\infty, B_\infty\right)$ during the recursion. For each newly emerged subproblem, either for $\left(A_0, B_0\right)$ or $\left(A_\infty, B_\infty\right)$, the spectral bounds for the scaled problems are given by:

$$\rho\left(\theta_0 A_0, B_0\right) \leq \theta_0, \quad \rho\left(B_0, \theta_0 A_0\right)^{-1} = \theta_0 \rho\left(B_0, A_0\right)^{-1} \tag{3.87}$$

and

$$\rho\left(\theta_\infty A_\infty, B_\infty\right) = \theta_\infty \rho\left(A_\infty, B_\infty\right), \quad \rho\left(B_\infty, \theta_\infty A_\infty\right)^{-1} \geq \theta_\infty. \tag{3.88}$$

Now, our initial divide-and-conquer framework from Section 3.1 together with the matrix disc function turns into the Divide-Scale-and-Conquer Algorithm shown in Algorithm 5. Regarding the recursion depth we, by construction, have:

Corollary 3.22:

Theorem 3.12 holds for the Divide-Scale-Conquer Algorithm 5, as well.

Algorithm 5 Divide-Scale-and-Conquer Algorithm

Input: $A \in \mathbb{C}^{m \times m}$ and $B \in \mathbb{C}^{m \times m}$ nonsingular without eigenvalues on the unit circle,
$\quad \theta_{low}, \theta_{up} \in \mathbb{R}$ lower and upper bound for the absolute magnitude of the eigenvalues, and
$\quad$ recursion depth d (all 0 at start).
Output: $Q \in \mathbb{C}^{n \times n}$ and $Z \in \mathbb{C}^{n \times n}$ unitary, such that $(Q^H AZ, Q^H BZ)$ as in (2.21).
1: **if** m small enough or $d \geq \tau_{rec} \log_2 \frac{m}{m_t}$ **then**
2: $\quad$ Compute Q, Z directly using QZ algorithm.
3: $\quad$ **return** $[Q, Z]$
4: **end if**
5: Compute W_∞ using Theorem 3.19 or $\hat{W}_\infty$ from Remark 3.21.
6: Compute Q_D and Z_D from W_∞ (or $\hat{W}_\infty$) using Algorithm 2.
7: Transform A and B:

$$(Q_D^H AZ, Q_D^H BZ) = \left(\begin{bmatrix} A_0 & * \\ & A_\infty \end{bmatrix}, \begin{bmatrix} B_0 & * \\ & B_\infty \end{bmatrix} \right)$$

$\quad$ and check the decoupling residual (3.19). If the spectral split fails, compute Q and Z using the
$\quad QZ$ algorithm and return.
8: **if** $d = 0$ **then**
9: $\quad$ Set $\theta_{low} = \frac{1}{1+\rho(B_0, A_0)^{-1}}$ and $\theta_{up} = \frac{1}{1+\rho(A_\infty, B_\infty)}$.
10: **else**
11: $\quad$ Set $\theta_0 = \frac{2}{1+\theta_{low}}$ and $\theta_\infty = \frac{2}{1+\theta_{up}}$.
12: **end if**
13: Compute Q_0 and Z_0 using Algorithm 5 for $(\theta_0 A_0, B_0)$ with $\theta_{low} \leftarrow \theta_0 \theta_{low}$, $\theta_{up} \leftarrow \theta_0$, and
$\quad d \leftarrow d + 1$.
14: Compute Q_∞ and Z_∞ using Algorithm 5 for $(\theta_\infty A_\infty, B_\infty)$ with $\theta_{low} \leftarrow \theta_\infty$, $\theta_{up} \leftarrow \theta_\infty \theta_{up}$, and
$\quad d \leftarrow d + 1$.
15: **return** $Q = Q_D \begin{bmatrix} Q_0 & \\ & Q_\infty \end{bmatrix}$ and $Z = Z_D \begin{bmatrix} Z_0 & \\ & Z_\infty \end{bmatrix}$.

Operation Count Finally, we present a naive best case flop count. Under the assumptions of Theorem 3.12 that in each step of the recursion the problems get split into equally sized subproblems and the inverse free iteration needs 15 steps to converge, one spectral division step in Algorithm 5 costs

$$\underbrace{195m^3}_{\text{Compute } W_\infty \text{ or } Y_\infty} + \underbrace{\left(\frac{16}{3}m^3 + 2m^3 + 4m^3 \right)}_{\text{Compute } Q \text{ and } Z.} + \underbrace{4m^3 + 2m^3}_{\text{Apply } Q \text{ and } Z} = \left(212 + \frac{1}{3} \right) m^3. \tag{3.89}$$

Together with the final setup of Q and Z in Algorithm 5 the overall flop count fulfills the recursion:

$$\text{flops}(m) = \left(212 + \frac{1}{3} \right) m^3 + m^3 + 2 \, \text{flops}\left(\frac{m}{2} \right). \tag{3.90}$$

Using the assumption that problems of size $m = 1$ are solved by definition, we set $\text{flops}(1) = 0$, setting $m = 2^M$, $M \in \mathbb{N}$, and neglecting lower order terms, we finally end up with

$$\text{flops}(m) = \frac{2590}{9}m^3 + O(m) = 287\frac{7}{9}m^3. \tag{3.91}$$

In comparison to the Divide-Shift-and-Conquer Algorithm 4 we need 130% more flops in the best case, but the algorithm does not need to solve linear systems and heavily relies on unitary transformations. Furthermore, the matrices containing the deflating subspaces are computed using unitary transformation in the inverse free iteration, which causes less round-off errors compared to the computation of the generalized matrix sign function.

Example 3.23:

Let (A, B) be as in Example 3.13. A problem is assumed to be trivial-to-solve if it is at most of the order 2. Then the first spectral division using the matrix disc function provides:

$$(A_0, B_0) = \left(\begin{bmatrix} 1.0755e+01 & 9.6253e+00 & -4.6375e+01 \\ -1.3185e+00 & 7.6405e-01 & 3.5600e+00 \\ 1.3312e+00 & 1.6078e+00 & 8.4535e-02 \end{bmatrix}, \right.$$
$$\left. \begin{bmatrix} 3.0921e+01 & 1.6835e+01 & -9.7704e+01 \\ 1.1502e+00 & -6.9710e+00 & 2.6604e-15 \\ -3.0845e+00 & -4.9385e-15 & -6.1098e-15 \end{bmatrix} \right) \quad (3.92)$$

and

$$(A_\infty, B_\infty) = \left(\begin{bmatrix} 3.5559e+01 & -2.6981e+00 \\ 8.2042e+00 & 6.1243e+00 \end{bmatrix}, \right.$$
$$\left. \begin{bmatrix} 1.0826e+01 & 5.5104e+00 \\ -3.4485e+00 & -3.8656e+00 \end{bmatrix} \right), \quad (3.93)$$

where (A_∞, B_∞) is solved directly. This gives:

$$\Lambda(A_\infty, B_\infty) = \{-1.5, 7\}.$$

For the matrix pair (A_0, B_0), we compute

$$\rho(B_0, A_0)^{-1} \approx \left\| A_0^{-1} B_0 \right\|_F^{-1} = 0.1192$$

and obtain

$$\theta_0^{(1)} = \frac{2}{1 + 0.1192} = 1.787$$

as scaling factor. Because this does not move any eigenvalue of $\left(\theta_0^{(1)} A_0, B_0 \right)$ outside the unit circle, we have to scale again. Therefore, we use Formula (3.88) to obtain

$$\rho\left(B_0, \theta_0^{(1)} A_0 \right)^{-1} = \theta_0^{(1)} \rho(B_0, A_0)^{-1} \approx 0.21298.$$

This yields $\theta_0^{(2)} = \frac{2}{1+0.21298}$ and we have

$$Q^H \left(\theta_0^{(1)} \theta_0^{(2)} A_0, B_0 \right) Z = \left(\begin{bmatrix} 5.3531e+00 & 1.0789e+02 & -2.3741e+00 \\ 0 & -9.3548e+01 & 4.9117e-01 \\ 0 & 1.3278e+01 & -6.7771e+00 \end{bmatrix}, \right.$$
$$\left. \begin{bmatrix} -7.2670e+00 & 8.1988e+01 & -4.6284e+00 \\ 0 & -6.3096e+01 & -1.2321e+00 \\ 0 & 6.6232e+00 & 4.7111e+00 \end{bmatrix} \right) \quad (3.94)$$

from which we get

$$\Lambda\left(\theta_0^{(1)} \theta_0^{(2)} A_0, B_0 \right) = \{-7.3663e-01, 1.4733e+00, -1.4733e+00\}$$

directly. Undoing the overall scaling $\theta_0^{(1)}\theta_0^{(2)}$ for this recursion path, we get

$$\Lambda\left(A_0, B_0\right) = \left(\theta_0^{(1)}\theta_0^{(2)}\right)^{-1} \cdot \Lambda\left(\theta_0^{(1)}\theta_0^{(2)}A_0, B_0\right) = \{-0.25, 0.5, -0.5\}.$$

Combined with the previously computed eigenvalues we obtain the spectrum

$$\Lambda\left(A, B\right) = \{7, 0.5, -0.25, -0.5, -1.5\}$$

for the generalized eigenvalue problem of the matrix pair (A, B).

Remark 3.24:

Due to the fact that the Divide-Shift-Conquer Algorithm 4 and the Divide-Scale-Conquer Algorithm 5 split the spectra in different ways, it is possible to switch both whenever one of the Algorithms fails in the spectral division step. The estimates for the spectral radius can be reused in this case.

IMPLEMENTATION ON CURRENT HARDWARE

Contents

In this chapter we focus on the efficient implementation of the Divide-Shift-and-Conquer and the Divide-Scale-and-Conquer algorithms including their building blocks.

The hard- and software setup for the motivating experiments and benchmarks appearing during this section are briefly described in Appendix A.

4.1. Divide-Shift/Scale-and-Conquer Algorithms

Divide-and-conquer algorithms are a standard strategy to shrink down the algorithmic complexity in many areas of computer science. The most popular ones designed for sorting and searching tasks in large data sets are Quicksort [92], Mergesort [110], or Quickselect [91]. Many of the theoretical results and implementation strategies developed in the last six decades can be applied to other divide-and-conquer algorithms without large modifications. Even for symmetric standard eigenvalue problems, a divide-and-conquer solver exists [80], but in contrast to our approach it works on the matrix reduced to symmetric tridiagonal form. This is not applicable for generalized eigenvalue problems.

In this section, we discuss the outline of the implementation of our main algorithms from Chapter 3. Additional building blocks are collected and analyzed later in this chapter.

Memory Requirements   Before we present details about the implementation, we analyze the memory consumption of Algorithms 4 and 5. On the one hand, this is necessary to minimize time-consuming memory management operations like allocation and reallocation. On the other hand, an a priori upper bound for the memory consumption allows to guarantee that algorithms work on a given hardware without running out of memory. The analysis does not include extra workspace required by the generalized matrix sign function, the inverse free iteration, or the computation of the deflating transformations. We will see later that the memory requirements for these algorithms can be, mostly, satisfied by the memory of the divide-and-conquer algorithm, which is not in use in the current execution step. Furthermore, we show that the recursion does not need additional auxiliary memory after the workspace is allocated for the initial problem. When we start with a matrix pair (A, B) of order m, we have to store A, B, Q, Z, and need auxiliary memory for the following:

1. Before a deflating subspace is computed, we need to back up the input matrices A and B. On the one hand, we allow the subspace computation to be destructive on their inputs and, on the other hand, we need to restore the original data if the spectral splitting fails.

   additional memory requirements: $2m^2$

2. The generalized matrix sign function, or the matrix disc function, computes a matrix X containing the deflating subspace X in Algorithms 4 or 5.

   additional memory requirements: m^2

3. The matrices Q_D and Z_D are computed from the deflating subspace X in Algorithms 4 or 5.

   additional memory requirements: $2m^2$

4. The matrix pair (A, B) is transformed using Q_D and Z_D using general matrix-matrix multiplications. Since this operation does not work in-place and consists of three operands, a temporary location is necessary during the computation. The result is stored afterwards in (A, B). As temporary location we use the space for the deflating subspace, here, since this is no longer in use at this point.

   satisfied memory requirements: m^2 from 2.

5. The computation of the shift or scaling parameters $\theta_{+,-,0,\infty}$ may require additional memory depending on the used computation strategy, compare Section 3.2.2. Since the backups of A, B, and the deflating subspace are no longer required, up to $3m^2$ memory can be used for this purpose.

   satisfied memory requirements: $3m^2$ from 1. and 2.

6. The updates of the matrices Q_D and Z_D are done using general matrix-matrix products or sophisticated routines if they are still represented as product of Householder transformations. Thereby, at most m^2 memory is required due to the out-of-place nature of this operation. Here, the memory from the deflating subspace matrix from point 2 or 1 can be used again.

   satisfied memory requirements: m^2 from 1. and 2.

Hence, beside eventually required memory inside the computation of the generalized matrix sign function or the matrix disc function, we need $5m^2$ memory. Since not the whole additional memory is in use the point where the deflating subspace is computed, we have up to $2m^2$ memory available in the computation of the generalized matrix sign function or the inverse free iteration, as shown later in Section 4.2 and Section 4.3. This allows allocating almost all memory at the beginning of Algorithm 4 or Algorithm 5.

Building Blocks   Beside the implementation of the recursive schemes, we regard the following building blocks, to obtain an efficient algorithm which fits to current computer hardware:

- The extraction of the deflating transformations Q_D and Z_D including common operations are shown in Section 4.1.5.

- The computationally hardest task in our algorithms is the computation of the matrix containing the deflating subspace. Both presented ways are analyzed separately, and their implementation details are shown in Sections 4.2 and 4.3.

For all building blocks, we show how accelerator devices, like GPUs, can be used to increase the performance. Without loss of generality, and due to the basic similarity of the heterogeneous programming concepts for accelerators, we restrict ourselves to GPUs with Nvidia® CUDA support here. The techniques and concepts can be easily transferred to other implementation strategies like OpenCL [157], HIP [2], or OpenMP offloading [138].

4.1.1. Recursion Free Variant

Algorithms 4 and 5 are recursive divide-and-conquer schemes. Without loss of generality, we use the more general Algorithm 1 to describe different ways to remove the recursion. This is necessary to handle the auxiliary workspace easier and include further parallelization. The key observation is that our framework splits the input problem of order m in two non-equally sized subproblems of order k and l, which are solved independently in the top-down part and combined afterwards to obtain the actual solution of the input problem in the bottom-up part. Figure 4.1 shows an example of a recursion tree appearing in our Divide-Shift/Scale-and-Conquer framework. Such recursion trees are well-known from the Quicksort algorithm [92]. As with Quicksort, there are not necessarily two new sub-problems. In contrast to Quicksort our algorithms perform operations in the bottom-up phase of the recursion, as well. We show next that this can be rearranged to become equivalent to the recursion in Quicksort. In this way, we are able to use the techniques to remove the recursion and use the memory bounds for the involved data structures known from Quicksort [147].

The updates of Q and Z need to be moved from the bottom-up to the top-down phase of the recursion. In this way, the memory used by Q_D and Z_D is available in the next recursion step. Furthermore, this is the main ingredient to make the recursion equivalent to Quicksort. Therefore, we recall the fact that the transformations performed on a subproblem can be expressed as multiplications with block diagonal matrices one level above, compare Equation (3.4) and (3.5). We consider the update of Q in Algorithm 1:

$$Q = Q_D \begin{bmatrix} Q_L & \\ & Q_R \end{bmatrix}. \tag{4.1}$$

In order to avoid this operation in the bottom-up phase of the recursion, we have to split Q into a Q_L and a Q_R part, where Q_L is of order l. Then we obtain

$$Q(:, 1 : l) = Q_D(:, 1 : l)Q_L \quad \text{and} \quad Q(:, l + 1 : m) = Q_D(:, l + 1 : m)Q_R, \tag{4.2}$$

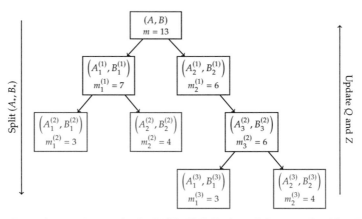

Figure 4.1.: Example recursion tree for the Divide-Shift/Scale-and-Conquer algorithm. Red nodes are trivially solvable and blue nodes are processed by divide-and-conquer again.

which can be computed independent of each other. Now these operations are moved to the step, where Q_L and Q_R become available. The updates on Z are handled analogously. In both cases the recursion has to keep track of the position l inside Q and Z to make both matrices shared over the whole procedure. This generates Algorithm 6 as a modified version of Algorithm 1. These modifications can be transferred directly to Algorithms 4 and 5.

Using this rearrangement of the operations during the recursion makes our algorithms, from the recursion point of view, equivalent to Quicksort. One can think of the deflation step with the subsequent update of Q and Z as the partition step in Quicksort.

In order to use strategies, known from Quicksort, to remove the recursion in our algorithms, we need to introduce the stack data structure.

Definition 4.1 ([110, 148]):
*A **stack** is a container of undetermined but positive size for arbitrary data, which provides two operations:*

- ***push**, which adds a new element to the container and*

- ***pop**, which retrieves the latest pushed element from the container.*

*Additionally, there is a property **isempty** indicating whether the are elements left in the container.*

Since the stack is an object of undetermined size, we see later in a theorem how we can limit its size in our application.

Remark 4.2:
Although the stack is an elementary data structure and used inside many algorithms it is not part of the standard library of programming languages like C or Fortran, which we use for implementation.

Now, we integrate the concept of the non-recursive Quicksort implementation [147] into the modified divide-and-conquer framework from Algorithm 6. Thereby, each entry on the stack consists of a matrix pair $\left(\hat{A}, \hat{B} \right)$, the column offset in Q and Z, and the dimension of the subproblem m_s. The algorithm starts with pushing $((A, B), 1, m)$ onto the stack and ends when the stack gets empty. Every time Algorithm 6 would be called by the recursion, we push the corresponding

---

Algorithm 6 Modified Divide-and-Conquer Framework for the Generalized Schur Decomposition

---

Input: $(A, B) \in \mathbb{C}^{m \times m} \times \mathbb{C}^{m \times m}$, Q and Z modified in-place, column offset l for Q and Z.
Output: Unitary $Q \in \mathbb{C}^{m \times m}$ and $Z \in \mathbb{C}^{m \times m}$ modified in-place such that $(Q^H A Z, Q^H B Z)$ is in generalized Schur decomposition.

1: **if** m is small enough **then**
2:     Compute Q_D and Z_D from (A, B) directly using the QZ algorithm [129].
3:     Update

$$Q(:, l : l + m - 1) \leftarrow Q(:, l : l + m - 1)Q_D$$

    and

$$Z(:, l : l + m - 1) \leftarrow Z(:, l : l + m - 1)Z_D.$$

4: **else**
5:     Determine a deflating subspace $\mathcal{X}$ for (A, B).
6:     Compute Q_D and Z_D from $\mathcal{X}$ using Equations (3.8) and (3.12) to obtain

$$(Q_D^H A Z_D, Q_D^H B Z_D) = \left(\begin{bmatrix} \hat{A}_{11} & \hat{A}_{12} \\ 0 & \hat{A}_{22} \end{bmatrix}, \begin{bmatrix} \hat{B}_{11} & \hat{B}_{12} \\ 0 & \hat{B}_{22} \end{bmatrix} \right).$$

7:     Update

$$Q(:, l : l + m - 1) \leftarrow Q(:, l : l + m - 1)Q_D$$

    and

$$Z(:, l : l + m - 1) \leftarrow Z(:, l : l + m - 1)Z_D.$$

8:     Rerun Algorithm 6 again for $\left(\hat{A}_{11}, \hat{B}_{11} \right)$ with offset $l = 1$ for Q and Z.
9:     Rerun Algorithm 6 again for $\left(\hat{A}_{22}, \hat{B}_{22} \right)$ with offset $l = \text{size} \left(\hat{A}_{11} \right) + 1$ for Q and Z.
10: **end if**

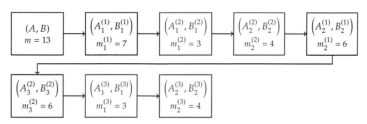

Figure 4.2.: Preorder traversal of the recursion tree from Figure 4.1. Red nodes are trivially solvable and blue nodes are pushed on the stack again.

subproblem onto the stack. Furthermore, we can implement the update 6 in Algorithm 6 such that the (A, B) is overwritten with the deflated version. Now, pushing only the offset and the size of the newly emerged subproblem to the stack provides enough information to let the algorithm work on the corresponding fraction of the initial locations of (A, B). Finally, this yields Algorithm 7.

Regarding the recursion tree in Figure 4.1, we obtain the preorder traversal of the tree [148], and Figure 4.2 shows the execution order if Algorithm 7 is applied to this example. Although the preorder traversal looks more sequential than the original recursion tree, some parallelization can

---

Algorithm 7 Recursion-Free Divide-and-Conquer Framework for the Generalized Schur Decomposition

---

Input: $(A, B) \in \mathbb{C}^{m \times m} \times \mathbb{C}^{m \times m}$.
Output: Unitary $Q \in \mathbb{C}^{m \times m}$ and $Z \in \mathbb{C}^{m \times m}$ modified in-place such that $\left(Q^H A Z, Q^H B Z\right)$ is in generalized Schur decomposition.

1: $Q = I_m, \quad Z = I_m$
2: $Stack \leftarrow \mathrm{push}\,(1, m)$.
3: **while** not isempty($Stack$) **do**
4:     $(l, m_s) \leftarrow \mathrm{pop}(Stack)$
5:     $b \leftarrow l : l + m_s - 1, \quad \hat{A} = A(b, b), \quad \hat{B} = B(b, b)$
6:     **if** m_s is small enough **then**
7:         Compute Q_D and Z_D from $\left(\hat{A}, \hat{B}\right)$ directly using the QZ algorithm [129].
8:         Update

$$Q(1 : m, b) \leftarrow Q(1 : m, b) Q_D$$

    and

$$Z(1 : m, b) \leftarrow Z(1 : m, b) Z_D.$$

9:     **else**
10:         Determine a deflating subspace $\mathcal{X}$ for $\left(\hat{A}, \hat{B}\right)$.
11:         Compute Q_D and Z_D from $\mathcal{X}$ using Equations (3.8) and (3.12) to obtain

$$(A(b, b), B(b, b)) \leftarrow \left(Q_D^H \hat{A} Z_D, Q_D^H \hat{B} Z_D\right) = \left(\begin{bmatrix} \hat{A}_{11} & \hat{A}_{12} \\ 0 & \hat{A}_{22} \end{bmatrix}, \begin{bmatrix} \hat{B}_{11} & \hat{B}_{12} \\ 0 & \hat{B}_{22} \end{bmatrix}\right),$$

    where $\left(\hat{A}_{11}, \hat{B}_{11}\right)$ is of order $\hat{m}$.
12:         Update

$$Q(1 : m, b) \leftarrow Q(1 : m, b) Q_D$$

    and

$$Z(1 : m, b) \leftarrow Z(1 : m, b) Z_D.$$

13:         $Stack \leftarrow \mathrm{push}\,(l, \hat{m}) \quad$ if $\hat{m} \leq 0$.
14:         $Stack \leftarrow \mathrm{push}\,(l + \hat{m}, m_s - \hat{m}) \quad$ if $\hat{m} < m_s$.
15:     **end if**
16: **end while**

---

also be included. Due to the Deflation Theorem 2.6, all leaves of the recursion tree can be handled independently. This enables us to neglect these nodes, i.e. the red ones, in the preorder traversal and handle them in parallel afterwards. Details about this approach are given in Section 4.1.3 later.

In the worst case, the recursion tree of Quicksort, and thus of our algorithms, can grow up to a depth of m. Since we do not want to extend the memory requirements too much, we have to find a way such that we can express such a recursion tree with a much smaller stack. Therefore, the following theorem gives an upper bound for the stack size if the newly emerged subproblems are pushed in a conscious order to the stack.

Theorem 4.3 ([147]):

We assume a Quicksort-like algorithm implemented recursion-free like Algorithm 7 and an input problem of size m. Then, if the larger one of the newly emerged subproblems is pushed onto the stack first, the number of elements on the stack never exceeds $\lceil \log_2 m \rceil$.

Corollary 4.4:

On computer systems with and address space of k bit, i.e. 2^k memory locations can be addressed by the processor, the maximum stack size in Algorithm 7 is limited to k.

The corollary shows that the recursion free variant can be implemented without causing an overflow through an infinitely increasing stack. Furthermore, most of the operations can be done in-place on the input data. Since all newly emerging subproblems never exceed order m, the auxiliary memory allocated of size $5m^2$ at the beginning for the initial problem is reused in each iteration. The memory requirements for the stack do not exceed more than 64 entries on a 64-bit computer architecture. In this way, the maximum size of the stack is known at compile time and is independent of the order m of the current problem, since m cannot exceed the maximum number of elements that can be addressed by the computer. Limiting the memory requirements in this way allows us to forecast if the algorithm will work without requiring the operating system to use the eventually existing swap memory of the computer.

4.1.2. Level-Set Traversing of the Recursion Tree

Since there exist several strategies to traverse the recursion tree, we present an alternative, which allows an easier parallelization and adjustment to modern hardware architectures. The main idea is switching from the preorder traversal scheme to level-set traversal. This means that all subproblems with the same recursion depth in Figure 4.1 are handled in the order of appearance. After finishing them, the depth is increased by one and the workflow starts on the next level again. To model this, we only have to replace the stack in Algorithm 7 by a queue:

Definition 4.5 ([110, 148]):

*A **queue** is a container of undetermined but positive size for arbitrary data, which implements the following two operations:*

- ***enqueue**, which adds a new element to the end of the existing elements and*

- ***dequeue**, which retrieves the first enqueued element from the container.*

*Additionally, there is a property **isempty** indicating whether the are elements left in the container.*

Remark 4.6:

Although the queue is an elementary data structure as the stack, it is not part of the C and Fortran standard library as well.

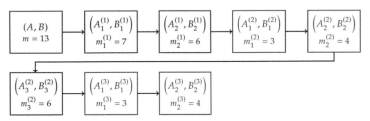

Figure 4.3.: Level-set traversal of the recursion tree from Figure 4.1. Red nodes are trivially solvable and blue nodes are enqueued again.

This change results in the execution order shown in Figure 4.3. Unfortunately, the small upper bound on the size of the stack given in Theorem 4.3 cannot be transferred to the queue-based recursion free approach. Here, a trivial upper bound is given by the following Theorem:

Theorem 4.7:
If Algorithm 7 uses a queue instead of a stack the maximum number of elements in the queue is limited by m, where m is the order of the input problem.

Proof. In the binary recursion tree, the number of elements on the same level is lower than m. Obviously, the smallest eigenvalue problem which emerges in the divide-and-conquer scheme is of order 1. Furthermore, this can only happen until the recursion tree has m leaves. This sets m as upper bound for the size of the queue, since the parent nodes are removed from it when the child nodes are added. To show that this bound is sharp, we assume $m = 2^M$, $M \in \mathbb{N}$ and the matrix pair (A, B) is split in two equally sized problems on each spectral division. Setting the level of the initial problem in the recursion tree to 0, we have 2^l subproblems of size $\frac{m}{2^l}$ on level l. For $l = M$, this results in m subproblems of order 1 in the queue at the begin of the level M. □

Remark 4.8:
Although the size of the queue is not invariant with respect to the problem dimension, the additional memory of size $O(m)$ is negligible. Hence, each element only requires $O(1)$ memory independent of the order of the input data.

For a practical implementation, we replace the abstract data structure of a queue using an integer vector q of dimension m. Using this vector we create a mapping from the start column in the transformation matrices Q and Z to the dimension m_s of the corresponding subproblem. For each index l, the entry q_l gives the size of the subproblem beginning in column l of Q and Z. At the beginning, we initialize $q_1 = m$. Every time a new subproblem emerges at column k in Q (and Z), the entry q_k is set to the size of the subproblem. Overall, this yields the reformulation of our general framework shown in Algorithm 8. Already solved subproblems are indicated by setting $q_k = -m_s$. In this way, a second information is stored in q. The procedure ends when the sum of all entries in q is $-m$. This enables us to implement the algorithm without using advanced abstract data types like a stack or a queue. Especially when we implement the divide-and-conquer schemes in Fortran we do not need any compound data type. Everything can be realized using an integer array.

In difference to the preorder traversal of the recursion tree, the level-set approach focuses on an earlier splitting of the large emerging subproblems. Furthermore, the level-set traversal provided additional properties that help us during the parallelization as we will see in the next section.

Algorithm 8 Level-Set Divide-and-Conquer Framework for the Generalized Schur Decomposition

---

Input: $(A, B) \in \mathbb{C}^{m \times m} \times \mathbb{C}^{m \times m}$
Output: Unitary $Q \in \mathbb{C}^{m \times m}$ and $Z \in \mathbb{C}^{m \times m}$ modified in-place such that $(Q^H A Z, Q^H B Z)$ is in generalized Schur decomposition.

1: $Q = I_m, \quad Z = I_m, \quad q = (0)_{l=1}^{m+1}$
2: $q_1 \leftarrow m$
3: **while** $\sum_{k=1}^{m} q_k \neq -m$ **do**
4:     $l \leftarrow 1$
5:     **while** $l \neq m + 1$ **do**
6:         $m_s \leftarrow q_l, \quad b \leftarrow l : l + m_s - 1, \quad \hat{A} = A(b, b), \quad \hat{B} = B(b, b)$
7:         **if** $m_s > 0$ and m_s is small enough **then**
8:             Compute Q_D and Z_D from $\left(\hat{A}, \hat{B}\right)$ directly using the QZ algorithm [129].
9:             Update
$$Q(1:m, b) \leftarrow Q(1:m, b)Q_D \text{ and } Z(1:m, b) \leftarrow Z(1:m, b)Z_D.$$

10:             $q_l \leftarrow -m_s, \quad l \leftarrow l + m_s$
11:         **else if** $m_s > 0$ **then**
12:             Determine a deflating subspace $\mathcal{X}$ for $\left(\hat{A}, \hat{B}\right)$.
13:             Compute Q_D and Z_D from $\mathcal{X}$ using Equations (3.8) and (3.12) to obtain
$$(A(b, b), B(b, b)) \leftarrow \left(Q_D^H \hat{A} Z_D, Q_D^H \hat{B} Z_d\right) = \left(\begin{bmatrix} \hat{A}_{11} & \hat{A}_{12} \\ 0 & \hat{A}_{22} \end{bmatrix}, \begin{bmatrix} \hat{B}_{11} & \hat{B}_{12} \\ 0 & \hat{B}_{22} \end{bmatrix} \right),$$
            where $\left(\hat{A}_{11}, \hat{B}_{11}\right)$ is of order $\hat{m}$.
14:             Update
$$Q(1:m, b) \leftarrow Q(1:m, b)Q_D \text{ and } Z(1:m, b) \leftarrow Z(1:m, b)Z_D.$$

15:             $q_l \leftarrow \hat{m}, \quad q_{l+\hat{m}} \leftarrow m_s - \hat{m}, \quad l \leftarrow l + m_s.$
16:         **else**
17:             $l \leftarrow l + |m_s|$ {Already solved.}
18:         **end if**
19:     **end while**
20: **end while**

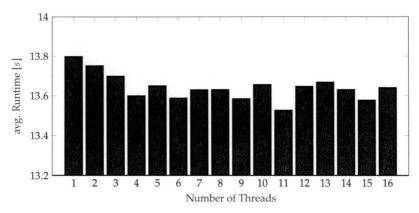

Figure 4.4.: Runtime of Intel® MKL's QZ Implementation on a 16 core Intel® Haswell System for $m = 1\,024$. (Benchmark setup see Appendix A)

4.1.3. Parallelization of the Recursion Tree

Regarding parallel capabilities in our algorithms, Theorem 2.6 is our biggest help. Using the level-set formulation (compare Algorithm 8), it states that all subproblems on the same level of the recursion tree can be handled independent of each other.

A natural idea to parallelize the algorithm would be to use common frameworks to handle all problems on the same level in parallel. Unfortunately, if we implement this using OpenMP [138] or PThreads [94], most of the BLAS implementations will detect that they are executed inside a parallel region. This causes them to switch to single thread operation mode [137, 96]. If this happens, even large problems only make use of a single CPU core during the computation of the deflating subspaces. Some BLAS libraries support nested parallelism, which needs to be adjusted for each level. Regarding the required flops to compute the generalized Schur decomposition, we need at least $117\frac{7}{9}m^3$ flops for the generalized matrix sign function or $270\frac{1}{9}m^3$ flops for the matrix disc function. Comparing this with the $66m^3$ flops [80] required for the QZ algorithm, we see that we need two to four CPU cores running perfectly in parallel in order to compete with single core runtime of the QZ algorithm. Hence, restricting the execution of our algorithms to a single thread will make them noncompetitive.

Another aspect to be aware of is that the current implementation of the QZ algorithm in LAPACK does not benefit from multi-threading. In order to illustrate this problem, we measure the influence multi-threading on the solution of ten random generalized eigenvalue problems of order $m = 1\,024$ using the current QZ algorithm in LAPACK. Figure 4.4 shows that utilizing many threads for computing the generalized Schur decomposition did not result in mentionable runtime savings. If the problems are smaller, even the remaining small speed up using many threads will vanish and a multi-core CPU is no longer necessary. This motivates to classify the recursion tree in two categories. First, we have the problems still handled by spectral splitting, which are rich in level-3 BLAS operations. Second, there are the trivial-to-solve subproblems only requiring a single CPU core. We rearrange Algorithm 8 such that the spectral splitting uses all CPU cores for the BLAS operations and the trivial-to-solve problems are processed afterwards in parallel using a single thread for each problem. In this way, all building blocks of Algorithm 8 benefit from the multi-core CPU. From the implementation point of view, this means that Algorithm 8 skips all small problems and afterwards they are solved in parallel. The benefit from running them in parallel is larger than running a single QZ algorithm with a parallel BLAS over all available CPU cores.

Example 4.9:

> *We consider the solution of 1 024 random generalized eigenvalue problems of order $m = 64$ on a 16 core Intel® Haswell system used. Solving them in parallel with only single threaded calls to BLAS and LAPACK takes 0.19 seconds. The sequential solution, i.e., solving all problems after each other but allowing the BLAS and LAPACK routines to use up to 16 CPU cores, which will never be done, takes 2.98 seconds. This accelerates the computation by a factor of 15.63.*

Translating the parallel loop over the remaining problems in Algorithm 8 into an OpenMP aware Fortran or C code causes some problems. Suppose the helper array q storing the subproblems only contains trivial-to-solve ones. Then the final step processing the trivial-to-solve subproblems would result in a loop with non-constant increments to get to the next subproblem. This type of loops cannot be processed using the parallel-do or parallel-for workshare constructs of OpenMP, since they require constant increments. In our case the increment to get to the next subproblem in the helper array q depends on the previous entries in q. Even without OpenMP, Fortran does not allow this in its do loop since its boundaries and its increment must be invariant during its execution. In contrast, C allows such behavior inside the three control statements of for loops, but OpenMP is not possible in this case.

We can think of two ways to solve this issue. On the one hand, the vector q can be compressed, such that it can be processed with a constant increment. We set

$$\hat{q}_1 = 1 \text{ and } \hat{q}_{k+1} = \hat{q}_k + q_{\hat{q}_k} \quad \forall \hat{q}_k \leq m. \tag{4.3}$$

Afterwards, $\hat{q}$ contains the start indices of the s remaining subproblems and the end marker $\hat{q}_{s+1} = m + 1$. The size of the k-th subproblem m_k is obtained by

$$m_k = \hat{q}_{k+1} - \hat{q}_k \quad \forall 1 \leq k \leq s. \tag{4.4}$$

The transformation from q to $\hat{q}$ can be done in-place on the location of q. Now, we iterate over $\hat{q}$ using a constant increment of one, which enables us to use OpenMP loop parallelization.

On the other hand, one can process the remaining subproblems in q using the while loop and create a new OpenMP task for each subproblem. In this way, $\hat{q}$ is not required. Since the creation of each task introduces an overhead, we prefer the loop based solution.

Regarding current computers with many CPU sockets and thus providing a non-uniform memory access scheme, one can think of even more sophisticated parallelization schemes. These require extended programming models and a complex memory management inside the algorithms. Since our target architectures only have two CPU sockets, we assume that the standard parallelization leads to decent results and do not focus on further approaches here.

4.1.4. Handling Trivial Problems

Until now, caused by the theory of computing roots of polynomials, we consider only eigenvalue problems of order up to four to be trivially solvable. Regarding the parallelization idea in the previous subsection, we can think of another definition. We first observed this from the parallelization point of view in [26] and [27]. This approach is extended and combined with an alternative motivation employing the different memory hierarchy levels. The idea was choosing the trivial-to-solve problems as large as possible such that they do not disturb a parallel ongoing solution on a different CPU core. This means, all data and additional memory required to execute the QZ algorithm must fit into the largest memory, which is exclusive to a single CPU core. This

CPU	excl. cache	C for double precision	$\frac{1}{2}\sqrt{C}$
Intel® Xeon® E5-2640v3 (Haswell)	level-2	32 768	≈ 88.5
Intel® Xeon® Silver 4110 (Skylake)	level-2	131 072	≈ 179
AMD Epyc 7501 (Zen)	level-2	65 536	≈ 126
IBM POWER 8	level-3	1 048 576	≈ 510
	level-2	65 536	≈ 126

Table 4.1.: Size of the largest exclusive CPU cache per core and the corresponding trivial-to-solve problem size using IEEE-754.2008 double precision.

enables fast access to all data required due to high speed of low level CPU caches. Even memory bandwidth bound operations, like Givens rotations [129, 80, 84], benefit from this. Depending on the hardware architecture, the largest exclusive memory to a CPU core is either the level-2 cache for Intel® Haswell-, Broadwell-, or Skylake-based Xeon® CPUs or the level-3 cache in case of IBM POWER 8 CPUs.

Besides $2m^2$ memory for the input matrix pair (A, B) and $2m^2$ memory for the matrices Q and Z, the QZ algorithm implemented in LAPACK needs at least auxiliary memory of size $8m + 16$ if it works in real arithmetic. For complex arithmetic, the necessary auxiliary memory needs to be at least $6m$ memory locations. Let $C > 0$ be the size of the largest exclusively used cache counted in elements of the desired precision, then the trivial-to-solve problem size $m_t > 0$ must fulfill in real arithmetic

$$C > 4m_t^2 + 8m_t + 16$$

$$\Rightarrow \quad m_t < \frac{1}{2}\left(\sqrt{C - 12} - 2\right) \quad (C > 12)$$

$$\Rightarrow \quad m_t < \frac{1}{2}\sqrt{C} - 2 \quad (C > 16) \tag{4.5}$$

or in complex arithmetic

$$C \geq 4m_t^2 + 6m_t$$

$$\Rightarrow \quad m_t \leq \frac{1}{4}\left(\sqrt{4C + 9} - 3\right)$$

$$\Rightarrow \quad m_t < \frac{1}{2}\sqrt{C}. \tag{4.6}$$

The requirements on C to be larger than 16 is not restrictive for practical applications because $C = 16$ and IEEE-754.2008 double precision data would result in a CPU cache of 128 bytes. In IEEE-754.2008 double precision, a real number takes 8 bytes in memory. Table 4.1 shows the cache sizes counted in double precision numbers and the trivial-to-solve problem sizes for several CPU architectures. The IBM POWER 8 CPU has also a level-3 cache exclusive to each CPU core, which is kept coherent such that other cores always have a correct view on the data in other cores' caches. Furthermore, the level-3 is used as cache for the memory controller, which makes the level-2 more favorable on this platform although the level-3 would be possible as well.

In order to verify this heuristic bound, we solve 100 different random generalized eigenvalues problems of order $m = 10, 20, \ldots, 300$ on each core of a 16 core Intel® Xeon® Haswell System. All 48 000 solves are done in parallel such that we force the CPU cores to disturb each other in the shared memory, once the problems do no longer fit into the exclusively owned memory. On each core, the QZ algorithm runs in a single thread without parallelization. In real double precision

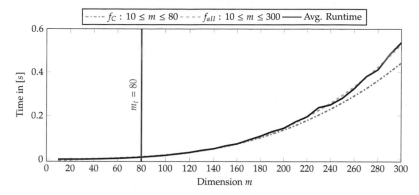

Figure 4.5.: Runtime of 16 QZ Algorithms running in parallel on a 16 core Intel® Haswell with varying problem dimension.

arithmetic and a 256kB exclusive level-2 cache, Formula (4.5) suggests

$$m_t < \frac{1}{2}\sqrt{\frac{256 \cdot 1024 Byte}{8 Byte}} - 2 = 88.5.$$

Since the level-2 cache is used for data and instructions of the algorithm we use a safety factor of 0.9 and set

$$m_t = 80$$

on this architecture. From the cubic operation count of the QZ algorithm [129], we interpolate a cubic polynomial $f(x) = a_3 x^2 + a_2 x^2 + a_1 x + a_0$ to connect the size of the input with the runtime. From the measurements with $m \leq m_t$ we obtain a polynomial f_C and from all measurements we get f_{all}. If the runtime gets influenced by the slower memory accesses on the non-exclusive cache or the disturbance of the parallel execution, the tangent of f_{all} would be steeper than f_C. Figure 4.5 shows exactly this behavior: the extrapolated runtime using f_C for problems larger than m_t is smaller than the measured one. This small experiment shows that the choice of m_t is reasonable for our purpose.

From this "trivial-to-solve" definition together with the initial memory requirement analysis in Section 4.1, we obtain for the maximum number of threads in real arithmetic:

$$\#\text{Threads} < \frac{5m^2}{2m_t^2 + 8m_t + 16}. \tag{4.7}$$

In the complex case, the QZ algorithm needs only $2m_t^2 + 6m_t$ additional memory for the trivial-to-solve problems. Thereby, m is the order of the input problem and m_t the trivial-to-solve problem size.

4.1.5. Computation of Deflating Subspaces and Common Operations

Before we present the efficient implementation of the two computationally most expensive operations, the generalized matrix sign function and the matrix disc function, details for matrix-matrix products and the subspace extraction are shown.

Matrix-Matrix Products Many operations, like the two-sided transformation of the matrix pair (A, B) by the matrices Q_D and Z_D (from Algorithm 4 or 5) or the right sided update of Q and Z are performed by general matrix-matrix (GEMM) products. The update $Q_D^H A Z_D$ (or $Q_D^H B Z_D$) decomposes into

$$M \leftarrow Q_D^H A \tag{4.8}$$

and

$$\hat{A} \leftarrow M Z_D, \tag{4.9}$$

where M is a temporary matrix of order m. The temporary matrix is necessary because the GEMM operation does not work in-place. Efficient implementations of the GEMM product for CPUs are available as single- and multi-threaded variants in all optimized BLAS libraries (Intel® MKL [96], OpenBLAS [137], IBM ESSL [97], BLIS [171]). Even if only single-thread optimal routines are available, a fast multi-threaded variant can be built with the help of OpenMP. On accelerator enabled architectures, the GEMM computations are accelerated by using automatic offloading to the accelerator devices. This can easily be implemented on top of the accelerator's BLAS library, or by using sophisticated vendor libraries. On Nvidia® CUDA architectures, the cuBLASXT extension of CUDA [132] provides this. On the IBM POWER platforms, the ESSL library [97] can make use of CUDA devices for acceleration of standard BLAS routines as well. Intel® introduced similar techniques for using Xeon® Phi accelerators in their MKL library. Since the optimization of the GEMM operation is a topic on its own and lies outside our scope, we assume that the given implementations are optimal.

Another variant of the GEMM operation required by our application is the multiplication of Q (or Z) with a square matrix from the right in Algorithm 1. Since an in-place multiplication is only possible if Q (or Z) is symmetric or triangular [66], which is in general not the case, we have to use a temporary matrix, here, as well. In this way, for Q and Q_D the following has to be implemented

$$M \leftarrow Q Q_D$$

and

$$Q \leftarrow M,$$

where M again is a temporary matrix of the same size as Q.

In both cases the memory required for the temporary matrix M is already covered by the memory requirements and allocations for the divide-and-conquer framework discussed at the beginning of this chapter.

Since we deal with unitary matrices in many parts of the algorithm, matrices partially exist as products of Householder transformations. In this case the matrix Q_D is represented by [80]

$$Q_D = \prod_{j=1}^{k} \left(I - \beta_j v_j v_j^H \right), \tag{4.10}$$

or with block transformations, i.e. resulting from a compact storage QR decomposition [40], by

$$Q_D = \prod_{j=1}^{K} \left(I - V_j T_j V_j^H \right). \tag{4.11}$$

Then the application of each single factor from the left to a matrix A of order m is performed as

$$\left(I - \beta_j v_j v_j^{\mathsf{H}}\right) A = A - \beta_j v_j \underbrace{v_j^{\mathsf{H}} A}_{w^{\mathsf{H}}} = A - \beta_j v_j w^{\mathsf{H}} \tag{4.12}$$

where w is only a vector of length m. In case of the block transformation with block size n_b ,this yields

$$\left(I - V_j T_j V_j^{\mathsf{H}}\right) = A - V_j T_j \underbrace{V_j^{\mathsf{H}} A}_{W} = A - V_j T_j W,$$

where W is a temporary matrix of size $m \times n_b$. Since T_j is upper triangular here, the product $T_j W$ is computed in-place. Both factorizations of the unitary matrix Q_D show a way to perform the matrix-matrix product with little temporary workspace nearly in-place on the matrix A. This is included in several variants in LAPACK, depending on the factorization of Q_D and its origin. In our case, this will mostly be the QR decomposition in the subspace extraction.

Deflating Subspaces The computation of the deflating transformations Q_D and Z_D from X is another important operation in our algorithms. The scheme proposed by Sun and Quintana-Ortí in [162] and recalled in Subsection 3.1.1 consists of two pivoted and/or rank revealing QR decompositions, together with some auxiliary operations. The following three variants of the QR decomposition are considered to compute Q_D and Z_D from X:

GEQPX/GEQPY The rank revealing QR decomposition implemented by Bischof and Quintana-Ortí [43, 42] is partly level-3 BLAS enabled and uses the τ_{ICE} tolerance to estimate the dimension of the deflating subspace. Since it does not support an initial guess for this estimation, like we obtain it from the properties in Corollary 3.10 or Theorem 3.16, the tolerance τ_{ICE} needs to be chosen carefully. The rank revealing procedure includes an expensive post processing step.

GEQP3 The column pivoted QR decomposition of LAPACK is the level-3 BLAS enabled variant of the Businger and Golub approach [49], developed by Quintana-Ortí, Sun, and Bischof [143]. Although the algorithm uses level-3 BLAS operations, the internal structure and Amdahl's law [9] limit the speed up on parallel architectures to approximately two. Still, it is faster than the GEQPX/GEQPY approach, since no post processing is required. For reliability and stability reasons, this implementation needs to be enhanced by the modified norm updates developed by Drmač and Bujanović [68]. In the case where the decomposition does not reveal the rank, the relative decoupling residual from Subsection 3.1.2 indicates a wrong spectral division.

HQRRP The column pivoted QR decomposition with random sampling is developed by Martinsson et al. [125] to overcome the level-3 limitations of the GEQP3 algorithm. By using random sampling, the algorithm nearly reaches the performance of a standard QR decomposition. Caused by the randomization it may fail, like GEQP3. Then the spectral division is incorrect. Again, the relative decoupling residual is used to detect this. The algorithm can act as a drop-in replacement for GEQP3 with compatible input and output. Similar randomization techniques are developed by Duersch and Gu in [69].

The GEQP3 and the HQRRP approach do not return the unitary or orthogonal matrix Q directly. Since the algorithms are based on Householder Transformations [40, 42, 125], the unitary/orthogonal matrix is returned as a product of those. Using the corresponding LAPACK routines they can either be applied to another matrix from the left or right, or can be used to set up the corresponding unitary/orthogonal matrix. Since the procedure shown in Algorithm 2 computes the column permuted matrix

$$\tilde{Z} = \begin{bmatrix} Z_2 & Z_1 \end{bmatrix}$$

instead of

$$Z = \begin{bmatrix} Z_1 & Z_2 \end{bmatrix},$$

we have to set up $\tilde{Z}$ and swap the column blocks afterwards. The second QR decomposition is required to compute the left part Q of the deflating transformation in Algorithm 2. There, the matrix Q stays in its Householder Transformation representation since it only needs to be applied from the left or right to other matrices. This avoids unnecessary flops to assemble Q first.

The most crucial point, in the computation of the deflating transformation, is the estimation of the rank p of $\tilde{Z}$ to decide which columns need to be swapped. On the one hand, we can use the Incremental Condition Estimation [41] from Subsection 3.1.1. On the other hand, the projection properties of the generalized matrix sign function (see Corollary 3.10) and of the matrix disc function (see Theorem 3.16) allow to compute the rank of Z_2 with the help of a few additional operations during the computation of the generalized matrix sign function, or matrix disc function.

As already mentioned, the trivial subproblems and the ones where the spectral splitting fails are solved using the QZ algorithm as implemented in LAPACK, namely GGES [129, 155], or its level-3 BLAS improved successor GGES3 [104]. More advanced algorithms, like the multi-shift aggressive deflation QZ by Kågström and Kressner [103], are not used, since they are not implemented in standard software yet and their runtime advantage compared to the LAPACK implementations strongly depends on the parameter, nontrivial, choice [103].

The overall procedure of computing the deflating transformations is implemented as a single subroutine:

SUBROUTINE GETQZ(M, A, LDA, B, LDB, Q, LDQ, S, LDS, RS, P1, P2, TAUICE, &
    & WORK, LWORK, IWORK, INFO)

Thereby, all matrices are passed including their leading dimensions, namely LDx. This enables to work in-place on submatrices. In order to save memory, the matrix S, containing the deflating subspace, is overwritten with $\tilde{Z}$. The column swap in $\tilde{Z}$ is done implicitly when Z is applied later on. The matrix Q contains its Householder representation, with the convention that the m-th column in the upper triangular part of Q contains the β_*, while the strict lower triangular part contains the Householder vectors, compare Equation (4.12). Since Q is constructed from $\begin{bmatrix} AZ_1 & BZ_1 \end{bmatrix}$, its memory location is at most of size $m \times 2m$. The RS argument provides the rank of the matrix S on input and P1 and P2 contain the sizes of the blocks in Z (or $\tilde{Z}$) on output. If RS is lower than zero on input, the incremental condition estimation from Subsection 3.1.1 is used with the given tolerance τ_{ICE} = TAUICE. The remaining parameters handle the workspace needed in addition to the input/output data. The required real or complex workspace is $8m + 1$ to compute the pivoted QR decompositions. For optimal performance, this should be larger, typically $70m + 32$. The pivoting vector stored in IWORK needs $2m$ integer elements. This workspace is additional to the estimations done in Section 4.1 since all the $5m^2$ memory locations are in use when deflating transformations are computed.

A final problem appears when the generalized Schur decomposition is computed by $Q^H(A, B)Z$ from the matrices Q and Z obtained by our algorithms. Since the algorithm computing Q and Z incorporates non-unitary operations and computers are only able to work in finite precision, the computation of $Q^H(A, B)Z$ will not lead to exact zero entries below the diagonal. The QZ algorithm uses the following heuristic to decide whether a subdiagonal entry of $Q^H(A, B)Z$ is zero or not [80, 129]:

$$a_{i,i-1}, b_{i,i-1} \leftarrow 0 \quad \text{if} \quad |a_{i,i-1}| \leq \varepsilon \left(|a_{i-1,i-1}| + |a_{i,i}| \right), \tag{4.13}$$

where $\varepsilon > 0$ is a tolerance factor. When we compute the generalized Schur decomposition $Q^H(A, B)Z$, we have to apply a similar heuristic. Thereby, a wrong choice of ε can influence the result dramatically. Values close to the machine precision have turned out as a good choice in practice.

Alternatively, the structure of our recursive scheme allows to track where zeros below the diagonal of $Q^H(A, B)Z$ have to appear. Therefore, we use the knowledge of the zero blocks introduced in each deflation step. Keeping track of them during the whole process allows us to set these zeros in the generalized Schur decomposition. The tracking is implemented using an integer vector of length m storing at the k-th position how many consecutive zero entries are inside column k, counted from the bottom of the matrix. Every time a deflating transformation was computed, the corresponding entries of the vector are updated using the obtained block structure. For the trivial subproblems, or the fallback to the QZ algorithm, this information is extracted from the generalized Schur decomposition of the subproblem, where the heuristic (4.13) is used.

4.2. Computation of the Generalized Matrix Sign Function

Most of the computational effort for the Divide-Shift-and-Conquer algorithm is required for the computation of the generalized matrix sign function

$$\text{sign}(A, B) = B \, \text{sign}\left(B^{-1}A\right) \tag{4.14}$$

using the Newton scheme presented in Subsection 3.2.1. Therefore, we recall the main aspects of Equation (3.27) for a matrix pair (A, B):

$$Y_0 = A$$
$$Y_{k+1} = \frac{1}{2}\left(\gamma_k Y_k + \gamma_k^{-1} B Y_k^{-1} B\right), \qquad k = 0, 1, \ldots, \tag{4.15}$$

where $\gamma_k = 1$, or γ_k is set to

$$\gamma_k = \left(\frac{|\det B|}{|\det Y_k|}\right)^{\frac{1}{m}} \tag{4.16}$$

to accelerate the convergence. Since the scheme converges globally quadratic, we use the relative change in Y_k as stopping criterion [18, 21, 161, 31]:

$$\|Y_{k+1} - Y_k\|_F \le \tau_S \|Y_k\|_F. \tag{4.17}$$

Thereby, τ_S is a tolerance factor usually set to $\tau_S = cm\varepsilon$, where m is the order of the matrix pair (A, B), ε is the machine precision, and c is a safety constant chosen as 10 or 100. In order to avoid stagnation, the quadratic convergence of the Newton Scheme (4.15) enables us to increase the tolerance factor further to $\tau_S = cm\sqrt{\varepsilon}$ at the cost of two or three additional steps after the criterion (4.17) is fulfilled. Other criteria, like checking the residual $\left\|I - \left((B^{-1}\, \text{sign}(A, B))^2\right)\right\|_F$, are computational more expensive.

Each iteration step in the Newton-Scheme (4.15) consists of the solution of a linear system

$$Y_k X = B \tag{4.18}$$

and a subsequent generalized matrix-matrix product

$$Y_{k+1} \leftarrow \frac{1}{2}\left(\gamma_k Y_k + \gamma_k^{-1} B X\right) \tag{4.19}$$

is required. Several strategies for an efficient implementation are discussed in the following subsections. The algorithms are annotated with the BLAS and LAPACK subroutines realizing the operations, to allow a straight forward implementation in higher programming languages like Fortran or C. The algorithms include eventually existing helper arrays and memory locations to focus more on the implementation aspects than on their mathematical formulation.

Algorithm 9 Naive LAPACK-based Newton-Iteration for sign (A, B)

Input: $(A, B) \in \mathbb{C}^{m \times m} \times \mathbb{C}^{m \times m}$, $\tau_s \in \mathbb{R}$ and maxit $\in \mathbb{N}$.
Output: A overwritten with sign (A, B)
1: $W_1 \leftarrow B$ {LAPACK: LACPY}
2: Overwrite W_1 with $[L^{(B)}, U^{(B)}, P^{(B)}] \leftarrow LU(W_1)$ {LAPACK: GETRF}
3: $d_B \leftarrow \prod_{i=1}^{m} \left(\left| U_{ii}^{(B)} \right| \right)^{\frac{1}{m}}, s \leftarrow 0$
4: **for** $k = 1, \ldots,$ maxit **do**
5:     $W_1 \leftarrow A, W_2 \leftarrow B$ {LAPACK: LACPY}
6:     W_1 overwritten with $[L^{(Y)}, U^{(Y)}, P^{(Y)}] \leftarrow LU(W_1)$ {LAPACK: GETRF}
7:     $W_2 \leftarrow W_1^{-1} W_2$ {LAPACK: GETRS}
8:     $d_B \leftarrow \prod_{i=1}^{m} \left(\left| U_{ii}^{(Y)} \right| \right)^{\frac{1}{m}}, \gamma \leftarrow \frac{d_B}{d_Y}$
9:     $W_1 \leftarrow A$ {LAPACK: LACPY}
10:    $A \leftarrow \frac{1}{2\gamma} B W_2 \cdot \frac{1}{2}\gamma A$ {BLAS: GEMM}
11:    **if** $||W_1 - A||_F < \tau_S \cdot ||W_1||_F$ **then**
12:        $s \leftarrow s + 1$, if $s \geq 3$ stop.
13:    **end if**
14: **end for**

4.2.1. Naive Implementation

Solving the linear system using the LU decomposition we end up with Algorithm 9. The implementation assumes that both input matrices A and B are allowed to be overwritten during the iteration. This does not conflict with our divide-and-conquer framework since we backup A and B every time we start the spectral splitting. The matrices W_1 and W_2 are both of order $m \times m$ and used as temporary workspace. The LU decomposition needs additional m integers to store its permutation $P^{(B)}$ or $P^{(Y)}$. On convergence, the matrix A is overwritten with the result sign (A, B).

Regarding the computational complexity of Algorithm 9 we have: Each LU decomposition takes $\frac{2}{3}m^3$ flops, $2m^3$ flops are required for solving with the LU decomposition, and another $2m^3$ flops to compute the final update from Equation (4.19) using the general matrix-matrix multiply (GEMM) operation from BLAS. The scaling parameter γ_k needs one LU decomposition of B before the first iteration to compute det B. Since we have the LU decomposition of Y_k in every iteration, we obtain det Y_k within $O(m)$ flops.

As the computation of the determinant of a matrix easily causes overflows, we rearrange Equation (4.16)

$$\gamma_k = \left(\frac{|\det B|}{|\det Y_k|} \right)^{\frac{1}{m}} = \left(\frac{|\det U^{(B)}|}{|\det U^{(Y)}|} \right)^{\frac{1}{m}} = \left(\frac{\prod_{i=1}^{m} \left| U_{ii}^{(B)} \right|}{\prod_{i=1}^{m} \left| U_{ii}^{(Y)} \right|} \right)^{\frac{1}{m}} = \frac{\prod_{i=1}^{m} \left(\left| U_{ii}^{(B)} \right|^{\frac{1}{m}} \right)}{\prod_{i=1}^{m} \left(\left| U_{ii}^{(Y)} \right|^{\frac{1}{m}} \right)} = \frac{d_B}{d_Y}, \quad (4.20)$$

where $U^{(B)}$ and $U^{(Y)}$ are the U factors of the LU decompositions of B and Y_k, respectively.

All auxiliary operations, like copying matrices and the evaluation of the convergence criterion, take $O(m^2)$ operations and can be neglected. Then, the overall process takes

$$\frac{2}{3}m^3 + it \cdot 4\frac{2}{3}m^3 \qquad (4.21)$$

flops, where it is the number of performed iterations.

```
!$omp parallel do reduction(+:DIFF,NRM) private(K)
DO L = 1, M
    !$omp simd reduction(+:NRM,DIFF)
    DO K = 1, M
        NRM = NRM + W1(K,L) ** 2
        DIFF = DIFF + (W1(K,L) - A(K,L))**2
    END DO
    !$omp end simd
END DO
!$omp end parallel do
```

Listing 4.1: OpenMP Parallelized and Vectorized Computation of the Cancelation Criterion for the Matrix Sign Function Computation

Since neither general LAPACK nor BLAS implementations provide a function to add or subtract matrices, the evaluation of the stopping criterion needs to be implemented directly. Thereby, two observations help to obtain an efficient implementation. First, $W_1 - A$ is not required afterwards, that means it is not required to be stored, and second the square of the Frobenius norm can be accumulated easily. Furthermore, the matrix W_1 is involved in both sides of the condition, i.e. every time we access an element in W_1, we can update the norm of the difference and the norm of W_1. Listing 4.1 shows how this is implemented in Fortran using OpenMP parallelization. Since the entries in each column are continuous in memory, the inner loop over the rows can be vectorized using the OpenMP simd statement. After execution, NRM contains $\|W_1\|_F^2$ and DIFF is set to $\|W_1 - A\|_F^2$. Both parallelization and vectorization are necessary to exploit the full memory bandwidth of current CPUs, since the execution speed of Listing 4.1 is bound by the de facto memory transfer rate.

Finally, we want to obtain the dimension of the deflating subspace corresponding to the eigenvalues inside the left or right half plane. In Sections 3.1.1 and 4.1.5, we already discussed the problem of estimating the rank from the pivoted QR decomposition. Corollary 3.10 shows that the rank with respect to the eigenvalues with positive real part is given by

$$p_+ = \frac{1}{2} \left(m + \mathrm{tr}\big(B^{-1} \operatorname{sign}(A, B)\big) \right). \tag{4.22}$$

The LU decomposition of B is already computed in order to obtain scaling parameter γ_k. Hence, it is only required to store this in an additional memory location W_3 of size m^2 until the iteration convergences. In this way we can compute the dimension of the deflating subspace without a heuristic at the cost of $2m^3$ flops for solving an additional linear system.

4.2.2. Reducing the Numerical Effort using One-Sided Transformations

The basic implementation, in the previous subsection, showed that the naive evaluation using the LU decomposition and general matrix-matrix products needs $4\frac{2}{3}m^3$ flops in each iteration. The Strassen based matrix-matrix products [158], which could be used to reduce the complexity from $O(m^3)$ to $O(m^{\log_2 7})$, are known to be inaccurate and increase the round-off errors and require additional memory. Our approach to minimize the number of flops uses the following lemma:

Lemma 4.10:

Let (A, B) be a matrix pair of order m with nonsingular B and $\operatorname{sign}(A, B)$ its generalized matrix sign function. Then for all decompositions $V\hat{B}W = B$, with nonsingular matrices V and W, it holds

$$\operatorname{sign}(A, B) = V \operatorname{sign}\left(V^{-1}AW^{-1}, \hat{B}\right) W. \tag{4.23}$$

Proof. We consider the Definition 3.8

$$\text{sign}(A, B) = B\,\text{sign}\left(B^{-1}A\right)$$

and insert the decomposition $V\hat{B}W = B$

$$\text{sign}(A, B) = V\hat{B}W\,\text{sign}\left(W^{-1}\hat{B}^{-1}V^{-1}A\right).$$

Then, Theorem 3.5 yields

$$\text{sign}(A, B) = V\hat{B}WW^{-1}\,\text{sign}\left(\hat{B}^{-1}V^{-1}AW^{-1}\right)W.$$

Using Definition 3.8 again, we obtain

$$\text{sign}(A, B) = V\,\text{sign}\left(V^{-1}AW^{-1}, \hat{B}\right)W. \qquad \square$$

The lemma helps us in the following way: If we find a decomposition $V\hat{B}W = B$ such that the multiplication with $\hat{B}$ is cheaper than with B, we can reduce the computational complexity of $\hat{B}Y_k^{-1}\hat{B}$. Especially, if V and W are unitary, this increases the costs of the overall algorithm by computing the decomposition and four matrix-matrix products. A first approach, where $\hat{B}$ is lower triangular and $W = I$, was introduced in the context of the solution of generalized Lyapunov equations by Benner and Quintana-Ortí [31]. Their idea uses a QL decomposition of B and the linear system (4.18) is solved by employing a pivoted LU decomposition with column interchanges. In our application this yields

$$Q_B L_B = B, \qquad (4.24)$$

where Q_B is unitary and L_B is lower triangular. The iteration (4.15) changes to

$$Y_0 = Q_B^H A \text{ and } Y_{k+1} = \frac{1}{2}\left(\gamma_k Y_k + \gamma_k^{-1} L_B Y_k^{-1} L_B\right) \qquad (4.25)$$

and we restore $\text{sign}(A, B)$ afterwards by

$$\text{sign}(A, B) = Q_B Y_\infty. \qquad (4.26)$$

The involved solution of a linear system with Y_k with the pivoted LU decomposition reads

$$X = \underbrace{P^\mathsf{T} U^{-1} L^{-1}}_{Y_k^{-1}} L_B. \qquad (4.27)$$

Thereby, the forward substitution $L^{-1}L_B$ can be done in $\frac{1}{3}m^3$ flops. Together with the backward substitution for U and the LU decomposition, the solution step costs $2m^3$ instead of $2\frac{2}{3}m^3$ flops. The multiplication with L_B from the left only costs m^3 flops. The overall cost for iteration step reduces from $4\frac{2}{3}m^3$ flops to $3m^3$. The whole iteration with the initial QL decomposition and two multiplications with Q_B and Q_B^H takes

$$\frac{4}{3}m^3 + 4m^3 + it \cdot 3m^3 \qquad (4.28)$$

flops, since γ_k is computed from L_B for free. After three iterations, this variant needs less flops than the naive approach.

This approach has two disadvantages from the implementation point of view. First, the QL decomposition is rarely maintained in many software packages. Furthermore, many optimizations like the Tile-QR approach [51] are not yet transferred to the QL decomposition. The second disadvantage is that the efficient forward substitution with L requires that the LU decomposition of Y_k is done without pivoting or with column interchanges. Third, all major software packages only implement the LU decomposition with row interchanges providing $Y_k = PLU$. To gain an advantage from the saved number of flops, we develop a modified version of (4.24) and (4.25), which uses the QR decomposition and the LU decomposition with row interchanges. We compute

$$Q_B R_B = B,\qquad(4.29)$$

where Q_B is unitary and R_B is upper triangular. Combined with the LU decomposition of Y_k this gives

$$Y_0 = Q_B^{\mathsf{H}} A \quad \text{and} \quad Y_{k+1} = \frac{1}{2}\left(\gamma_k Y_k + \gamma_k^{-1} R_B \left(P_k L_k U_k\right)^{-1} R_B\right),\qquad(4.30)$$

where P_k is the row permutation matrix. In contrast to Equation (4.18) and (4.19), we evaluate the product from left to right and obtain

$$X = R_B U_k^{-1} L_k^{-1} P_k^{\mathsf{T}}\qquad(4.31)$$

and

$$Y_{k+1} = \frac{1}{2}\left(\gamma_k Y_k + \gamma_k^{-1} X R_B\right).\qquad(4.32)$$

This scheme takes the same number of flops as using the QL-decomposition but can be implemented with optimized routines from LAPACK. Still, the available routines in BLAS and LAPACK do not allow a concise and efficient implementation of this scheme. The two critical operations are the multiplication with a triangular matrix and the solution of a triangular linear system with a triangular left-hand side of the same kind. The first one, known as triangular matrix multiply (TRMM) in BLAS works in-place on the non-triangular matrix, which yields additional operations and memory transfers to finalize the update (4.32). The second one is neither implemented in BLAS/LAPACK nor full-featured in other common software packages. These issues are briefly discussed before we present the final optimized implementation of this idea.

Three Operand Triangular Matrix-Matrix Multiplication   The classic implementation of the triangular matrix-matrix multiplication computes the product of a triangular matrix with a general matrix working in-place on the later one. In BLAS [66], the required operation reads

$$X \leftarrow \alpha L_B X \quad \text{or} \quad X \leftarrow \alpha X R_B,$$

where L_B is a lower triangular, R_B an upper triangular matrix, and α a scaling factor. We assume that all matrices are square and of order m. Many applications require this operation in the same style as the general matrix-matrix multiplication:

$$Y \leftarrow \alpha L_B X + \beta Y \text{ or } Y \leftarrow \alpha X R_B + \beta Y.\qquad(4.33)$$

This can be realized in a three step algorithm using low-level operations. First, the matrix X is saved to another memory location and afterwards multiplied by L_B (or R_B) and α. Finally, the temporary matrix is added to the scaled Y.

We build a parallelized variant, which uses the OpenMP 4 task dependencies for efficient parallelization and load balancing across the CPU cores without copying the whole matrix X

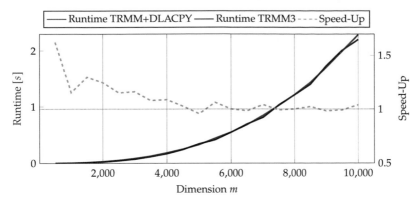

Figure 4.6.: Runtime of TRMM with LACPY and TRMM3 on a 16-core Intel® Skylake system.

before. A crucial point is that the memory locations of the unused triangle are not allowed to be modified. In many cases, like in the QR decomposition, these locations are used to store data independent of the triangular matrix for subsequent use. Hence, zeroing the unwanted part of the matrix storage and applying a general matrix-matrix multiplication, afterwards, is not an option.

We consider the partitioning of (4.33) in $p \times q$ blocks of size $N_B \times N_B$:

$$\begin{bmatrix} Y_{11} & \cdots & Y_{1q} \\ \vdots & & \vdots \\ Y_{p1} & \cdots & Y_{pq} \end{bmatrix} \leftarrow \alpha \begin{bmatrix} X_{11} & \cdots & X_{1q} \\ \vdots & & \vdots \\ X_{p1} & \cdots & X_{pq} \end{bmatrix} \begin{bmatrix} R_{11} & \cdots & R_{1q} \\ & \ddots & \vdots \\ & & R_{pq} \end{bmatrix} + \beta \begin{bmatrix} Y_{11} & \cdots & Y_{1q} \\ \vdots & & \vdots \\ Y_{p1} & \cdots & Y_{pq} \end{bmatrix}. \quad (4.34)$$

Then Y_{kl}, $k = 1, \ldots, p$, and $l = 1, \ldots, q$ is given by

$$Y_{kl} \leftarrow \beta Y_{kl} + \sum_{j=1}^{l-1} \alpha X_{kj} R_{jl} + \alpha X_{kl} R_{ll}, \quad (4.35)$$

where the last product is only a small triangular matrix-matrix multiplication. Since this is only of size $N_B \times N_B$, we can create a copy of R_{ll} and zero the lower part to use a general multiply operation. Alternatively, we create a copy of X_{kl} and use the TRMM operation with a subsequent matrix addition. The scheme is parallelized using one task for each block Y_{kl} in (4.35). This avoids race conditions by design and no task dependencies appear. Listing 4.2 shows a simplified Fortran-like implementation for this procedure performing the update (4.33), which is referred to as TRMM3 in the following. The variant for lower triangular matrices and multiplication from the left works in the same way.

The advantage over using TRMM with an extra copy and scaling is, that we only need a constant amount of memory independent of the size of the inputs. The only requirement is that each thread needs its own auxiliary memory. This is guaranteed by including the TMP array into the list of private variables for each task. Since the number of floating- point operations is essentially the same in the TRMM and the TRMM3 case, an acceleration can only be expected if we work on inputs of moderate size. For increasing dimensions both should result in the same runtime but with a smaller memory footprint for the TRMM3 case. In Figure 4.6, we see exactly this behavior. For the small problems we obtain a speed up from avoiding the additional copy. By implementing a TRMM3 kernel routine in the style of the high performance GEMM implementations [81], the additional memory can be removed as well.

```
SUBROUTINE TRMM3(M,N,NB,ALPHA,X,R,BETA,Y)
  DOUBLE PRECISION TMP(NB,NB)              ! Temporary Memory
  !$omp parallel
  !$omp master
  DO L=1,N,NB
    LB = MIN(NB,N-L+1)
    DO K=1,M,NB
      KB = MIN(NB,M-K+1)
      !$omp task firstprivate(K,KB,L,LB) private(TMP,REMAIN)
      REMAIN = M - (K+KB) + 1
      IF (REMAIN .GT.0) THEN
        GEMM(KB,LB,REMAIN,ALPHA,X(K,K+KB),R(K+KB,L),BETA,Y(K,L))
      END IF
      TMP(1:KB,1:LB) = B(K:K+KB-1,L:L+LB-1)
      TRMM(KB, LB, ALPHA, A(K,K), TMP(1,1))
      IF (REMAIN .EQ. 0) THEN
        Y(K:K+KB-1,L:L+LB-1) = BETA*Y(K:K+KB-1,L:L+LB-1)
                             + TMP(1:KB,1:LB)
      ELSE
        Y(K:K+KB-1,L:L+LB-1) = Y(K:K+KB-1,L:L+LB-1)
                             + TMP(1:KB,1:LB)
      END IF
      !$omp end task
    END DO
  END DO
  $omp end master
  $omp end parallel
END SUBROUTINE
```

Listing 4.2: Implementation of the TRMM3 Operation with OpenMP for $\alpha XR + \beta Y$. Y and X are of size $M \times N$, R of order N and NB the block size.

Remark 4.11:

The BLIS library [171] contains an implementation of the TRMM3 routine. Nevertheless, the routine is not available in BLAS and the BLAS wrapper does not contain the interface. This does not permit a painless and portable usage in Fortran Codes.

Solution of Triangular Systems with Triangular Right-Hand Sides   BLAS and LAPACK, as well as other libraries like BLIS, only support the solution of triangular linear systems with a generic right-hand side without paying attention to eventually existing structures. Even new developments, like no longer maintained ELEMENTAL [141], only support this operation for some special cases, e.g. only lower triangular matrices are supported.

Our algorithm needs the solution of either

$$XU = R_B \quad \text{or} \quad LX = L_B, \tag{4.36}$$

where U and R are upper triangular matrices and L and L_B are lower triangular matrices. Since triangular matrices form a group with respect to multiplication [80], the solution X is an upper triangular, or a lower triangular matrix, respectively. Without loss of generality we only consider $XU = R$, since we require this case for our implementation. As usual for linear system solves, the right-hand side will be overwritten by the solution.

The naive approach would solve for X row-wise using the BLAS routine TRSV. In this case, we

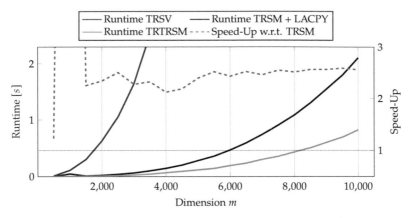

Figure 4.7.: Runtime of TRTRSM and its naive replacements on a 16-core Intel® Skylake system.

have to solve

$$X(k, k : m)U(k : m, k : m) = R_B(k, k : m) \quad \forall k = 1, \ldots, m \quad (4.37)$$

for each row k of X and R. Although this can be done in parallel, accessing the rows of a column-major stored matrix is expensive from a memory access point of view. The problem is that in column-major storage two consecutive elements in the same row have a distance of at least m elements in the memory. In order to avoid this problem, we create a copy of the actual row before solving the subproblem and copy it back afterwards.

Alternatively, one can create a temporary copy with the correct zero block, use TRSM, and copy the triangular part back to its original memory location. This approach needs $m \times m$ auxiliary memory locations and requires more floating-point operations than necessary.

To overcome these issues, we consider the block partitioning of $XU = R$

$$\begin{bmatrix} X_{11} & X_{12} \\ & X_{22} \end{bmatrix} \begin{bmatrix} U_{11} & U_{12} \\ & U_{22} \end{bmatrix} = \begin{bmatrix} R_{11} & R_{12} \\ & R_{22} \end{bmatrix}, \quad (4.38)$$

where X_{11}, U_{11}, and R_{11} are of size $N_B \times N_B$. Then, we solve $X_{11}U_{11} = R_{11}$ using the naive approach (4.37). Finally, we solve for X_{12} using the TRMM3 operation and a standard triangular solve:

$$X_{12}U_{22} = (R_{12} - X_{11}U_{12}).$$

Applying the same scheme again to a block partitioning of X_{22} provides the solution. Like the naive case, this can be implemented using easy loop parallelization over the block rows of X and R_B. Since the length of the rows is decreasing by N_B from block to block, a loop based parallelization will create an unbalanced parallel execution. Again, we create a single task for each loop as in the TRMM3 case to avoid this issue. In Listing 4.3, a simplified Fortran implementation with OpenMP is shown. According to the BLAS/LAPACK naming scheme, this routine is referred as TRTRSM (Triangular Solve on a Triangular Matrix). Figure 4.7 shows the runtime and the speed-up for this implementation on the Intel® Skylake architecture. We see that using the TRSV routine, although it preserves structure, leads to unacceptable runtime and will slow down the overall process dramatically. Comparing our approach with the one based on TRSM we obtain a speed up of 2.5 for almost all problem sizes. Only for small problems the TRSM approach is extremly slow in some cases causing a speed-up of 21.

```
SUBROUTINE TRTRSM(M,NB,U,R)
  !$omp parallel taskloop private(KB,MKB)
  DO K = 1, M, NB
    KB = MIN(NB,M-K+1)
    MKB = M-(K+KB)+1
    ! Solve the small problem for X(K:KB,K:KB)
    CALL TRTRSM_SMALL(KB,U(K,K),X(K,K))
    ! Solve the rest of the row X(K,K+KB:M)
    CALL TRMM3(KB,MKB,NB,-1.0, X(K,K),U(K,K+KB),1.0,X(K,K+KB))
    CALL TRSM(KB,MKB,1.0,U(K+KB,K+KB),X(K,K+KB))
  END DO
  !$omp end parallel taskloop
END SUBROUTINE
```

Listing 4.3: Implementation of the TRTRSM Operation with OpenMP for $XU = R$. X, U, and R of order M with block size NB.

The variants for $LX = L_B$ and other possible combinations are obtained in a similar manner.

Optimized Algorithm Using the routines for triangular matrices, we extend Algorithm 9 such that the flop count minimizing optimizations are included. For the already explained reasons, we follow the QR based approach, here, instead of the QL based approach by Benner and Quintana-Ortí [31].

First, the matrix B is overwritten with its QR decomposition and A with $Q_B^H A$. Therefore, we only need $m \cdot N_{QR}$ workspace, where N_{QR} is the block size in the blocked QR decomposition [40, 72]. Since the QR decomposition represents the matrix Q as Householder vectors, compare Equation (4.10), we overwrite B with Q_B and R_B. Only m or $m \cdot N_{QR}$ additional memory is required to save the Householder factors β_j from (4.10) or the matrices T_j from (4.11), respectively. Additionally, two temporary matrices W_1 and W_2 of order $m \times m$ are required, as in the naive implementation in Algorithm 9. Without the TRMM3 and TRTRSM routine, further copy operations are required. After convergence, the result is multiplied from the left with Q to restore sign (A, B).

The availability of the QR decomposition changes the computation of the dimension p_+ of X to

$$p_+ = \frac{1}{2}\left(m + \text{tr}\left((Q_B R_B)^{-1} \text{sign}(A, B)\right)\right) = \frac{1}{2}\left(m + \text{tr}\left(R_B^{-1} Q_B^H \text{sign}(A, B)\right)\right).$$

Together with Equation (4.26) this yields

$$p_+ = \frac{1}{2}\left(m + \text{tr}\left(R_B^{-1} Q_B^H Q_B Y_\infty\right)\right)$$
$$= \frac{1}{2}\left(m + \text{tr}\left(R_B^{-1} Y_\infty\right)\right). \tag{4.39}$$

The evaluation of the trace costs $m^3 + m$ flops if $T = R_B^{-1} Y_\infty$ is fully computed by backward substitution. The definition of the trace

$$\text{tr}(T) = \sum_{k=1}^{m} T_{kk}$$

shows that only the m entries of the diagonal of T are required. Computing R_B^{-1} allows switching to

$$\text{tr}(T) = \sum_{k=1}^{m} R_B^{-1}(k, :)^\mathsf{T} Y_\infty(:, k). \tag{4.40}$$

Algorithm 10 LAPACK-based Newton-Iteration for sign (A, B) with one-sided transformations.

Input: $(A, B) \in \mathbb{C}^{m \times m} \times \mathbb{C}^{m \times m}$, $\tau_s \in \mathbb{R}$ and maxit $\in \mathbb{N}$.

Output: A overwritten with sign (A, B), p_+ rank of $P_+ = \frac{1}{2} \left(I + B^{-1} \text{sign} (A, B) \right)$

1: Overwrite B with $Q^B R^B$ {LAPACK: GEQRT}

2: $A \leftarrow Q_B^H A$ {LAPACK: GEMQRT}

3: $d_B \leftarrow \prod\limits_{i=1}^{m} (|R_{Bii}|)^{\frac{1}{m}}, s \leftarrow 0$

4: **for** $k = 1, \ldots,$ maxit **do**

5:     $W_1 \leftarrow A, W_2 \leftarrow R_B$ {LAPACK: LACPY}

6:     W_1 overwritten with $[L^{(k)}, U^{(k)}, P^{(k)}] \leftarrow LU(W_1)$ {LAPACK: GETRF}

7:     $W_2 \leftarrow W_2 L^{(Y)^{-1}}$ {TRTRSM}

8:     $W_2 \leftarrow W_2 U^{(Y)^{-1}}$ {BLAS: TRSM}

9:     $W_2 \leftarrow W_2 P^{(Y)^{-T}}$ {LAPACK:LASWP}

10:     $d_Y \leftarrow \prod\limits_{i=1}^{m} \left(\left| U_{ii}^{(k)} \right| \right)^{\frac{1}{m}}, \gamma \leftarrow \frac{d_B}{d_Y}$

11:     $W_1 \leftarrow A$ {LAPACK: LACPY}

12:     $A \leftarrow \frac{1}{2}\gamma A + \frac{1}{2\gamma}W_2 R_B$ {TRMM3}

13:     **if** $||W_1 - A||_F < \tau_S \cdot ||W_1||_F$ **then**

14:        $s \leftarrow s + 1$, if $s \geq 3$ stop.

15:     **end if**

16: **end for**

17: $R_B \leftarrow R_B^{-1}$ {LAPACK: TRTRI}

18: $p_+ \leftarrow \frac{1}{2} \left(m + \sum_{k=1}^{m} R_B(k,:)^T A(:,k) \right)$

19: $A \leftarrow Q_B^H A$ {LAPACK GEMQRT}

Since R_B^{-1} is upper triangular, this costs $m^2 + m$ flops in addition to the triangular matrix inversion. The inversion of R_B is done in-place using TRTRI from LAPACK. In this way, we reduce the complexity for estimating the rank p_+ from $m^3 + m$ to $\frac{1}{3}m^3 + m^2 + m$ flops. Algorithm 10 shows the overall procedure for the generalized matrix sign function computation with all optimizations presented in this section. Beside the input data, the algorithm needs $2m^2 + m + mN_{QR}$ additional memory and m integers for storing the permutation of the LU decomposition. Thereby, $2m^2$ memory locations are already available in the divide-and-conquer scheme at the point, where the subspace is computed.

4.2.3. Reducing the Numerical Effort using Two-Sided Transformations

In the previous subsection we used Lemma 4.10 to reduce the number of required flops in the generalized matrix sign function iteration by reducing B to an upper triangular matrix. Thereby, we use only a one-sided transformation but Lemma 4.10 allows using two-sided ones as well.

In [161] Sun and Quintana-Ortí proposed an enhanced version of the scheme shown above, where B is reduced to bidiagonal form using unitary transformations from the left and the right. This can be achieved using Householder transformations [80]. Then we have

$$B_{\text{BD}} = \begin{bmatrix} b_1 & \hat{b}_1 & & \\ & \ddots & \ddots & \\ & & b_{m-1} & \hat{b}_{m-1} \\ & & & b_m \end{bmatrix} = X^H B W, \tag{4.41}$$

where X and W are unitary matrices of order m. Using Lemma 4.10, we set Y_0 to

$$Y_0 = X^H A W \tag{4.42}$$

and use B_{BD} instead of B in the iterations. Finally, we recover $\text{sign}(A, B)$ from Y_∞ using Lemma 4.10:

$$\text{sign}(A, B) = X Y_\infty W^H. \tag{4.43}$$

During the iteration, the computation of the update

$$Y_{k+1} = \frac{1}{2} \left(\gamma_k Y_k + \gamma_k^{-1} B_{BD} Y_k^{-1} B_{BD} \right)$$

now only costs $2m^3 + O(m^2)$ flops, which is basically the cost of the matrix inversion [161]. In this way the computation of the generalized matrix sign function gets nearly as cheap as the matrix sign function $\text{sign}(A)$ using the Newton iteration. Including the costs for the bidiagonalization, the initial transform, and the final transform, we end up with

$$\frac{8}{3}m^3 + 8m^3 + it \cdot 2m^3 \tag{4.44}$$

flops for the whole iteration. If we need at least four iterations to converge, this gets cheaper than the naive approach, compare Subsection 4.2.1.

Although the bidiagonalization is included in LAPACK as level-3 enabled algorithm, namely GEBRD, it does not scale well on multi-core CPUs. In Table 4.2, the runtimes for the pre- and post-processing, namely computing B_{BD}, Y_0, and $\text{sign}(A, B) = X Y_\infty W^H$, and one iteration step of the bidiagonal approach are compared to the duration of one iteration step of the naive implementation. Remember, the naive implementation does not require an expensive pre or post processing. Thereby, the turnaround presents the number of iteration steps required, in order to compensate the costs of the pre- and post-processing by the cost savings in each iteration step. Since the iteration normally needs around 15 to 25 steps to converge, we do not accelerate the algorithm by using the bidiagonal approach at all, in most cases. Another problem is the implementation of the matrix inversion in the MKL [96] library on the Intel® architectures, which does not achieve a high performance compared to the GEMM operation or the LU decomposition in the case of small matrices. In Table 4.2, we see that one iteration step in the bidiagonal case, which is mostly the matrix inversion, takes more than half of the time spent for the evaluation of one naively implemented iteration step, although it takes less than half of the flops. The implementation contained in the IBM ESSL [97] library on contrary shows nearly a perfect behavior. Section 4.2.4 will introduce a replacement for the LAPACK matrix inversion procedure to overcome this problem on the affected architectures.

Even if a fast matrix inversion is available, the computation of the bidiagonal matrix B_{BD} is too expensive. Thus, our goal is the replacement of reduction (4.41) by a more efficient one. Thereby the multiplication with the reduced matrix from the left or from the right in the iteration step should be cheaper than $2m^3$ flops. This can be achieved if B is reduced to an upper band matrix. Since this can be interpreted as block variant of the bidiagonalization, we can obtain unitary matrices X and W such that for $B \in \mathbb{R}^{pN_{BW} \times pN_{BW}}$ (or $B \in \mathbb{C}^{pN_{BW} \times pN_{BW}}$), $p \in \mathbb{N}$, it holds

$$B_{BAND} = \begin{bmatrix} \boxed{} & & \\ & \ddots & \ddots & \\ & & & \boxed{} \\ & & & \end{bmatrix} = \begin{bmatrix} U_1 \, L_1 & & & \\ & \ddots & \ddots & \\ & & U_{p-1} \, L_{p-1} & \\ & & & U_p \end{bmatrix} = X^H B W, \tag{4.45}$$

	Dimension m	2 000	4 000	6 000	8 000	10 000
Intel® Haswell	Bidiag. Pre + Post	0.44s	9.44s	45.34s	101.97s	190.16s
	Bidiag. Iter. Step	0.12s	0.56s	1.64s	3.78s	7.10s
	Naive Iter. Step	0.11s	0.81s	2.53s	5.69s	10.22s
	turnaround (#Iter.)	–	38	52	54	61
Intel® Skylake	Bidiag. Pre + Post	0.55s	5.47s	20.93s	47.82s	103.10s
	Bidiag. Iter. Step	0.15s	0.55s	1.60s	3.41s	6.31s
	Naive Iter. Step	0.20s	0.75s	2.29s	5.26s	10.03s
	turnaround (#Iter.)	12	28	31	26	28
IBM POWER 8	Bidiag. Pre + Post	0.88s	5.20s	23.71s	63.51s	115.68s
	Bidiag. Iter. Step	0.06s	0.38s	1.16s	2.63s	5.00s
	Naive Iter. Step	0.10s	0.76s	2.46s	5.66s	10.88s
	turnaround (#Iter.)	24	14	19	21	20

Table 4.2.: Runtime of the bidiagonal reduction and a single iteration step compared to the naive iteration step.

where U_* and L_* are upper and lower triangular matrices of order N_{BW}, respectively. If the order of B is not a multiple of N_{BW} the matrix U_p is a smaller upper triangular one and L_{p-1} is a lower trapezoidal one. The multiplication with B_{BAND} from the left (or the right) costs

$$\underbrace{p \cdot m N_{BW}^2}_{\text{multiplication with } U_*} + \underbrace{(p-1) \cdot m N_{BW}^2}_{\text{multiplication with } L_*} \overset{N_{BW}=\frac{m}{p}}{=} 2\frac{m^3}{p} - \frac{m^3}{p^2}$$

flops. For moderate sized bandwidths N_{BW}, p is larger than 1 and hence this is much smaller than $2m^3$. In the worst case, $p = 1$, we obtain the same algorithm scheme and the same flop count as presented in Subsection 4.2.2.

The upper band structure shown in (4.45) can be achieved using unitary transformations. First, the leading column block is reduced to upper triangular form using a QR decomposition. This yields

$$X_1^H B = \begin{bmatrix} \boxed{} & \boxed{} & \cdots & \boxed{} \\ & \boxed{} & \cdots & \boxed{} \\ & \vdots & \vdots & \vdots \\ & \boxed{} & \cdots & \boxed{} \end{bmatrix} = \tilde{B}_1. \tag{4.46}$$

Using an LQ factorization on the first block row, we generate a lower triangular matrix on the top of the second block column. We get

$$\tilde{B}_1 W_1 = \begin{bmatrix} \boxed{} & \boxed{} & & \\ & \boxed{} & \cdots & \boxed{} \\ & \vdots & \vdots & \vdots \\ & \boxed{} & \cdots & \boxed{} \end{bmatrix} = \hat{B}_1. \tag{4.47}$$

	Dimension m	2 000	4 000	6 000	8 000	10 000
Intel® Haswell	Band Pre + Post	0.23s	2.46s	9.15s	17.15s	32.10s
	Band Iter. Step	0.12s	0.61s	1.74s	4.16s	7.48s
	Naive Iter. Step	0.11s	0.81s	2.53s	5.69s	10.22s
	turnaround (#Iter.)	–	12	12	12	12
Intel® Skylake	Band Pre + Post	0.33s	1.95s	6.21s	14.18s	27.32s
	Band Iter. Step	0.12s	0.59s	1.76s	3.59s	6.69s
	Naive Iter. Step	0.20s	0.75s	2.29s	5.26s	10.03s
	turnaround (#Iter.)	5	13	12	9	9
IBM POWER 8	Band Pre + Post	0.61s	3.58s	12.00s	28.37s	48.48s
	Band Iter. Step	0.08s	0.42s	1.32s	2.84s	5.25s
	Naive Iter. Step	0.10s	0.76s	2.46s	5.66s	10.88s
	turnaround (#Iter.)	25	11	11	11	9

Table 4.3.: Runtime of the upper band reduction and a single iteration step compared to the single naive iteration step, $N_{BW} = 256$.

Repeating this procedure recursively on the remaining square beginning at the $(2, 2)$ block of $\hat{B}_1$ leads to the desired decomposition. The procedure is easily implemented using the LAPACK routines GEQRF and GELQF. Since the QR and the LQ decomposition store their Householder vectors in-place in the input matrix, this procedure only requires to store $2m - N_{BW}$ scalars for the β_* values of the Householder transformations. This is the same demand as for the bidiagonalization procedure. Using $N_{BW} = 256$, we obtain the runtimes presented in Table 4.3. We can see that using the upper band matrix with a moderate bandwidth leads to a huge performance increase in the pre and post processing. Further, using a band matrix instead of a bidiagonal matrix does not lead to a large increase on the time taken per iteration. Algorithm 11 shows the generalized matrix sign function computation with the two-sided iteration. If the computation of the next iterate in Step 10 and the computation of the convergence criterion are combined, the second helper matrix W_2 can be annihilated. In this way, Algorithm 11 only requires $m^2 + O(m)$ workspace to compute the generalized matrix sign function. If the bandwidth N_{BW} stays moderate, the overall number of flops is given by:

$$\frac{8}{3}m^3 + 8m^3 + it \cdot \left(2m^3 + O(m^2)\right). \tag{4.48}$$

The bandwidth $N_{BW} = 256$ seems to be selected arbitrary but in coincidence of the typical block sizes found in other algorithms from LAPACK [10]. An empirical study about its optimal value would minimize the turnaround point shown in Tables 4.2 and 4.3 and is subject to future work.

Especially, we see that the problem of computing the generalized matrix sign function of a matrix pair using the Newton approach reduces to the computation of the inverse of a matrix.

4.2.4. Gauss-Jordan Elimination and GPU acceleration

The optimizations presented above already use level-3 enabled operations, but some of them are not well suited for GPUs, especially if multiple accelerator devices come into play. The massive parallel architecture of the GPUs is well suited for operations like the general matrix-matrix multiplies but it does not reach high performance if the algorithm has an inherent sequential structure like the forward/backward substitution. Furthermore, some algorithms, like the general matrix-matrix multiply or the LU decomposition, can easily be distributed across several GPUs. Other operations, like the triangular solve, have a sequential character and are communication intensive. Hence, these are critical from this point of view. Therefore, in this section the algorithms to compute the

---

Algorithm 11 LAPACK-based Newton-Iteration for sign (A, B) with two-sided transformations

---

Input: $(A, B) \in \mathbb{C}^{m \times m} \times \mathbb{C}^{m \times m}$, $\tau_s \in \mathbb{R}$ and maxit $\in \mathbb{N}$.
Output: A overwritten with sign (A, B), p_+ rank of $P_+ = \frac{1}{2}\left(I + B^{-1}\text{sign}(A, B)\right)$

1: Overwrite B with $B_{\text{BAND}} = X^H B W$ and update $A \leftarrow X^H A W$

2: $d_B \leftarrow \prod\limits_{i=1}^{m}(|B_{\text{BAND}}(i, i)|)^{\frac{1}{m}}, s \leftarrow 0$

3: **for** $k = 1, \ldots, $ maxit **do**

4:     $W_1 \leftarrow A$ {LAPACK: LACPY}

5:     $W_1 \leftarrow W_1^{-1}$ {LAPACK: GETRF+GETRI}

6:     $W_1 \leftarrow W_1 B_{\text{BAND}}$

7:     $W_1 \leftarrow B_{\text{BAND}} W_1$

8:     $d_Y \leftarrow \prod\limits_{i=1}^{m}\left(\left|U_{ii}^{(k)}\right|\right)^{\frac{1}{m}}, \gamma \leftarrow \frac{d_B}{d_Y}$

9:     $W_2 \leftarrow A$ {LAPACK: LACPY}

10:    $A \leftarrow \frac{1}{2}\gamma A + \frac{1}{2\gamma}W_1$

11:    **if** $||W_2 - A||_F < \tau_S \cdot ||W_2||_F$ **then**

12:       $s \leftarrow s + 1$, if $s \geq 3$ stop.

13:    **end if**

14: **end for**

15: $B_{\text{BAND}} \leftarrow B_{\text{BAND}}^{-1}$

16: $p_+ \leftarrow \frac{1}{2}\left(m + \sum_{k=1}^{m} B_{\text{BAND}}(k, :)^{\mathsf{T}} A(:, k)\right)$

17: $A \leftarrow X^H A W$

---

generalized matrix sign function are rearranged to integrate multi-GPU aware operations in all major steps. Due to the reuse of the developed components and the limited memory on GPUs, we do not focus on a pure GPU implementation of the Divide-Shift-and-Conquer algorithm, here. The main effort is to identify computationally expensive tasks and find proper GPU-enabled replacements.

Regarding Algorithms 9, 10, and 11, a multi-GPU aware solution of a linear system with many right-hand sides, a computation of a matrix inverse, and the matrix-matrix product are key ingredients for fast execution. The MAGMA library [166, 167, 65] provides a multi-GPU aware LU decomposition but the subsequent solution step does only work using a single accelerator. This motivates to use an alternative algorithm. Following from previous experiments [111], the Gauss-Jordan-Elimination [142] and its GPU implementation [23, 111] are promising candidates, since it handles both cases appearing in our algorithms:

Inversion of a Matrix The inversion procedure performed by the Gauss-Jordan Elimination is equivalent to using the LU decomposition based inversion in LAPACK [142]. In this way we use it as drop-in replacement in Algorithm 11 to compute the inverse. Like the LU decomposition based approach, this works in-place taking $2m^3$ flops. Especially the dissatisfying performance of the Intel® MKL library, which already showed up in Subsection 4.2.3, motivates to develop a more efficient algorithm.

Solution of Linear Systems Algorithms 9 and 10 require the solution of a linear system with m right-hand sides. The Gauss-Jordan Elimination overwrites the right-hand side B with the solution X of the linear system $AX = B$ with $3m^3$ flops. This amounts to 12.5% more flops than using the LU decomposition. Previous experiments [111] showed that the performance gain of a GPU accelerated Gauss-Jordan-Elimination is more than 12.5% faster than the LU decomposition implemented in MAGMA.

Both steps, the inversion process and the solution of linear systems, can be combined but this is not required in our application. If we replace the solution of the linear system in Algorithm 9 with the Gauss-Jordan Elimination, the cost for one iteration increases from $4\frac{2}{3}m^3$ flops to $5m^3$ flops. However, the solution and the subsequent GEMM operation can be performed on multiple GPUs.

The flop reducing strategy from Subsection 4.2.2 cannot be implemented by switching to the Gauss-Jordan Elimination since no triangular factors are produced. Having the triangular factor R_B of B in Algorithm 10, we replace the solution step (4.31)

$$X = R_B U_k^{-1} L_k^{-1} P_k^\mathsf{T}$$

by

$$X = R_B Y_k^{-1},$$

where Y_k^{-1} is computed directly and the multiplication with R_B is done using the TRMM operation. This costs $3m^3$ flops instead of the $2m^3$ in the LU decomposition case. In the case of the second flop minimizing technique with initial band reduction, the computation of the inverse is replaced straight forward without further adjustments and does not increase the flop count.

Beside the mathematical reformulation to obtain a proper GPU accelerated scheme of the Gauss-Jordan-Elimination and the generalized matrix sign function, we focus on hardware related aspects, like hidden unnecessary memory accesses.

Gauss-Jordan Elimination on the CPU   Our implementation of the Gauss-Jordan Elimination is based on the work by Quintana-Ortí et al. [142] and Benner et al. [23] for the computation of the matrix inverse. The extension to linear systems with multiple right-hand sides is straight forward, see Köhler et al. [111]. We consider a general linear system $AX = B$, where no special properties of the matrices, except of A being nonsingular, are assumed. The main goal in the Gauss-Jordan Elimination is to find a sequence of transformations G_i such that

$$G_q \cdot G_{q-1} \cdots G_2 \cdot G_1 \begin{bmatrix} A & B \end{bmatrix} = \begin{bmatrix} I & X \end{bmatrix} \tag{4.49}$$

is fulfilled. Then, we have $A^{-1} = G_q \cdot G_{q-1} \cdots G_2 G_1$. We choose G_i, $i = 1, \ldots, q = m$, as

$$G_i = \frac{1}{\hat{a}_{ii}^{(i-1)}} \begin{bmatrix} 1 & & & -\hat{a}_{1i}^{(i-1)} & & \\ & \ddots & & \vdots & & \\ & & 1 & -\hat{a}_{(i-1)i}^{(i-1)} & & \\ & & & 1 & & \\ & & & -\hat{a}_{(i+1)i}^{(i-1)} & 1 & \\ & & & \vdots & & \ddots \\ & & & -\hat{a}_{mi}^{(i-1)} & & 1 \end{bmatrix}, \tag{4.50}$$

where $\hat{a}_*^{(i-1)}$ are the elements of $A^{(i-1)} = \left(\prod_{k=1}^{i-1} G_k \right) A$. This creates a one on the (i, i) position of the matrix $A^{(i)}$ by zeroing all remaining elements in the i-th column.

Since a naive multiplication with G_i is inefficient, we want to obtain a block algorithm. The algorithm should work in-place to save memory. That means, A is overwritten by A^{-1} and B is overwritten by the solution X. This is achieved by storing G_i in the i-th column of A since it is known that it has to contain a one on its diagonal position.

The derivation of a block algorithm works as follows. We rewrite the application of a single transformation G_i, to A and B, from the left into a rank-1 update [111] working in-place on the memory of A and B. This yields

$$\begin{bmatrix} A & B \end{bmatrix} \leftarrow -\frac{1}{\hat{a}_{ii}} \left(\hat{a}_{1i}, \cdots, \hat{a}_{(i-1)i}, 0, \hat{a}_{(i+1)i}, \ldots, \hat{a}_{mi} \right)^{\mathsf{T}} \begin{bmatrix} A_{i,\cdot} & B_{i,\cdot} \end{bmatrix} \tag{4.51}$$

with the subsequent correction of the i-th row

$$\begin{bmatrix} A_{i,\cdot} & B_{i,\cdot} \end{bmatrix} \leftarrow \frac{1}{\hat{a}_{ii}} \begin{bmatrix} A_{i,\cdot} & B_{i,\cdot} \end{bmatrix} . \tag{4.52}$$

By turning the rank-1 update into a rank-N_B update, we obtain the block algorithm, easily. Thereby, N_B is the block size of the algorithm. We partition the current augmented matrix $\begin{bmatrix} A & B \end{bmatrix}$ into

$$\begin{bmatrix} A & B \end{bmatrix} \rightarrow \begin{bmatrix} A_{11} & A_{12} & A_{13} & B_1 \\ A_{21} & A_{22} & A_{23} & B_2 \\ A_{31} & A_{32} & A_{33} & B_3 \end{bmatrix} , \tag{4.53}$$

where A_{22} is of dimension $N_B \times N_B$ and begins at the (i, i) position in A. Then, the updates (4.51) and (4.52) read

$$\begin{bmatrix} A & B \end{bmatrix} \leftarrow \begin{bmatrix} A_{11} & 0 & A_{13} & B_1 \\ 0 & 0 & 0 & 0 \\ A_{31} & 0 & A_{33} & B_3 \end{bmatrix} + \underbrace{\begin{bmatrix} -A_{12}A_{22}^{-1} \\ A_{22}^{-1} \\ -A_{32}A_{22}^{-1} \end{bmatrix}}_{H} \begin{bmatrix} A_{21} & I_{N_B} & A_{23} & B_2 \end{bmatrix} . \tag{4.54}$$

By performing this for $i = 1, N_B + 1, 2N_B + 1, \ldots, (q-1)N_B, q = \lceil m/N_B \rceil$, we obtain the block version of (4.49). It is easy to see that the update is now a rank-N_B update which is a general matrix-matrix multiply. In this way the idea of applying Gauss-Transforms to invert a matrix or solve a linear system is realized by the inverse of a small matrix A_{22} and one large matrix-matrix multiply in each of the q steps.

Remark 4.12:
In order to guarantee that $\hat{a}_{ii}^{(i-1)} \neq 0$ and to improve the numerical stability, we introduce pivoting as known from the LU decomposition. Each transformation G_i precedes a row permutation P_i which swaps the current row i with the row $p = \text{argmax}_{l \geq i} |\hat{a}_{li}^{(i-1)}|$, where $\hat{a}_{\bullet}^{(i-1)}$ are the elements of $\left(\prod_{k=1}^{i-1} G_k P_k \right) A$. In this way, we obtain

$$G_m \cdot P_m \cdot G_{m-1} \cdot P_{m-1} \cdots G_2 \cdot P_2 \cdot G_1 \cdot P_1 \begin{bmatrix} A & B \end{bmatrix} = \begin{bmatrix} I & X \end{bmatrix} . \tag{4.55}$$

Using the fact that P_i are permutation matrices and the scheme works like the LU decomposition, we rewrite (4.55) into [142, 80]

$$\tilde{G}_m \cdot \tilde{G}_{m-1} \cdots \tilde{G}_2 \cdot \tilde{G}_1 \cdot P \begin{bmatrix} A & B \end{bmatrix} = \begin{bmatrix} I & X \end{bmatrix} , \tag{4.56}$$

with $P = P_{m-1} \cdots P_1$. Now, the product $\prod_{k=1}^{m} \tilde{G}_k$ gives $A^{-1}P^{\mathsf{T}}$.
In the case of the block algorithm, this needs to be integrated into the computation of the update matrix H, as well. We either achieve this by using a pivoted Gauss-Jordan-Elimination scheme or an LU decomposition on the column $\begin{bmatrix} A_{12}^{\mathsf{T}} & A_{22}^{\mathsf{T}} & A_{32}^{\mathsf{T}} \end{bmatrix}^{\mathsf{T}}$. Since the LU decomposition is optimized in almost all

LAPACK implementations, such as the ones from OpenBLAS [137], Intel® MKL [96], or IBM ESSL [97], we prefer this approach, here. Then the computation of H works in the following way:

$$\begin{bmatrix} L_1 & L_2 \end{bmatrix} U_1 = \hat{P} \begin{bmatrix} A_{22} \\ A_{23} \end{bmatrix} \quad and \quad H := \begin{bmatrix} -A_{12}U_1^{-1}L_1^{-1} \\ U_1^{-1}L_1^{-1} \\ -L_2L_1^{-1} \end{bmatrix}.$$

Finally, the row permutation matrix $\hat{P}$ needs to be applied to

$$\begin{bmatrix} A_{21} \\ A_{32} \end{bmatrix}, \quad \begin{bmatrix} A_{23} \\ A_{33} \end{bmatrix}, \quad and \quad \begin{bmatrix} B_2 \\ B_3 \end{bmatrix}$$

before the update (4.54) *is performed.*

The overall procedure inverts A and solves $AX = B$, with B of order m, taking $4m^3$ flops. These are 50% more flops than using the LU decomposition if only the solution of the linear system is required. The number of flops can be reduced to $3m^3$ flops in this case if the update on the first block column in (4.54) is neglected. Then the procedure costs 12.5% more flops than the LU decomposition. If the performance of the rank-N_B update is much higher than the performance of the triangular solves on an architecture, it is still beneficial to use the Gauss-Jordan Elimination instead of the LU decomposition. Furthermore, the rank-N_B update is easily distributable across multiple accelerators.

The update (4.54) is the main operation for the algorithm. Regarding an efficient realization on the CPU, as well as on the GPU, we have to rearrange (4.54) into four disjoint updates:

$$\begin{bmatrix} A_{11} \\ A_{21} \\ A_{31} \end{bmatrix} \leftarrow \begin{bmatrix} A_{11} \\ 0 \\ A_{31} \end{bmatrix} + HA_{21}, \tag{4.57a}$$

$$\begin{bmatrix} A_{12} \\ A_{22} \\ A_{32} \end{bmatrix} \leftarrow \begin{bmatrix} 0 \\ 0 \\ 0 \end{bmatrix} + HI_{N_B}, \tag{4.57b}$$

$$\begin{bmatrix} A_{13} \\ A_{23} \\ A_{33} \end{bmatrix} \leftarrow \begin{bmatrix} A_{13} \\ 0 \\ A_{33} \end{bmatrix} + HA_{23}, \tag{4.57c}$$

and

$$\begin{bmatrix} B_1 \\ B_2 \\ B_3 \end{bmatrix} \leftarrow \begin{bmatrix} B_1 \\ 0 \\ B_3 \end{bmatrix} + HB_2. \tag{4.57d}$$

The first observation from (4.57b) is that the second block column is overwritten by H. That means the computation of H can be done in-place on the data and no additional memory is required for H. The second observation is that the remaining three updates (4.57a), (4.57c), and (4.57d) need the second block row zeroed out before overwriting it. Regarding that the standard matrix-matrix product in BLAS realizes $D \leftarrow \beta D + \alpha EF$, we have to split this further into three products. In the case of (4.57a) this gives:

$$A_{11} \leftarrow A_{11} + H_1 A_{21}, \quad A_{21} \leftarrow 0 + H_2 A_{21}, \text{ and } A_{31} \leftarrow A_{31} + H_3 A_{21}. \tag{4.58}$$

Thereby A_{21} needs to be copied to a temporary location before, because, on the one hand, the result of a matrix-matrix product is not allowed to share its memory location with one of the two factors.

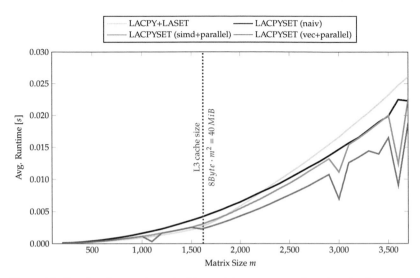

Figure 4.8.: Runtime for copying a double precision $m \times m$ matrix and setting it to zero.

On the other hand, if they are implemented in the presented order, A_{21} is overwritten although its old value is required to update A_{31}. The main disadvantage of this approach is the splitting of large matrix-matrix products into three smaller ones, where only one has a constant number of rows. The other two are varying in the number of rows depending on the diagonal position i. Since we need to back up the A_{21} block anyway, we zero its original location and use a single large matrix-matrix multiply. This yields the following scheme:

$$A_{21} \rightarrow W, \quad 0 \rightarrow A_{21}, \text{ and } \begin{bmatrix} A_{11} \\ A_{21} \\ A_{31} \end{bmatrix} \leftarrow \begin{bmatrix} A_{11} \\ A_{21} \\ A_{31} \end{bmatrix} + HW. \tag{4.59}$$

Although the number of flops for both approaches are the same, the approach using the large matrix-matrix product allows a better utilization of the parallel features inside the BLAS implementation.

The backup and set-to-zero operations can get a critical point in the implementation depending on the hardware capabilities. On almost all hardware architectures, writing to a variable causes the variable to be transferred from the memory to the cache, where its value is updated, and only afterwards written back to the memory [84]. This procedure is called *write-allocation*. Simplified, this means setting a double precision variable to zero causes 16 bytes to be transferred instead of eight. This effect is increased if $A_{21} \rightarrow W$ and $0 \rightarrow A_{21}$ are realized using the LAPACK routines:

```
CALL LACPY(NB, N, A(I,1), LDA, W, NB)
CALL LASET(NB, N, 0.0, 0.0, A(I,1), LDA)
```

where N is the number of columns of A_{21}. The LACPY operation fetches the $2N_B N$ elements of A_{21} and W from the memory and writes the $N_B N$ elements of W back. The subsequent LASET operation reads the $N_B N$ elements of A_{21} from the memory although it needs only a write-only access to the memory. One possibility to remove the unnecessary $2N_B N$ memory transfers is to use non-temporal store operations [84] in LACPY and LASET, if the hardware architecture allows this. A non-temporal store operation allows a direct write operation from the CPU to the memory bypassing the CPU cache and the write-allocation. Since this avoids to store data in the cache, this can have a negative influence on the performance for small data sets but is beneficial for large ones [84]. A better implementation, which is invariant under the availability of the non-temporal store operation, is to combine both operations and reuse the already fetched elements out of A_{21} for the write-allocation of the set operation. A concise implementation copying A to W and setting A to α afterwards would be:

```
DO L = 1, N
  DO K = 1, M
    W(K,L) = A(K,L)
    A(K,L) = ALPHA
  END
END DO
```

This is further optimized using OpenMP SIMD statements, Fortran vector notation, or both variants enhanced with OpenMP parallelization. Figure 4.8 shows the difference between the LAPACK approach, the concise loop based implementation, and the parallel ones with vectorization. Thereby, we see that the classic version using LASET and LACPY is fast as long as the data fits into the CPU cache since this avoids the write allocation problem. If the size of the matrices exceeds the cache, our combined version gets faster. We see that the variant using Fortran vector notation together with OpenMP achieves the best performance. Even the explicit usage of the OpenMP SIMD statement is not that fast at all. Regarding a practical implementation, we switch between the LAPACK approach and the combined version with Fortran vectorization and OpenMP parallelization depending on the size of the data and the accumulated size of the CPU caches. The same problem appears in GPUs, but their much smaller caches let us use the combined version for all problem sized there.

GPU Implementation The previously presented optimization and rearrangement of the matrix-matrix multiply and the copy-and-set operation are useful for both CPUs and GPUs. Now, we create a GPU based implementation of the Gauss-Jordan Elimination, which integrates into the update formula (4.15) of the Newton iteration.

Basically, each step of the Gauss-Jordan Elimination decomposes into two parts. The first one is the computation of H, which is done inside the current block column of A. The second one are the updates of the remaining parts of the matrix A and of the right-hand side B. By employing a hybrid CPU-GPU approach, we leave the computation of the matrix H on the CPU and perform only the matrix-matrix multiply to update A and B on the GPUs. If more than one GPU is available we distribute A and B using the Column Block Cyclic (CBC) distribution [111, 177]. Two dimensional distributions as used in ScaLAPACK [44] cause an organizational overhead, here, since our formulation of the Gauss-Jordan Elimination works mostly on block columns. If a two-dimensional data distribution is used, the transfers of the block columns between CPU and GPU will lead to many fine-grained transfers. This increases the number of messages to be sent between the CPU and the GPUs and thus the latency of initiating a transfer gets more influence on the runtime. The structure of the rank-N_B updates (4.59) allows an easy parallel execution by only distributing the matrix H to all GPUs in each iteration. Figure 4.9 visualizes this for two GPUs.

Putting these ideas together, we obtain an easy CPU-GPU implementation of the Gauss-Jordan Elimination. This naive approach lets either the CPU or the GPU work, but not both in parallel. In order to have the CPU and the GPU working in parallel, we extend the approach using a look-ahead scheme. The ability of the GPUs to transfer data between their memory and the CPU's memory while they are computing enables asynchronous data transfers. This capability of the GPUs is

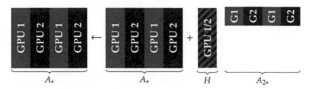

Figure 4.9.: Data Distribution and computational distribution of rank-N_B update in the Gauss-Jordan Elimination scheme on two GPUs

used by partitioning the rightmost block column of A from (4.54) into

$$\begin{bmatrix} A_{13} \\ A_{23} \\ A_{33} \end{bmatrix} := \begin{bmatrix} \hat{A}_{13} & \bar{A}_{13} \\ \hat{A}_{23} & \bar{A}_{23} \\ \hat{A}_{33} & \bar{A}_{33} \end{bmatrix}, \tag{4.60}$$

where $\hat{A}_{13}$, $\hat{A}_{23}$, and $\hat{A}_{33}$ have at most N_B columns and $\bar{A}_{*3}$ is possibly empty. Then the GPUs update

$$\begin{bmatrix} \hat{A}_{13} \\ \hat{A}_{13} \\ \hat{A}_{23} \end{bmatrix} \leftarrow \begin{bmatrix} \hat{A}_{13} \\ 0 \\ \hat{A}_{23} \end{bmatrix} + H\hat{A}_{23} \tag{4.61}$$

first, before they perform the remaining parts of the update from (4.54). This enables us to transfer the updated block column $\begin{bmatrix} \hat{A}_{13}^\mathsf{T} & \hat{A}_{13}^\mathsf{T} & \hat{A}_{23}^\mathsf{T} \end{bmatrix}^\mathsf{T}$ back to the CPU and prepare the matrix H for the next subproblem. Since all function calls to the GPUs can be done asynchronous, i.e. they do not interrupt the CPU program flow, the CPUs work in parallel, as well. The non-blocking behavior of this execution model requires synchronization points between CPU and GPU to avoid race conditions while computing and transferring data. Here, the synchronization takes place after the update matrix H is copied to the GPUs and before the next look-ahead (4.61) is computed on the GPUs. If only a linear system needs to be solved and the inverse of A is not required to be available in the end, the update of the left-hand side blocks A_{*1} can be neglected. Since the workload on the GPU reduces in each step, we further accelerate this asynchronous working concept by adjusting the block size N_B depending on the time how long the CPU or the GPU wait at the synchronization point. This dynamic block size adjustment was described by Catalán et al. in [53] for the LU decomposition. If the matrix inverse A^{-1} is computed, the strategy proposed in [53] may not be beneficial, since the workload on the GPU is nearly constant in each step. Furthermore, the dynamic block size may conflict with the column block cyclic data distribution.

In our case, the large right-hand side B in the linear system causes enough computational workload such that we can stick to a classical a priori selection of the optimal block size N_B by running a set of benchmarks. The optimal block size, on the one hand, should minimize the waiting time for the CPUs/GPUs at the synchronization point and, on the other hand, should increase the performance of the rank-N_B update on the GPUs.

Algorithm 12 shows the hybrid CPU-GPU algorithm with asynchronous operations. The different parallel execution paths on the GPU are referred as *CtoG* (CPU to CPU) for the interaction of the CPU and the GPU and *GPU-Compute* for the computations on the GPU. In the Nvidia® CUDA terminology [131], these execution paths are called *streams*. Although current accelerators allow at least three streams running in parallel (one for computational tasks, one copying data to the GPU and one copying data back to the CPU), we can restrict ourselves to two of them in our implementation. A stream handles the operations queued in its execution pipeline in their order of submission independent of computations or data transfers in other streams. This allows us to overlap the data transfers with the computations done in the GPU.

Data Layout Observations   As already described in the previous paragraph, we use a column block cyclic data distribution to obtain a multi-GPU aware algorithm. Although BLAS, LAPACK, and Nvidia® cuBLAS [132] use the column major storage known from Fortran, we have to check if this is optimal for our algorithm on GPUs. The main reason for this additional analysis is that the memory controller of the GPUs and the execution model are highly influenced by the way we access the memory. Data can only be fetched in chunks of 128 or 256 bytes, which are called cache-lines, and it can only be accessed fast if the access is aligned to 128 or 256 bytes. That means the fetch of a single double precision value, i.e. 8 byte, from the GPU's memory transfers at least 16 times more data than required. Hence, we have to adjust our algorithm to minimize this overhead.

Remark 4.13:
CPUs have a similar memory access scheme as GPUs but, on the one hand, they transfer up to a page counting 4096 bytes. This increases the probability that two required elements lie in the same page. On the other hand, the cache per computational unit is much larger than on GPUs. This increases the probability that already fetched data can be reused later without being transferred from memory again. Since the matrix H will be relatively small, i.e. many parts of it will fit into the CPU cache, the influence of this behavior is negligible on the CPU.

As long as no pivoting is integrated in the Gauss-Jordan Elimination, we can ignore this memory access property since the GEMM operation in cuBLAS is optimized to access the memory in the correct way. Integrating the pivoting procedure, i.e. adding the necessity to swap rows on the GPUs, makes it a crucial point. In the work by Köhler et al. [111], experiments on Nvidia® Kepler K20 GPUs showed that between 20% and 37.5% of the runtime of the Gauss-Jordan Elimination with pivoting is spent on row exchanges.

The column-major matrix storage scheme in combination with the row swap operation results in one of the following two cases (assuming double precision entries and a cache line length of 128 bytes):

1. Both row entries in one column are inside the same cache-line. In this case, two out of sixteen double precision entries of the cache-line are used in the swap operation. That means 87.5% of the fetched and stored data are not used.

2. The row entries are in different cache-lines. That means for exchanging the two elements in one column two cache-lines need to be fetched and stored. This means 93.5% of the transferred data is useless for the swap operation.

Figure 4.10 illustrates the second using a cache line length of 4 entries, for explanatory reasons. This overhead in the data transfers during the application of the pivoting wastes time, energy, and slows down the whole algorithm. By changing the matrix storage scheme to row-major, we can use all double precision entries of all cache-lines during the row exchange, compare Figure 4.10. For optimal performance the leading dimension of the row-major matrix should be chosen as a multiple of the cache line length. In this way swapping two arbitrary rows of length 16 requires only the transfer of 256 bytes instead of 4 kB in the worst case. This motivates the switch to row major storage on the GPU in contrast to the column major scheme used by LAPACK on the CPU. It is easy to see that the row-major storage of a matrix is equivalent to the column major storage of its transpose. Hence, we only have to transpose all input data of the Gauss-Jordan Elimination when it gets copied to the GPUs, but interpret the data now in row-major. In this way the transposition

---

Algorithm 12 Multi-GPU Gauss-Jordan-Elimination for Inversion and Solution of Linear Systems

---

Input: $A \in \mathbb{C}^{m \times m}$ and optionally $B \in \mathbb{C}^{m \times n}$
Output: A^{-1} and/or X solving $AX = B$ depending on the requirements.

1: Asynchronous distribution and row-major storage conversion of A and B across all GPUs with block width N_B.
2: **for** $J := 1, 1 + N_B, 1 + 2N_B, \ldots, m$ **do**
3:     $J_B := \min(N_B, m - J + 1)$
4:     **if** $J = 1$ **then**
5:         **CPU:** Compute H and pivoting from $A(1 : m, 1 : J_B)$
6:     **else**
7:         **CPU:** Wait for the download of $\hat{A}_{13}$, $\hat{A}_{23}$, and $\hat{A}_{33}$ in stream *CtoG* and compute H and pivoting from them.
8:     **end if**
9:     **CPU:** Upload H to all GPUs in stream *CtoG*.
10:     **GPU:** Convert H to row-major storage in stream *CtoG* and apply the pivoting to A in stream *GPU-Compute*.
11:     **GPU:** Synchronize stream *CtoG* and stream *GPU-Compute*.
12:     **GPU:** Partition $A := \left[\begin{array}{cc|c|c} A_{11} & A_{12} & \hat{A}_{13} & \bar{A}_{13} \\ \hline A_{21} & A_{22} & \hat{A}_{23} & \bar{A}_{23} \\ \hline A_{31} & A_{32} & \hat{A}_{33} & \bar{A}_{33} \end{array} \right]$, where $A_{11} \in \mathbb{C}^{J-1 \times J-1}$, $A_{22} \in \mathbb{C}^{J_B \times J_B}$, and $\hat{A}_{\cdot 3}$ has $m - (J + J_B) + 1$ columns.
13:     **GPU:** Perform Look-ahead update $\begin{bmatrix} \hat{A}_{13} \\ \hat{A}_{13} \\ \hat{A}_{23} \end{bmatrix} \leftarrow \begin{bmatrix} \hat{A}_{13} \\ 0 \\ \hat{A}_{23} \end{bmatrix} + H\hat{A}_{23}$ and convert the result to column-major storage in stream GPU-Compute.
14:     **CPU:** Synchronize stream *GPU-Compute* and download $\hat{A}_{13}$, $\hat{A}_{23}$, and $\hat{A}_{33}$ in stream *CtoG*.
15:     **GPU:** Compute $\begin{bmatrix} A_{11} \\ A_{21} \\ A_{31} \end{bmatrix} \leftarrow \begin{bmatrix} A_{11} \\ 0 \\ A_{31} \end{bmatrix} + HA_{21}, \begin{bmatrix} A_{12} \\ A_{22} \\ A_{32} \end{bmatrix} \leftarrow H,$

$\begin{bmatrix} \bar{A}_{13} \\ \bar{A}_{23} \\ \bar{A}_{33} \end{bmatrix} \leftarrow \begin{bmatrix} \bar{A}_{13} \\ 0 \\ \bar{A}_{33} \end{bmatrix} + H\bar{A}_{23}$, and $\begin{bmatrix} B_1 \\ B_2 \\ B_3 \end{bmatrix} \leftarrow \begin{bmatrix} B_1 \\ 0 \\ B_3 \end{bmatrix} + HB_2$ in stream *GPU-Compute*.
16: **end for**
17: **GPU:** Synchronize all streams and convert A to column-major storage.
18: **CPU:** Download A and/or B and revert the column pivoting.

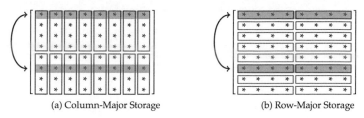

(a) Column-Major Storage   (b) Row-Major Storage

Figure 4.10.: Cache line access for swapping two rows in a matrix.

is only done from a data management point of view and not in a mathematical sense. Studies on the Nvidia® Kepler architecture [111] already showed that the overhead of combining the CPU to GPU transfers (as well as the GPU to CPU transfers) with a transposition operation can be neglected compared to the runtime increase by badly styled memory accesses during the row swap operation.

Changing the storage scheme on the GPUs, we have to adjust all BLAS like operations on the GPU without reimplementing them to keep their optimization and benefit from improved implementations, easily. A typical call to the general matrix-matrix multiply GEMM(α,A,B,β,C), as implemented in cuBLAS or clBLAS, computes: $C := \alpha AB + \beta C$, where $A \in \mathbb{R}^{m \times k}$, $B \in \mathbb{R}^{k \times n}$, and $C \in \mathbb{R}^{m \times n}$ are stored in column-major storage. Recalling the fact that the row-major storage is corresponding to the column major storage of the transpose, we have A^T, B^T, and C^T in column-major storage on the GPU. Applying this knowledge we see that GEMM(α,B,A,β,C) computes

$$C^\mathsf{T} \longleftarrow := \alpha B^\mathsf{T} A^\mathsf{T} + \beta C^\mathsf{T} \tag{4.62}$$

if A, B, and C are stored in row-major storage, although the operation works on column-major storage. That means, exchanging the roles of A and B (and the corresponding dimension arguments) in the call to the GEMM routine allows us to use the optimized implementation, for the row-major storage as well, without reimplementing this routine. In complex arithmetic, this works exactly the same way, i.e. with the complex transpose rather than the Hermitian transpose.

Remark 4.14:

If the pivoting is enabled during the Gauss-Jordan Elimination, $A^{-1}P^\mathsf{T}$ is computed instead of A^{-1}. That means we have to apply a column permutation to the matrix now stored in row-major storage. This yields the same memory access problem as for the swap operation during pivoting. Furthermore, if we have a multi-GPU system, the matrix is distributed across the GPUs in a column cyclic way. In this way the application of the permutation to retrieve A^{-1} either causes an undesirable memory access pattern and/or requires a lot of communication between all GPUs. Even on modern platforms supporting Nvidia® NVLink, the interconnection between the GPUs is slower than the interconnection between the CPU and its memory. In the case of an IBM POWER 8 system, the CPU's total memory bandwidth is 230GB/s and the interconnection between the two NVLink enabled GPUs reaches up to 40GB/s. Furthermore, for each transfer of $O(m)$ data, the latency for initiating the transfer must be taken into account. This will slow down the matrix inversion in its final step. Thus, we copy the matrix back to the CPU to permute it and if we need it afterwards on the GPU, we copy it back to the GPUs.

On systems with slower CPU–GPU interconnections like PCIExpress, this problem has even a higher influence on the overall process. Also the number of CPU sockets plays a role, since on current computer architectures each CPU socket has its own memory controller. In this way an increasing number of CPU sockets also enhances the CPU's total memory bandwidth.

4.2.5. Integrating Gauss-Jordan Elimination and Generalized Matrix Sign Function

The hardware capabilities and the performance of the provided BLAS implementation are the main factors to decide how the Gauss-Jordan Elimination is integrated in the Newton scheme. Since GPUs require a critical amount of data to achieve their high performance, we can switch between a GPU and CPU implementation depending on the size of the input data. Using an a priori benchmark, we select one of the following variants.

Naively Implemented Newton Scheme If the Newton scheme is implemented naively, i.e. the matrix B is a square matrix without any special structure, the solution of a linear system with many right-hand sides is one of the main operations. On CPU-only hardware, the replacement of a standard LU decomposition based solver in favor of the Gauss-Jordan Elimination requires an a priori benchmark to select when this replacement is worth to be done. In most cases, the LU decomposition will be selected, since the Gauss-Jordan Elimination scheme needs 12.5% more operations and this cannot be compensated by an improved CPU utilization. On GPU accelerated systems, however, it was shown by Köhler et al. [111] that the higher flop count of the Gauss-Jordan Elimination scheme can indeed be compensated. Additionally, it overcomes the parallelization issues in the forward/backward substitution phase of a multi-GPU aware LU decomposition.

Newton Scheme with One Sided Transformations The reduction of B to upper triangular form and the adjustment of the forward substitution step, as shown in Subsection 4.2.2, leads to a large reduction of the operation count. Since, the used techniques rely on the triangular structure of the involved matrices, the Gauss-Jordan Elimination will destroy most of the benefits. If only CPUs are available, it is not worth to replace the LU decomposition based scheme. If GPUs are available, we alter the scheme, such that Y_k^{-1} is computed and multiplied from the left and the right with the upper triangular matrix B using multiple GPUs. This increases the number of flops per iteration from $3m^3$ to $4m^3$, but we obtain a multi-GPU aware algorithm. Since the GPUs have a much higher performance than the CPUs, the increased number of flops is compensated.

Newton Scheme with Two-Sided Transformations Reducing the matrix B to an upper band structure leads to the largest flop reduction. The inversion of the current iterate Y_k is the only operation with a cubic flop count. During Subsection 4.2.3, we saw that the performance of the inversion, as implemented in LAPACK, strongly depends on the underlying BLAS library. On a CPU-only system, an a priori benchmark has to decide whether LAPACK or the Gauss-Jordan Elimination scheme is used. Especially on Intel® architectures with the LAPACK implementation of the Intel® MKL library, it seems to be beneficial to replace the LU decomposition based matrix inversion, compare Tables 4.2 and 4.3. On GPU accelerated systems, the Gauss-Jordan Elimination provides a multi-GPU aware way to compute the inverse. The update with the upper band matrix B can be done on the GPUs, as well as on the CPUs.

4.2.6. Summary

Summarizing this section, Table 4.4 shows the number of flops and the additional memory required by our implementation of the Newton iteration scheme. The naive CPU implementation uses the LU decomposition with forward/backward substitution since switching to the Gauss-Jordan Elimination scheme will not lead to a better performance in most cases. The one-sided transformed variant uses the LU decomposition as well because the flop reduction strongly depends on the involved triangular structures. In case of the two-sided variant, we can switch between LU decomposition and Gauss-Jordan Elimination depending on a priori benchmarks. The flop count is not influenced by this change. In case of the GPU accelerated variants, the whole iteration

Variant	Number of Flops			Extra Memory	
	Before	Iteration	After	on CPU	on GPUs
Naive CPU	$\frac{2}{3}m^3$	$4\frac{2}{3}m^3$	$2m^3$	$3m^2$	$-$
One Sided CPU	$3\frac{1}{3}m^3$	$3m^3$	$2\frac{2}{3}m^3$	$2m^2 + N_B^{QR}m$	$-$
Two-Sided CPU	$\frac{20}{3}m^3$	$2m^3$	$4m^3$	$m^2 + N_B^{QR}m$	$-$
Naive GPU	$\frac{2}{3}m^3$	$5m^3$	$2m^3$	$2m^2$	$2m\left(m + N_B^{GJE}\right)$
One Sided GPU	$3\frac{1}{3}m^3$	$4m^3$	$2\frac{2}{3}m^3$	$m^2 + N_B^{QR}m$	$2m\left(m + N_B^{GJE}\right)$
Two-Sided GPU	$\frac{20}{3}m^3$	$2m^3$	$4m^3$	$m^2 + N_B^{QR}m$	$m\left(m + N_B^{GJE}\right)$

Table 4.4.: Comparison of Newton scheme implementations with determinantal scaling and estimation of $p_\pm$ neglecting lower order flop counts.

step is done on the GPUs and in this way the required memory on the CPU reduces. All GPU implementations use the Gauss-Jordan Elimination scheme instead of the LU decomposition due to the better performance and a better scalability on multi-GPU systems. For the usage inside our Divide-Shift-and-Conquer scheme, we use a look-up-table to decide which variant of the Newton iteration is used depending on the input dimension. This look-up-table is determined by an auto-tuning procedure. Therefore, a set of benchmark problems need to be executed on each hardware platform. From the obtained data, the auto-tuning procedure sets up the corresponding boundaries for which size of the input data which of the presented algorithms is used. With a properly defined benchmark, this allows us to select the fasted algorithm in each situation a priori.

Remark 4.15:
Since GPUs have a very limited amount of memory available compared to the CPUs, the maximum problem size handled by the Gauss-Jordan Elimination on the GPU is restricted. Regarding a system with two Nvidia® P100 accelerators, having 16 GB memory each, we can invert matrices up to a dimension of $m = 65\,000$ in real double precision arithmetic. Compared to the performance of the QZ algorithm this exceeds the size of eigenvalue problems that can be solved on a single computer in a reasonable amount of time. Since this is the case on our target systems, we do not cover out-of-core algorithms to overcome the limitations of the GPUs memory here.

4.3. Computation of the Matrix Disc Function

The Divide-Scale-and-Conquer algorithm presented in Section 3.3 uses the matrix disc function instead of the generalized matrix sign function. Thereby, the proposed algorithm – the inverse free iteration – relies on the QR decomposition in contrast to the linear system solves used above. The iterative scheme reads:

$$W_0 = A, \quad Y_0 = B$$
$$W_{k+1} = Q_{k,12}^H W_k, \quad Y_{k+1} = Q_{k,22}^H Y_k, \tag{4.63}$$

with the QR decomposition

$$\begin{bmatrix} Q_{k,11} & Q_{k,12} \\ Q_{k,21} & Q_{k,22} \end{bmatrix} \begin{bmatrix} R_k \\ 0 \end{bmatrix} = \begin{bmatrix} Y_k \\ -W_k \end{bmatrix}. \tag{4.64}$$

This only requires the computation of the QR decomposition and two stacked generalized matrix-matrix products.

4.3.1. CPU Implementation

During the last four decades, different variants of the QR decomposition were developed and efficient implementations were found [40, 51, 65]. A common property of nearly all implementations is that the matrix Q is never set up explicitly but represented in terms of Householder reflections [80]:

$$Q = (I - \beta_1 v_1 v_1^H) \cdots (I - \beta_j v_j v_j^H), \qquad (4.65)$$

where v_ℓ, for each $\ell \in \{1, \ldots, j\}$, are the Householder vectors and β_ℓ are the corresponding scalar factors. Since the iterative scheme (4.63) needs the blocks $Q_{k,12}$ and $Q_{k,22}$ of Q_k, the factorized representation of Q_k is applied to

$$\begin{bmatrix} 0 \\ I \end{bmatrix}.$$

Only a few of the factorization variants allow direct access to the blocks of Q, see [139, 25, 24]. Afterwards, the matrices W_{k+1} and V_{k+1} are computed using the standard matrix-matrix product. The missing level-3 capabilities in the standard Householder vector implementation of the QR decomposition prevents the efficient usage on current computers. To overcome this issue, Bischof and Van Loan developed the storage efficient WY-QR decomposition [40]. Thereby, the unitary matrix Q is factored as

$$Q = (I - V_1 T_1 V_1^T) \cdots (I - V_j T_j V_j^T), \qquad (4.66)$$

where V_ℓ, for each $\ell \in \{1, \ldots, j\}$, contains the Householder vectors and T_ℓ is an upper triangular matrix. In both cases (4.65) and (4.66), the available implementations overwrite the input matrix with the upper triangular R factor and the lower triangular part is used to store v_* or V_*. This is possible since the top entry of v_* is always one and V_* has only ones on the diagonal, which do not need to be stored. Only the scalars β_* or the small upper triangular matrices T_* need to be stored separately or be recomputed every time Q is applied. The standard QR decompositions in LAPACK, GEQRF and GEQRT, represent either Q as Householder vectors (4.65) or in compact storage representation (4.66), respectively. In case of GEQRF, the implementation internally uses the compact storage representation but does not provide access to T_* afterwards. For this reason, we use GEQRT in our implementations since it provides the matrices T_* and we do not need to recompute them every time on demand when Q is applied. This is the behavior of the application routines belonging to GEQRF [10].

In order to check the convergence of the inverse free iteration, the relative change in the matrix R of the QR factorization is used [161].

Regarding the additional memory to implement the inverse free iteration, we need $2m^2$ memory to build $\begin{bmatrix} V_k^T & -W_k^T \end{bmatrix}^T$, $2m^2$ elements to set up the part of Q containing Q_{12} and Q_{22} and $2N_B m$ memory to store the matrices T_* together with the workspace for the QR decomposition. Thereby, N_B is the block size used in the QR decomposition. Finally, we have to store the absolute values of the upper triangular part of R to check convergence. This requires further $\frac{m^2-m}{2} + m$ elements. These additional memory locations are denoted by E_R, E_Q, E_T, E_W and E_{ROLD} in Algorithm 13. The algorithm is destructive on its input data and overwrites (A, B) with the result $(W, V) = \text{disc}_R(A, B)$.

The presented procedure needs $\frac{40}{3} m^3$ flops per iteration. Compared to the most efficient generalized matrix sign function iteration from the previous section, this scheme is $6\frac{2}{3}$ times more expensive. Although the QR decomposition exists in highly optimized variants inside LAPACK, it does not reach the same performance as the LU decomposition. Figure 4.11 illustrates this problem, comparing their performances. This loss of performance on multi-core architectures lead to the development of the Tile-QR algorithms presented by Buttari et al. [51]. The main difference to the purely Householder based QR decomposition is that inside a block column, a specialized

---

Algorithm 13 Inverse free iteration using LAPACK

Input: (A, B) of order m
Output: (A, B) of order m overwritten with $(W, V) = \mathrm{disc}_R (A, B)$

1: **for** $k := 1, \ldots$ **do**
2:      $E_R \leftarrow \begin{bmatrix} B \\ -A \end{bmatrix}$ {LACPY, LASCL}
3:      $E_R, E_T \leftarrow \mathrm{QR}(E_R)$ using E_W as workspace.   {GEQRT}
4:      $E_Q \leftarrow \begin{bmatrix} 0 \\ I \end{bmatrix}$ {LASET}
5:      Apply Q contained in E_R and E_T to E_Q to obtain $\begin{bmatrix} Q_{12} \\ Q_{22} \end{bmatrix}$ {GEMQRT}
6:      $A \leftarrow Q_{12}^H A$ {GEMM}
7:      $B \leftarrow Q_{22}^H B$ {GEMM}
8:      **if** $k > 1$ **then**
9:          Check convergence using $\||R| - E_{ROLD}\|_F / \|E_{ROLD}\|_F$
10:     **end if**
11:     $E_{ROLD} \leftarrow |R|$
12: **end for**

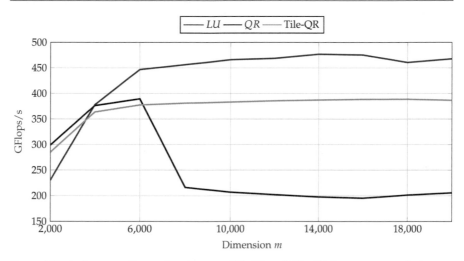

Figure 4.11.: Performance Comparison between LU, QR, and Tile-QR decomposition of a $2m \times m$ matrix on a 16-core Intel® Skylake architecture.

QR decomposition for tall-skinny matrices is used and the updates on the remaining matrix are done as soon as the necessary data gets available. This yields a highly parallelizable scheme, using Directed-Acyclic-Graph scheduling. First implemented in the PLASMA library [51] employing the QUARK library [178] as parallel scheduler, the OpenMP 4 standard now allows to implement such algorithms requiring only compiler features. The runtime libraries of the compiler then take care of the dependencies and the scheduling. The performance of the Tile-QR is shown in Figure 4.11 as well.

Since the Tile-QR needs a different storage scheme for the matrices, where each matrix is stored in $p \times q$ blocks of size $N_B \times N_B$, we have to adjust the implementation of the inverse free iteration. The matrix $E_R := \begin{bmatrix} B^\mathsf{T} & -A^\mathsf{T} \end{bmatrix}^\mathsf{T}$ needs to be set up any way at the beginning of each iteration. Thereby, the change between Fortran storage and the tile storage scheme is done. The application of the factored matrix Q to $\begin{bmatrix} 0 & I \end{bmatrix}^\mathsf{T}$ is implemented having the factors of Q in tile storage from the left to an array Fortran column major storage. This leads to the highest performance, compare Figure 4.11. Especially for large problems, we benefit from the better data locality of the tile storage scheme. The performance of the non-tile storage based approaches decreases for larger problems, since the way the QR decomposition accesses its data is not cache friendly [72]. As a consequence, it is only worth to implement the inverse free iteration using the Tile-QR approach.

Although we change the way the QR decomposition is computed in Algorithm 13, its complexity does not change.

4.3.2. CPU Implementation with Directed Acyclic Graph Scheduling

The previous section showed that the inverse free iteration can be implemented with the help of tuned QR decompositions. For the generalized matrix sign function we have seen in Subsections 4.2.2 and 4.2.3 that the numerical effort can be reduced by optimizing the structure in the matrices A and B before the iteration begins. Lemma 4.10 is adjusted to the matrix disc function as follows, and extends a Lemma given by Benner in [21]:

Lemma 4.16:
Let (A, B) be a matrix pair of order m, further let K and P be nonsingular $m \times m$ matrices. Then the right matrix disc function $\mathrm{disc}_R (KAP, KBP)$ is given by

$$\mathrm{disc}_R (KAP, KBP) = P^{-1} \mathrm{disc}_R (A, B) P. \tag{4.67}$$

For the left matrix disc function it holds

$$\mathrm{disc}_L (KAP, KBP) = K \mathrm{disc}_L (A, B) K^{-1}. \tag{4.68}$$

Proof. Using the definition of the Weierstrass/Kronecker canonical form (3.60) we obtain

$$(KAP, KBP) = \left(KV \begin{bmatrix} J_0 & \\ & J_\infty \end{bmatrix} U^{-1}P, KV \begin{bmatrix} I & \\ & N \end{bmatrix} U^{-1}P \right)$$

which, for the right matrix disc function, yields

$$\begin{aligned}
\mathrm{disc}_R (KAP, KBP) &= \left(P^{-1}U \begin{bmatrix} I & \\ & 0 \end{bmatrix} U^{-1}P, P^{-1}U \begin{bmatrix} 0 & \\ & I \end{bmatrix} U^{-1}P \right) \\
&= P^{-1} \left(U \begin{bmatrix} I & \\ & 0 \end{bmatrix} U^{-1}, U \begin{bmatrix} 0 & \\ & I \end{bmatrix} U^{-1} \right) P \\
&= P^{-1} \mathrm{disc}_R (A, B) P.
\end{aligned}$$

The result for the left matrix disc function is obtained analogously. $\qquad\square$

Corollary 4.17:
Lemma 4.16 shows that the right matrix disc function is invariant under left sided transformations and the left matrix disc function is invariant under right sided transformations.

This was used by Marqués et al. [124] to reduce the number of flops required to perform one iteration step in Algorithm 13 by 32% to $9m^3$ flops. Additionally, only one QR decomposition of the matrix A is required before the iteration starts. Since this idea does not result in a high performance on current computers by its self, we want to improve it to make use of OpenMP 4. Hence, we recall it here.

Employing Lemma 4.16 we can use a QR decomposition of A to make the lower matrix in the inverse free iteration upper triangular. Having $A = Q_A R_A$, we apply the inverse free iteration to $\left(Q_A^H A, Q_A^H B\right) = \left(R_A, \tilde{B}\right)$. Thus the algorithm has to decompose

$$E_R := \begin{bmatrix} \tilde{B} \\ -R_A \end{bmatrix} = \begin{bmatrix} \Box \\ \diagdown \end{bmatrix} \tag{4.69}$$

in the first iteration. If this structure is kept over all iterations, the number of operations required per iteration can be considerably reduced. The update $A \leftarrow Q_{12}^H A$ in Algorithm 13 turns into

$$R_A \leftarrow Q_{12}^H R_A, \tag{4.70}$$

which only results in an upper triangular R_A again, if and only if Q_{12} is lower triangular. Marqués et al. [124] showed that this is achieved with specialized QR decompositions. Their first approach using Givens-rotations showed, the matrix Q_{12} is lower triangular if R_A is upper triangular. Although the theoretical operation count reduces to $9m^3$ flops per iteration, this approach is not useful to reach the goal to obtain an algorithm, which is rich in level-3 BLAS operations. A more convenient approach using block operations on top of Householder transformations was presented in [124] and works as follows: For simplicity, we partition E_R into $2p \times p$ blocks of size $N_B \times N_B$ and obtain

$$\begin{bmatrix} \tilde{B} \\ -R_A \end{bmatrix} := \begin{bmatrix} E_R^{(1,1)} & E_R^{(1,2)} & \cdots & E_R^{(1,p)} \\ E_R^{(2,1)} & E_R^{(2,2)} & \cdots & E_R^{(2,p)} \\ \vdots & \vdots & \cdots & \vdots \\ E_R^{(p,1)} & E_R^{(p,2)} & \cdots & E_R^{(p,p)} \\ E_R^{(p+1,1)} & E_R^{(p+1,2)} & \cdots & E_R^{(p+1,p)} \\ 0 & E_R^{(p+2,2)} & \cdots & E_R^{(p+2,p)} \\ \vdots & \vdots & \ddots & \vdots \\ 0 & 0 & \cdots & E_R^{(2p,p)} \end{bmatrix}. \tag{4.71}$$

Now, the QR decomposition of E_R is computed block by block from bottom to top and from the first to the last column. Beginning with

$$\tilde{Q}_{(p,1)} \tilde{R} = \begin{bmatrix} E_R^{(p,1)} \\ E_R^{(p+1,1)} \end{bmatrix} \tag{4.72}$$

and updating

$$\begin{bmatrix} E_R^{(p,2)} & \cdots & E_R^{(p,p)} \\ E_R^{(p+1,2)} & \cdots & E_R^{(p+1,p)} \end{bmatrix} \leftarrow \tilde{Q}_{(p,1)}^H \begin{bmatrix} E_R^{(p,2)} & \cdots & E_R^{(p,p)} \\ E_R^{(p+1,2)} & \cdots & E_R^{(p+1,p)} \end{bmatrix}.$$

The matrix $\tilde{R}$ overwrites the block $E_R^{(p,1)}$ and we compute $\tilde{Q}_{k,1}$ for $k = p, \ldots, 1$. After finishing the first block column, this goes on with $\tilde{Q}_{(k,2)}$, $k = p + 1, \ldots, 2$ until $\tilde{Q}_{(k,p)}$, $k = 2p - 1, \ldots, p$. The matrices Q_{12} and Q_{22} are now obtained by applying $\tilde{Q}_{(\bullet,\bullet)}$ in reverse order to $\begin{bmatrix} 0 & I \end{bmatrix}^H$. In [124], it is shown that Q_{12} becomes block lower triangular or strictly lower triangular if $N_B = 1$. The fact that Q_{12} is block lower triangular leads to $Q_{12}^H R_A$ being block upper triangular. This fits the block partitioning from Equation (4.71) and the structure is kept during the inverse free iteration. Following [124], the QR decomposition now takes $3m^3$ flops. The same amount of operations is required to assemble Q_{12} and Q_{22} from the single blocks. Together with the update of $R_A \leftarrow Q_{12}^H R_A$ and $\tilde{B} \leftarrow Q_{22}^H \tilde{B}$, this sums up to $9m^3$ flops per iteration.

A practical implementation of the QR decomposition of E_R consists of two different types of smaller QR decompositions. First, we consider the compact storage QR decomposition [40] of the bottom block $E_R^{(p+k,k)}$ in column k, compare (4.72):

$$\tilde{Q}_{(p+k-1,k)} \begin{bmatrix} \tilde{R} \\ 0 \end{bmatrix} := \left(I - V_{(p+k-1,k)} T_{(p+k-1,k)} V_{(p+k-1,k)}^H \right) \tilde{R} = \begin{bmatrix} E_R^{(p,k)} \\ E_R^{(p+k,k)} \end{bmatrix} \tag{4.73a}$$

with the structure

$$\tag{4.73b}$$

The zeros below the upper triangular block $\tilde{R}$ are used to store the unit lower triangular matrix $V_{(p+k-1,k)}$ containing the Householder vectors. Only the matrix $T_{(p+k-1,k)}$ containing the coefficients between the Householder vectors need to be stored separately. This requires $N_Q \times N_B$ memory for each QR decomposition, where N_Q is the blocking factor inside the QR decomposition.

Second, we regard the remaining blocks of each column from bottom to top. These consist of a lower triangular matrix $E_R^{(j+1,k)}, j = k+p-2, \ldots, k$, stemming from the previous QR decomposition of the block below, and a square matrix $E_R^{(j,k)}$. This yields the QR decomposition

$$\tilde{Q}_{j,k} \tilde{R} := \left(I - V_{(j,k)} T_{(j,k)} V_{(j,k)}^H \right) \begin{bmatrix} E_R^{(j,k)} \\ E_R^{(j+1,k)} \end{bmatrix} \tag{4.74a}$$

similar to (4.73a), but with the changed structure

$$\tag{4.74b}$$

Since the bottom block in $\begin{bmatrix} E_R^{(j,k)} \\ E_R^{(j+1,k)} \end{bmatrix}$ was upper triangular before, the unit lower triangular matrix $V_{(j,k)}$ is now a band matrix. Moreover, it is an upper triangular matrix stacked with a unit lower triangular one. In this way the matrices $V_{(j,k)}$ can be stored in the annihilated part of $E_R^{(j,k)}$ and $E^{(j+1,k)}$ without overwriting the values in the bottom lower triangle. This already contains the Householder vectors of the QR decomposition of the block below. Consequently, we are able to store all Householder vectors inside the matrix (4.71).

For storing the triangular factors $T_{(\bullet,\bullet)}$, we have an upper bound for memory requirements using the following fact: Since the block size N_Q in the QR decomposition (4.73a) and (4.74a) fulfills $N_Q \leq N_B$, the required memory to store the matrices $T_{(\bullet,\bullet)}$ does not exceed $p^2 (N_B \times N_B) = m^2$. The memory is mostly provided by allocating E_R in column major storage. Since R_A stays block upper triangular and Q_{12} is block lower triangular, we can reuse the unused memory locations in the bottom of E_R and on top of Q_{12} to store $p^2 - p$ Householder coefficients $T_{(\bullet,\bullet)}$ and only need $p (N_Q \times N_B)$ memory in addition.

Additional details about the structure preserving QR decomposition, including a Givens-Rotation based approach, are shown by Marqués et al. [124].

Remark 4.18:
If the block size in (4.71) does not match the dimension of the matrices R_A and $\tilde{B}$ perfectly, the first QR decomposition (4.73a) combines two blocks such that the resulting upper triangular matrix fits in an equally sized block structure.

Although the algorithm is rich in level-3 BLAS and LAPACK operations, it performs poorly on multi-core systems as shown in Table 4.5. This is in contrast to the time, the algorithm was presented since there, it outperforms the naive approach. The reason is that although it is rich in level-3 BLAS operations, its design conflicts with the parallelization capabilities of current BLAS and LAPACK implementations. In our case, the block layout (4.71) yields many small matrix-matrix products which are not capable of efficient parallel computation. Here, we have to move the parallelization from BLAS and LAPACK into our algorithm to execute many small operations in parallel. Since the reduction scheme is obviously sequential inside a block column, easy loop-based parallelization techniques are not promising here. Regarding the dependencies between all blocks and operations during the QR decomposition we obtain, with

$$M_{p,k} = \begin{bmatrix} E_R^{(p,k)} \\ E_R^{(p+1,k)} \end{bmatrix}, \tag{4.75}$$

the following properties:

1. The block $M_{p,k}$ can be decomposed after the QR decomposition of $M_{p+1,k}$ is ready and all updates from QR decompositions in block columns $l, 1 \leq l \leq k - 1$ are done.

2. The blocks right of $M_{p,k}$ ($M_{p,l}, k < l \leq p$) can be updated once $M_{p,k}$ is decomposed. This update can be done in parallel on all affected blocks.

3. The blocks $M_{p+k-1,l}$ and $M_{p+k-2,l}, k \leq l \leq p$ overlap. Especially, the block $M_{p+k-1,k}$ of column k can be processed when the update on using $M_{p+k-2,k-1}$ is done.

Using the first two properties, we define a directed acyclic graph $\mathcal{G}$ having edges from node (p, k) to $(p - 1, k)$ and from node (p, k) to nodes $(p, l), k < l \leq p$. The third property introduces an edge from $(p + k - 2, l)$ to $(p + k - 1, l), 1 \leq k \leq p$, since $M_{p+k-1,l}$ and $M_{p+k-2,l}$ overlap in the memory. For the first block column this edge is not required. Figure 4.12 visualizes this property for a matrix with $p = 3$. Regarding the parallelization, all nodes in the graph, which have no parents, i.e. no incoming edges, can be computed independently of each other. Once a computation, like the QR decomposition of $M_{p,k}$ finishes, the outgoing edges are removed from $\mathcal{G}$. If a node looses all its

```
SUBROUTINE DISC_QR(N, NB, A)
P = (N+NB-1) / NB
!$omp parallel
!$omp master
DO J = 1, N, NB
  K = NB * (P - 1) + 1, JB = MIN ( NB, N - J + 1)
  KB = MIN (N+J+NB-1-K+1, 2*N - K +1)

  !$omp task depend(inout: A(K,J))
  CALL DGEQRT(A(K:K+KB-1,J:J+JB-1))
  !$omp end task

  DO L = J+JB, N, NB
    LB = MIN(NB, N - L +1)
    !$omp task depend(in:A(K,J)) depend(inout: A(K,L)) &
    !$omp&  depend(out:A(K+NB,L))
    CALL DGEMQRT(A(K:K+KB-1,J:J+JB-1), A(K:K+KB-1,L:L+LB-1))
    !$omp end task
  END DO

  DO WHILE (K .GT. J)
    K = K - NB
    KB = MIN(JB+NB, 2*NB)

    !$omp task depend(in:A(K+NB,J)) depend(inout: A(K,J))
    CALL BAND_QR(A(K:K+KB-1,J:J+JB-1))
    !$omp end task

    DO L = J+JB, N, NB
      LB = MIN(NB, N - L +1)
      !$omp task depend(in: A(K,J), A(K+NB,L)) &
      !$omp& depend(inout:A(K,L))
      CALL BAND_MQR(A(K:K+KB-1,J:J+JB-1),A(K:K+KB-1,L:L+LB-1))
      !$omp end task
    END DO
  END DO
END DO
!$omp end master
!$omp end parallel
END SUBROUTINE
```

Listing 4.4: OpenMP accelerated trapezoidal QR (TRQR)

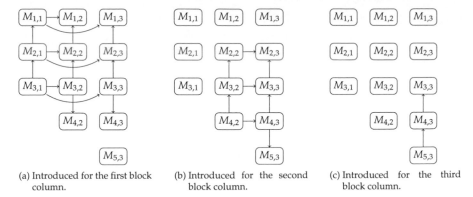

(a) Introduced for the first block column.

(b) Introduced for the second block column.

(c) Introduced for the third block column.

Figure 4.12.: Edges introduced in the Directed-Acyclic-Graph of the QR decomposition.

parents, by this operation, it is ready to be executed as well. The OpenMP 4 programming scheme allows to describe the graph $\mathcal{G}$ directly in the program code. This approach is already used by the Tile-QR [138, 51]. Since the matrix-matrix product is not commutative, we have to apply the unitary transformations $\tilde{Q}_*$ in the correct order. Therefore, we introduce a set of dummy edges in the graph pointing from a node to itself. These *inout* edges are removed in the order they were created, when the associated operation is done. In this way, we integrate the order of the $\tilde{Q}_*$ into the graph $\mathcal{G}$, as well. A simplified pseudo code algorithm enhanced with the OpenMP statements to generate the graph is shown in Listing 4.4. The algorithm has similar data locality problems, as those known from other variants of the QR decomposition, like the tall-skinny QR [59, 17]. Mainly, the column major layout causes the CPU's prefetching to load unnecessary data if a block $E_R^{(p,k)}$ is accessed, since two consecutive pages in memory are usually not in the same block. Switching to the tiled data layout [151, 51] enables the CPU's prefetcher to only pick additional data from the memory which belongs to the current block, too. This increases the probability that additionally prefetched data can be used as well. In contrast to the Tile-QR, where the combination of two blocks only requires the diagonal entry for the computation of the Householder vectors, our structure preserving approach needs full access to both blocks in $M_{p,k}$. One could copy both blocks to a temporary memory location, perform the operations, and copy them back to the tiled layout. This trivial idea allows optimal memory transfers with respect to the CPU's prefetcher, but consumes more memory and introduces too many additional copy operations.

The optimal solution is to adjust the QR decompositions and the corresponding application of $\tilde{Q}_*$ from (4.73a) and (4.74a) to the tiled layout. If the dimension of the matrices fits exactly an equally sized block structure, the standard QR decomposition of the bottom block works in a single tile. The application of $\tilde{Q}_*$ to the remaining block columns fits as well, in this case. In this way, the LAPACK routines GEQRT and GEMQRT can be used straight forward on the tiles. If the last block in a column of $E_R^{(p+k,k)}$ has $h < N_B$ rows, we compute the QR decomposition of

$$M_{p+k-1,k} = \begin{bmatrix} E_R^{(p+k-1,k)} \\ E_R^{(p+k,k)} \end{bmatrix}$$

without consecutive column access in E_R. One could extend all routines of the standard QR operations to handle this problem. This yields many function calls on small data sets. To use the strategy known from the tall-skinny QR decomposition [59, 17] and perform the calculation in two steps is more favorable. First, we decompose $E_R^{(p+k-1,k)}$ and obtain

$$\left[\;\Box\;\right] = E_R^{(p+k-1,k)} = \tilde{Q}_1 \tilde{E}_R^{(p+k-1,k)} = \left[\;\Box\;\right]\left[\;\searrow\;\right], \tag{4.76}$$

where $\tilde{E}_R^{(p+k-1,k)}$ is upper triangular. This operation is performed inside a single block. Afterwards, we annihilate the $E_R^{(p+k,k)}$ block with the help of the triangle-pentagonal QR decomposition known from the tall-skinny QR [59]. This computes

$$\begin{bmatrix}\searrow\\\Box\end{bmatrix} = \begin{bmatrix}\tilde{E}_R^{(p+k-1,k)}\\E_R^{(p+k,k)}\end{bmatrix} = \tilde{Q}_2 \hat{E}_R^{(p+k-1,k)} = \left[\;\Box\;\right]\left[\;\searrow\;\right], \tag{4.77}$$

where the Householder representation of $\tilde{Q}_2$ is stored in the memory location of $E_R^{(p+k,k)}$. Thus, we can implement the QR decomposition of the bottom block (4.73a) without consecutive column access in E_R. Applying the matrices $\tilde{Q}_1$ and $\tilde{Q}_2$ to the remaining columns of the matrix E_R is done in the same way to preserve the tiled layout.

Beside this, most of the QR decompositions are structured as shown in Equation (4.74b), where the upper tile contains a full matrix and the bottom tile contains an upper triangular one. After the QR decomposition, the matrices $V_{(p+k-1,k)}$ are split into an upper and a lower triangular part as well. For the j-th Householder vectors in $V_{(p+k-1,k)}$, it holds that it has $N_B - j$ elements in the upper tile and j elements in the lower tile. Naively, one can use the same strategy from the bottom block of the column again to obtain the decomposition in two steps. This gets too inefficient for p^2 blocks since the values in the upper triangle of the top tile get accessed twice and the application of the corresponding Q to the remaining columns inherits the same problem. Because this update is the most used operation in the whole procedure, we have to derive a strategy for the QR decomposition of $M_{p,k}$ which respects the data layout.

The matrix $\tilde{Q}_*$ of the QR decomposition of $M_{p,k}$ consists of the product of block Householder transformations:

$$M_{p,k} = QR_{p,k} = \left(\prod_{l=1}^{q}\left(I - V_l T_l V_l^H\right)\right)R_{p,k}, \tag{4.78}$$

where $\lceil N_B/q\rceil$ is the block size of the QR decomposition. The matrix $\begin{bmatrix}V_1 & \cdots & V_q\end{bmatrix}$ has the lower-upper triangular structure as presented in (4.74b). Regarding the tiled layout this yields the following structure for each V_l:

$$V_l = \begin{bmatrix}0\\\searrow\\\hline\Box\\\searrow\\0\end{bmatrix} = \begin{bmatrix}0\\V_l^{(U,L)}\\V_l^{(U,F)}\\\hline V_l^{(L,F)}\\V_l^{(L,U)}\\0\end{bmatrix}, \tag{4.79}$$

where the blocks $V_l^{(U,*)}$ and $V_l^{(L,*)}$ belong to the same tile. In this way, the application of one factor of $\tilde{Q}_*$ (or $\tilde{Q}_*^H$) to a matrix C, consisting of two tiles $C^{(U)}$ and $C^{(L)}$, is written as

$$\left(I - V_l T_l V_l^H\right) C = \left(I - V_l T_l V_l^H\right) \begin{bmatrix} C^{(U)} \\ C^{(L)} \end{bmatrix} = \left(I - \begin{bmatrix} 0 \\ V_l^{(U,L)} \\ V_l^{(U,F)} \\ V_l^{(L,F)} \\ V_l^{(L,U)} \\ 0 \end{bmatrix} T_l \begin{bmatrix} 0 \\ V_l^{(U,L)} \\ V_l^{(U,F)} \\ V_l^{(L,F)} \\ V_l^{(L,U)} \\ 0 \end{bmatrix}^H \right) \begin{bmatrix} C^{(U,1)} \\ C^{(U,2)} \\ C^{(U,3)} \\ \hline C^{(L,1)} \\ C^{(L,2)} \\ C^{(L,3)} \end{bmatrix}$$

$$= \begin{bmatrix} C^{(U,1)} \\ C^{(U,2)} \\ C^{(U,3)} \\ \hline C^{(L,1)} \\ C^{(L,2)} \\ C^{(L,3)} \end{bmatrix} - \begin{bmatrix} 0 \\ V_l^{(U,L)} \\ V_l^{(U,F)} \\ \hline V_l^{(L,F)} \\ V_l^{(L,U)} \\ 0 \end{bmatrix} T_l \underbrace{\left(V_l^{(U,L)^H} C^{(U,2)} + V_l^{(U,F)^H} C^{(U,3)} + V_l^{(L,F)^H} C^{(L,1)} + V_l^{(L,U)^H} C^{(L,2)} \right)}_{\tilde{C}}$$

$$= \begin{bmatrix} C^{(U,1)} \\ C^{(U,2)} - V_l^{(U,L)} T_l \tilde{C} \\ C^{(U,3)} - V_l^{(U,F)} T_l \tilde{C} \\ \hline C^{(L,1)} - V_l^{(L,F)} T_l \tilde{C} \\ C^{(L,2)} - V_l^{(L,U)} T_l \tilde{C} \\ C^{(L,3)} \end{bmatrix}. \quad (4.80)$$

This enables us to apply the factorization of $\tilde{Q}_*$ with operations that do not cross the tile boundaries and use, as in the standard QR decomposition, only a single temporary location $\tilde{C}$. With this knowledge the implementation of the QR decomposition of $M_{(p,k)}$ uses the tile-enabled formulation (4.80) for its internal updates and for the update of the remaining block columns in (4.71).

The last problem is to find the QR decomposition of a column block inside $M_{(p,k)}$ of width $N_B/q = N_Q$. Typically, for all current practical implementations we have $N_Q = 32$ or $N_Q = 64$. This means we only have a QR decomposition of at most $2N_B \times 32$ or $2N_B \times 64$ elements which crosses the tile boundary. Due to its small size, we avoid the overhead introduced by splitting the function calls to operate on two tiles and copy the column blocks of both tiles into a single temporary matrix of size $2N_B \times N_Q$. On this temporary location we use the algorithm for the column major layout from LAPACK. Afterwards, the block column is copied back into the tiled layout. This small auxiliary memory location can be statically allocated at the compile-time and hence does not influence the overall memory requirements of the inverse free iteration.

Together with the parallelization, the tiled layout leads to continuously high performance independent of the problem size compared to the previous approaches. Table 4.5 shows this comparison including the subsequent assembly of Q_{12} and Q_{22} on a test system.

Remark 4.19:
Since we already used the tile layout to compute the QR decomposition, we can use it to construct or apply Q in Algorithm 13, as well. The graph based scheduling using OpenMP and the tile-enabled operations as in the update (4.80) can be used in the same way to compute QC or $Q^H C$, where Q represents the overall matrix Q of the QR decomposition of E_R. The only prerequisite here is, that the matrix C is stored in tiled layout, too. Otherwise, we get the same memory access problems and the corresponding slow down as for the QR decomposition.

4.3.3. Block Upper Triangularization using Generalized Householder Transformations

Finally, we have the problem that the structure preservation of the inverse free iteration depends on the block size N_B, i.e. R_A will only be a block upper triangular matrix after the first iteration

Dimension m	Runtime without OpenMP 4	Runtime with OpenMP 4 and Tiled Layout	Speedup
2000	1.73s	0.19s	9.25
4000	12.26s	1.12s	10.94
6000	42.53s	3.60s	11.83
8000	88.30s	8.31s	10.62
10000	187.12s	16.31s	11.48
12000	310.47s	27.68s	11.21
14000	526.82s	44.15s	11.93
16000	674.00s	66.07s	10.20
18000	1065.45s	93.99s	11.34
20000	1380.55s	127.92s	10.79

Table 4.5.: Runtime and speedup of the structure preserving QR decomposition including assembling Q_{12} and Q_{22} on a 16-core Intel® Haswell platform.

although it was upper triangular before. Therefore, making the initial matrix R_A in (4.69) upper triangular is a waste of effort. Using generalized Householder transformations, as proposed by Schreiber and Parlett [146], we obtain the block upper triangular representation directly. Seeing that this algorithm is not used often, we briefly introduce it here.

Since we only have to guarantee that R_A is block upper triangular, we only need a transformation Q_A such that

$$\begin{bmatrix} A_{11} & \dots & A_{1p} \\ \vdots & & \vdots \\ A_{p1} & \dots & A_{pp} \end{bmatrix} = Q_A \begin{bmatrix} R_{11} & \dots & R_{1p} \\ 0 & \ddots & \vdots \\ 0 & 0 & R_{pp} \end{bmatrix} = Q_A R_A, \tag{4.81}$$

where $R_{l,l}$, $l = 1, \dots, p$, are not necessarily upper triangular matrices. This can be rewritten in similarity to the standard QR decomposition as a product of transformations Q_l, where each Q_l fulfills

$$Q_l^H \begin{bmatrix} R_{1l} \\ \vdots \\ A_{ll} \\ A_{l+1l} \\ \vdots \\ A_{pl} \end{bmatrix} = \begin{bmatrix} R_{11} \\ \vdots \\ R_{ll} \\ 0 \\ \vdots \\ 0 \end{bmatrix}. \tag{4.82}$$

Then, we obtain Q_A and R_A from

$$A = \underbrace{Q_1 Q_2 \cdots Q_p}_{Q_A} R_A. \tag{4.83}$$

Dietrich showed in [64] that this can be considered as a block version of the Householder QR decomposition and each Q_l can be written, in analogy to the vector valued Householder transformations, as

$$Q_l = I - 2 G_l G_l^H. \tag{4.84}$$

Regarding our global goal to use as many matrix-matrix operations as possible, this factorization of Q_l fits perfectly into our scheme. Properties of the generalized Householder transformation (4.84) and efficient computational schemes were provided by Schreiber and Parlett in [146]. Algorithm 14 shows an approach only requiring one small polar decomposition besides a set of

---

Algorithm 14 Computation of Generalized Householder Transformations

---

Input: $E \in \mathbb{C}^{m \times N_B}$

Output: $H = I - 2GG^H$ such that $HE = \begin{bmatrix} F \\ 0 \end{bmatrix}$, where $F \in \mathbb{C}^{N_B \times N_B}$

1: Compute a QR decomposition $PT = E$, with $P \in \mathbb{C}^{m \times N_B}$.

2: Compute the polar decomposition $Q_1 M_1 = P_1$, with $P = \begin{bmatrix} P_1 \\ P_2 \end{bmatrix}$ and P_1 is of size $N_B \times N_B$.

3: Set $F := -Q_1 T s$

4: Compute R, $R^H R = I + M_1$ using the Cholesky Decomposition.

5: Set $G := \begin{bmatrix} Q_1 R^H \\ P_2 R^{-1} \end{bmatrix}$.

---

standard BLAS and LAPACK level-3 enabled operations. A detailed derivation including variants based on matrix square-roots can be found in [146]. The algorithm is implemented straight forward using the techniques provided by BLAS and LAPACK. The only crucial operation is the computation of the polar decomposition of the block P_1 in Algorithm 14. Since P_1 is of dimension $N_B \times N_B$ and N_B is small in comparison to the size of the matrix A, it has a negligible influence on the runtime. Several ways to compute the polar decomposition exist. Higham [88] showed an iterative scheme similar to the computation of the matrix sign function. In order to obtain a high accuracy and the fact that we have only small matrices, we use the singular value decomposition based approach here [80]. Furthermore, this approach poses no restrictions on the input matrix, unlike the iterative approach.

Let P_1 have the singular value decomposition

$$P_1 = U \Sigma V^H, \tag{4.85}$$

then we obtain the polar decomposition $P_1 = Q_1 M_1$ by

$$Q_1 = U V^H \quad \text{and} \quad M_1 = V \Sigma V^H. \tag{4.86}$$

The overall procedure needs $O(N_B^2)$ additional memory mostly for the computation of the polar decomposition and $O(N_B^3)$ flops. Since we have $N_B \ll m$, this can be neglected in the flop count.

The whole procedure to reduce the matrix A of the matrix pair (A, B) to block upper triangular form is shown in Algorithm 15. Since Lemma 4.16 showed that the right matrix disc function is invariant under left-sided transformations, we do not require the transformation Q_A afterwards any longer. If we apply the transformations simultaneously to A and B, the matrices G_l are only required once, and their memory location can be reused. In this way, Algorithm 15 reduces the matrix pair (A, B) to $(R_A, Q_A^H B)$. Although we already computed a QR decomposition of each block column during the computation of G_l, the application of $(I - 2G_l G_l^H)$ to the remaining columns is more efficient on current computer architectures than using the factorization of the storage efficient QR decomposition. Using $N_B = 256$ for the blocks in E_R in (4.71), Figure 4.13 shows the runtime and the speed up compared to the standard QR decomposition. It shows that for increasing matrix dimensions, we accelerate the initial transformation by a factor of up to 4.5. Thereby, we benefit from the higher performance of the two matrix-matrix products with G_l compared to the possible triangular ones in the compact storage WY representation. In real world applications we perform an a priori benchmark on each machine to determine the switching point between both strategies.

Algorithm 15 Initial transformation for the inverse free iteration

---

Input: Matrix pair (A, B) of order m, N_B block size
Output: Matrix pair $(R_A, Q_A^H B)$, with R_A block upper triangular and $\Lambda(A, B) = \Lambda(R_A, Q^H B)$.
1: **for** $J = 1, 1 + N_B, 1 + 2N_B, \ldots, m$ **do**
2:     $J_B = \min N_B, m - J + 1$
3:     Compute G and F from $A(J : m, J : J + J_B - 1)$ using Algorithm 14.
4:     Set $A(J : J + J_B - 1, J + J_B - 1) := F$.
5:     Update $A(J : m, J + J_B : m) = (I - GG^H) A(J : m, J + J_B : m)$.
6:     Update $B(J : m, 1 : m) = (I - GG^H) A(J : m, 1 : m)$.
7: **end for**

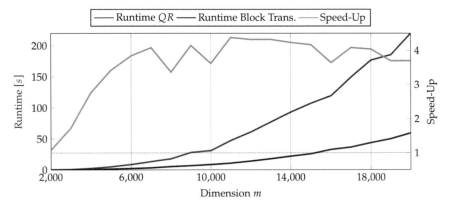

Figure 4.13.: Runtime and speedup of the initial reduction for the inverse free iteration on the Haswell architecture

4.3.4. GPU Implementation with Storage Efficient QR Decomposition

In the previous section we presented how to accelerate the QR decomposition for the matrix disc function on multi-core systems, which benefits from the directed acyclic graph parallelization using OpenMP. Although OpenMP supports accelerator devices, the execution model does not support tasks on them. Furthermore, accelerators require large datasets to achieve their peak performance and therefore the parallelization and implementation strategies presented before are not well suited for this purpose. The algorithm derived in this section covers the QR decomposition of an $m \times m$ matrix, which is required in the initial transformation of the generalized matrix sign function in Subsection 4.2.2, and the $2m \times m$ case from the inverse free iteration.

Hence, we focus on the GPU accelerated compact storage QR decomposition of a matrix A following the approach of Bischof and Van Loan [40]. Considerable work has already been done in this area, especially in the MAGMA library [3], but the experiments in Figure 4.14 show that their implementation performs good but not yet optimal on a single GPU. Regarding the two GPU case, the performance does not even reach 10% of the theoretical peak performance, which renders the multi-GPU implementation unusable for practical purposes. Other implementations like the ones provided by the cuSolver library [133] or the ArrayFire software package [176] do not provide multi-GPU variants of the algorithm at all. Our goal is to adjust the compact storage QR decomposition to multiple GPUs without being slowed down by the bad QR decomposition

---

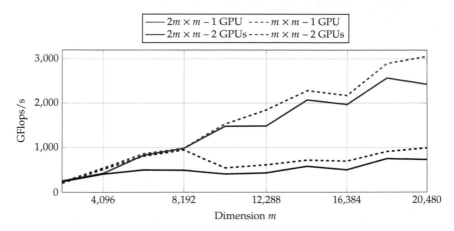

Figure 4.14.: Performance of the MAGMA QR decomposition on a GPU accelerated 20-core IBM POWER 8 System.

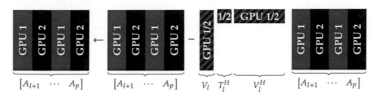

Figure 4.15.: Column Block Cyclic data distribution on two GPUs for the compact storage QR update.

performance of the CPUs as shown in Figure 4.11 and overcome the bad scaling properties of the MAGMA implementation. Similar to the hybrid CPU-GPU approach for the Gauss-Jordan Elimination in Subsection 4.2.4 the computation is split into a CPU part and GPU part. The CPU factorizes a block column A_l of size $m \times N_B$ to obtain

$$Q_l R_l = \left(I - V_l T_l V_l^{\mathsf{H}}\right) \begin{bmatrix} \tilde{R}_l \\ 0 \end{bmatrix} = A_l. \tag{4.87}$$

The GPU afterwards updates the remaining column blocks $A_k, l < k \le p$, using

$$\begin{bmatrix} A_{l+1} & \dots & A_p \end{bmatrix} \leftarrow \left(I - V_l T_l^{\mathsf{H}} V_l^{\mathsf{H}}\right) \begin{bmatrix} A_{l+1} & \dots & A_p \end{bmatrix}. \tag{4.88}$$

The look-ahead and the asynchronous operation is implemented employing the same strategy as in the Gauss-Jordan Elimination, compare Subsection 4.2.4. The update of A_{l+1} in (4.88) is performed first, such that the CPU can already prepare V_{l+1} and T_{l+1}, while the GPU updates the remaining columns of A. A multi-GPU-aware implementation, as in the Gauss-Jordan Elimination, uses the fact that the update (4.88) can be easily distributed across multiple GPUs. If the matrix A is distributed in a block column cyclic way, Figure 4.15 shows how the work is split over two GPUs. Algorithm 16 shows the workflow of the GPU accelerated QR decomposition, which is only a slight modification of the GPU accelerated Gauss-Jordan Elimination in Algorithm 12. Again, the block size N_B is a tuning parameter to obtain a well performing algorithm, which is determined with a set of benchmarks.

Remark 4.20:
Since the QR decomposition does not involve pivoting as in the Gauss-Jordan Elimination scheme, we use the column major storage on the CPU and the GPU. An adjustment is not necessary.

Although the QR decomposition is implemented similar to the Gauss-Jordan Elimination, additional optimizations need to be done:

Data transfer and management: For our application, we are interested in the factors of Q and upper triangular matrix R. Thereby, we have to check which datasets need to be transferred back and forth to the GPUs and which ones are already available on the CPU, since they are an outcome of the CPU part of Algorithm 16.

Optimization of the CPU part: The main CPU work is the decomposition of possibly tall-skinny matrices, which typically leads to a bad performance [59], due to its limited parallelization capabilities and their unfavorable data access patterns.

Handling the decreasing GPU workload: In contrast to the matrix inversion using the Gauss-Jordan Elimination, the number of flops performed on the GPUs decreases in every step.

Data Management: Although modern architectures have a fast interconnect between accelerators and CPUs, like PCIexpress Gen3 or NVLink, the communication between them is still the bottleneck of many algorithms. This is partly handled by the communication hiding the CPU-GPU data transfers in the asynchronously working algorithm. In the QR algorithm, we observe that once the CPU computed V_J and T_J in Algorithm 16, they stay untouched, i.e. they contain their final value. Thus, we already can store them to their final location on the CPU and do not need to download them from the GPUs memory at the end. That means we only need to fetch the upper triangle of A from the GPUs to obtain R in the end. Thereby, it is easy to see that once the diagonal block of R_{JJ} is computed on the CPU during the computation of V_J and T_J, it is not updated anymore. In order to obtain the full R, we only need to copy the strict upper block triangle of A back to the CPU at the end of Algorithm 16.

Alternatively, one can set up R on the CPU during Algorithm 16 if in step J the full column block $A(1 : m, J : J + J_B - 1)$ is copied. This hides the transfer of R, back to the CPUm behind the computation on the GPUs.

Like Algorithm 12, we need two parallel execution paths on the GPUs. Again these are *CtoG* for the CPU to GPU communication and *GPU-Compute* for the computations on the GPUs.

Beside the amount of data transferred between CPU and GPUs, the number of transfers plays a role because each transfer has an initial delay until the first byte is transferred. This latency is independent of the amount of data. We see in Algorithm 16 that in each step J, two data sets V_J and T_J need to be copied in a naive implementation and in this way the latency appears twice. Regarding the properties of both matrices we can reduce the transfer to a single one of size $(m - J - 1) \times J_B$. Thereby, we use that V_J is lower trapezoidal with a unit diagonal and T_J is upper triangular. Using the unit diagonal of V_J implicitly in every application of V_J, we can store both matrices in a rectangular matrix of size $(m - J + 1) \times J_B$, which is transferred to the GPUs in a single step.

Algorithm 16 GPU accelerated QR decomposition

Input: $A \in \mathbb{C}^{m \times n}$, block size N_B

Output: $QR := A$ with Q as in (4.66) and A overwritten by R.

1: Asynchronous distribution of A across all GPUs with block width N_B.
2: **for** $J := 1, 1 + N_B, 1 + 2N_B, \ldots, m$ **do**
3:     $J_B := \min(N_B, m - J + 1)$ and $l = \lfloor J - 1/N_B \rfloor + 1$.
4:     **if** $J = 1$ **then**
5:         **CPU:** Compute V_1 and T_1 from $A(1 : m, 1 : J_B)$
6:     **else**
7:         **CPU:** Wait for the download of $A(J : m, J : J + J_B - 1)$ in stream $CtoG$ and compute V_l and T_l from it.
8:     **end if**
9:     **CPU:** Upload V_l and T_l to all GPUs in stream $CtoG$.
10:     **GPU:** Synchronize stream $CtoG$ and stream $GPU\text{-}Compute$.
11:     **GPU:** Perform look-ahead update

$$A(J : m, J + J_B + 1 : J + J_B + \hat{J}_B - 1) =$$
$$\left(I - V_l T_l^H V_l^H\right) A(J : m, J + J_B + 1 : J + J_B + \hat{J}_B - 1)$$

    in stream $GPU\text{-}Compute$, with $\hat{J}_B = \min(N_B, n - (J + J_B) + 1)$.
12:     **CPU:** Synchronize stream $GPU\text{-}Compute$ and download $A(J + J_B : m, J + J_B : J + J_B + \hat{J}_B - 1)$ in stream $CtoG$.
13:     **GPU:** Update A using

$$A(J : m, J + J_B + \hat{J}_B : n) \leftarrow \left(I - V_l T_l^H V_l^H\right)(J : m, J + J_B + \hat{J}_B : n)$$

    in stream 2.
14: **end for**
15: **GPU:** Synchronize all streams.
16: **CPU:** Download the upper triangle of A containing R.

CPU Code Optimization: The benchmarks in the previous section showed that the performance of standard QR decompositions does not reach a high performance on current hardware. Especially, when the matrices are getting tall-skinny, like the column blocks in the GPU accelerated variant, the performance is far below the theoretical peak performance of the CPU. The QR decomposition used inside the inverse free iteration amplifies this problem even further, since the input matrix is already of size $2m \times m$ and the block columns become even more skinny. This results in the fact that the GPUs have to wait for the computations on the CPU and thus the CPU computation gets the limiting factor for the overall performance of the algorithm.

The first idea to overcome this problem is to use the tall-skinny QR decomposition (TSQR) on the CPU, which yields a better performance than the standard QR decomposition for this type of matrices [59, 17]. We already used building blocks of the algorithm in Section 4.3.2 to annihilate a block below a triangular matrix. Repeating this procedure subsequently for the blocks in a single block column, we obtain the TSQR algorithm. For a matrix $\tilde{A} \in \mathbb{C}^{m \times n}$, $m \ggg n$, we first compute the (standard) QR decomposition of the upper $n \times n$ block:

$$\tilde{A} = \begin{bmatrix} \tilde{Q}_1 \\ & 1 \\ & & \searrow \\ & & & 1 \end{bmatrix} \begin{bmatrix} \boxed{\searrow} \\ \boxed{} \\ \boxed{} \end{bmatrix} = Q_1 R_1. \tag{4.89}$$

Then we combine the next full block in R_1 with the upper triangular one using Q_2:

$$Q_1^H \tilde{A} = R_1 = \begin{bmatrix} 1 \\ & \searrow \\ & & 1 \\ & & & \tilde{Q}_2 \\ & & & & 1 \\ & & & & & \searrow \\ & & & & & & 1 \end{bmatrix} \begin{bmatrix} \boxed{\searrow} \\ \boxed{\searrow} \\ \boxed{} \end{bmatrix} = Q_2 R_2. \tag{4.90}$$

This continues until the bottom block in R_* is merged to the upper triangular block. The procedure is implemented in LAPACK as LATSQR or can be realized using TPQRT to compute the single block transformations on one's own. Although this scheme improves data locality [59], it still works sequentially and the parallelization of the BLAS library cannot be used very well in the single building blocks.

An extended but parallelizable scheme is the communication avoiding QR decomposition (CAQR) by Demmel et al. [59]. There, all $n \times n$ blocks in $\tilde{A}$ are decomposed using the standard QR decomposition first:

$$\tilde{A} = \begin{bmatrix} \tilde{Q}_1^{(1)} \\ & \tilde{Q}_1^{(2)} \\ & & \tilde{Q}_1^{(3)} \\ & & & \tilde{Q}_1^{(4)} \end{bmatrix} \begin{bmatrix} \boxed{\searrow} \\ \boxed{\searrow} \\ \boxed{\searrow} \\ \boxed{\searrow} \end{bmatrix} = Q_1 R_1, \tag{4.91}$$

which can be done independently and in parallel for each block. Afterwards, using the same building blocks as in the TSQR case, two consecutive triangles are merged to a single one until only one upper triangle is left. This results in

$$Q_1^H \tilde{A} = \begin{bmatrix} \tilde{Q}_2^{(1)} \\ & \tilde{Q}_2^{(2)} \end{bmatrix} \begin{bmatrix} \boxed{\searrow} \\ \boxed{\searrow} \end{bmatrix} = Q_2 R_2 \tag{4.92}$$

in the first step and leads to the final reduction

$$Q_2^H Q_1^H \tilde{A} = Q_3 \begin{bmatrix} \boxed{\searrow} \end{bmatrix} = Q_3 R. \tag{4.93}$$

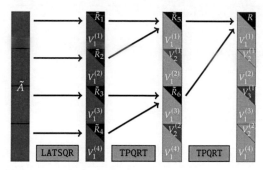

Figure 4.16.: LAPACK correspondence and storage scheme for the PTSQR of a matrix with four initial blocks.

This results in an easily parallelizable scheme where one either does all computations of $Q_j^{(i)}$ for a fixed j in parallel or one uses the underlying binary tree and the directed acyclic graph scheduling options of OpenMP 4. Both strategies can be implemented straight forward and chosen by the capabilities of the compiler. Although the scheme seems to allow good parallelization, the $n \times n$ blocks in the beginning are to small such that the parallelization overhead takes more time than we gain from the large amount of parallel tasks. To overcome this problem, we combine the TSQR approach with the CAQR approach to the parallel TSQR (PTSQR) in the following way: In the first step the matrix $\tilde{A}$ is split into P row blocks of nearly the same size, where P is the number of CPU cores working in parallel. Then each block is decomposed using the TSQR algorithm and afterwards the resulting upper triangular matrices are combined using the CAQR approach as shown in (4.92) and (4.93). Since the TSQR and the CAQR approach use the compact storage representation of $Q_j^{(i)}$, we have to store the corresponding $V_j^{(i)}$ and $T_j^{(i)}$. Thereby we use again that the matrix $V_j^{(i)}$ is unit diagonal and has the same shape as the annihilated block in the input matrix. In this way, we only have to store the sequence of $T_j^{(i)}$ in a separate location. Figure 4.16 illustrates this.

In contrast to the standard QR decomposition and the TSQR implemented in LAPACK we now have a highly parallelizable and efficient scheme to compute the QR decomposition of a block column on the CPU. The final problem now is that Algorithm 16 requires the factorization of a single matrix Q_l in each step as

$$Q_l = \left(I - V_l T_l V_l^H\right) \tag{4.94}$$

instead of the fine-grained decomposition the TSQR, CAQR, or PTSQR schemes produce. Therefore, we have to use the Householder reconstruction developed by Ballard et al. [17] with the goal of finding the compact storage representation (4.94) of an arbitrary factorization of a unitary matrix Q_l. Thereby, we only need to handle the effective part of Q_l, i.e. the first n columns for a column block or $A_l \in \mathbb{C}^{m \times n}$, $m \geq n$. Then equation (4.94) transforms into

$$Q_l = \left(I - V_l T_l \hat{V}_l^H\right), \tag{4.95}$$

where $\hat{V}_l$ is the top $n \times n$ block of V_l. The QR decomposition of Q_l reads [17]

$$Q_l = \left(I - V_l T_l \hat{V}_l^H\right) S, \tag{4.96}$$

Algorithm 17 Modified LU decomposition for Householder Reconstruction [17]

Input: $Q \in \mathbb{R}^{m \times n}$ or $Q \in \mathbb{C}^{m \times n}$ unitary, $n < m$.

Output: L lower trapezoidal, U upper triangular, and S diagonal such that $LU = Q - \begin{bmatrix} S \\ 0 \end{bmatrix}$

1: $S = 0$
2: **for** $j = 1, \dots, n$ **do**
3:     $S(j, j) = -\text{sign}\left(Q(j, j)\right)$
4:     $Q(j, j) = Q(j, j) - S(j, j)$
5:     $Q(j + 1 : m, j) = \frac{1}{Q(j,j)} Q(j + 1 : m, m)$
6:     $Q(j + 1 : m, j + 1 : n) = Q(j + 1 : m, j + 1 : n) - Q(j + 1 : m, j)Q(j, j + 1 : m)$
7: **end for**

Algorithm 18 Level-3 enabled LU decomposition for Householder Reconstruction

---

Input: $Q \in \mathbb{R}^{m \times n}$ or $Q \in \mathbb{C}^{m \times n}$ unitary with $n < m$.

Output: L lower trapezoidal, U upper triangular, and S diagonal such that $LU = Q - \begin{bmatrix} S \\ 0 \end{bmatrix}$

1: Compute L_1, U, and S from $Q(1 : n, 1 : n)$ using Algorithm 17.
2: **for** $j = n + 1, 2n + 1, \dots, m$ **do**
3:     $L(j : j + n - 1, 1 : n) = L(j : j + n - 1, 1 : n)U^{-1}$ {LAPACK: TRSM}
4: **end for**

---

where S is a diagonal matrix with ± 1 on its diagonal. The matrix S contains the sign changes performed during the computation of the Householder vectors in the standard QR algorithm [80]. Now, we obtain

$$Q_l - \begin{bmatrix} S \\ 0 \end{bmatrix} = -V_l T_l \hat{V}_l^H S \tag{4.97}$$

and by setting $L := V_l$ and $U = -T_l \hat{V}_l^H S$, we are able to compute V_l and T_l from the LU decomposition of $Q_l - \begin{bmatrix} S^H & 0 \end{bmatrix}^H$. Ballard et al. [17] proposed the modified LU decomposition to compute $LU = Q - \begin{bmatrix} S^H & 0 \end{bmatrix}^H$ including the determination of S during the computation shown in Algorithm 17. Additionally, they showed that V_l contains the same Householder vectors as the standard Householder QR decomposition would produce from the matrix A in exact arithmetic. Pivoting is not necessary in the LU decomposition since the current diagonal entry (j, j) in Q during step j has at least a magnitude of 1 by construction. The remaining entries in the j-th row or j-th column have an absolute value less than one since Q is unitary. Hence, the pivoting strategy of the LU decomposition would always select the diagonal entry.

Obviously, on the one hand, Algorithm 17 only works with BLAS level-2 operations which does not lead to a good memory reuse. On the other hand, there are accesses to long vectors and tall and skinny rank-1 updates. Using a common strategy, we derive Algorithm 18 as level-3 enabled variant of Algorithm 17, still using the assumption that $n \lll m$. If $m \approx n$ other strategies will be required. Another advantage of Algorithm 18 is the easy parallelization of the level-3 updates in terms of the triangular solves using the parallel loops in OpenMP. Furthermore, if the matrix is stored in tile representation with a tile size of $n \times n$, the data locality is improved. Since n can be made equal to the block size in the TSQR, CAQR, or PTSQR algorithm, we can easily implement the approach in tiled layout.

Finally, we have to solve

$$T_l \overset{S = S^{-1}}{=} -US\hat{V}_l^{-H} \tag{4.98}$$

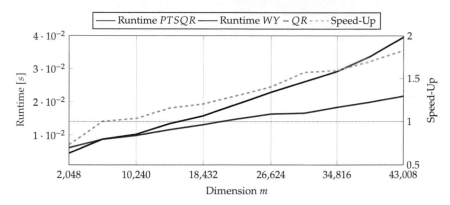

Figure 4.17.: Runtime of PTSQR and compact storage QR decomposition for a matrix of size $m \times 128$ on a 20-core IBM POWER 8 system.

to obtain T_l. Thereby, the solution with $\hat{V}_l$ will not cause problems since it is a triangular matrix with a unit diagonal and has full rank, as Q had full rank before. If the matrix S differs from the identity, the matrix R of the QR decomposition needs to be adjusted by

$$R \leftarrow SR. \tag{4.99}$$

Since PTSQR and the reconstruction of the Householder vectors are more expensive than the classical QR regarding the number of necessary operations, we set up a benchmark to figure out the favorable CPU algorithm depending on the size of the matrix. This is done by using an a priori benchmark. Thereby, for a fixed number of columns, we vary the number of rows to determine the critical point when it is worth to switch to the PTSQR algorithm. Figure 4.17 visualizes the runtimes for both approaches using a matrix with 128 columns and an increasing number of rows. We see that using 20 CPU cores, it is worth choosing the computational more expensive algorithm for large problems to obtain a better performance from the CPU.

GPU Workload Optimization   The last problem presented here is based on the decreasing number of flops performed on the GPUs during the QR decomposition. On the one hand, the CPU switches between the QR decomposition algorithms to minimize the time until the next Householder vectors are computed. On the other hand, once the GPU switched back to compact storage QR decomposition, the preparation time for the next step can be decreased only if the number of processed columns is decreased. Therefore, we develop an approach similar to the one presented by Catalán et al. [53] to adjust the block size N_B in Algorithm 16 dynamically. In practice that means, if the GPUs have to wait for the CPUs to finish the factorization of the next column block, the block size N_B is decreased to $N_B^{(w)}$ in order to obtain the result faster for the next column block.

In order to properly deal with a multi-GPU aware implementation of Algorithm 16, the reduced block size $N_B^{(w)}$ has to fulfill

$$2^k \cdot N_B^{(w)} = N_B, \quad k \in \mathbb{N} \tag{4.100}$$

to fit the data distribution across the participating GPUs easily for an initial data distribution with the block size N_B. This guarantees that a column block which is used to compute the next step on the CPU is never distributed across several GPUs. Now, every time the GPUs have to wait for the CPUs to finish their work, the block size is divided by two.

Since the CPU's performance is influenced by many factors, like different scheduling of parallel tasks, operating system interactions, etc., a safety factor is introduced such that the GPUs have to wait r times until the working block size N_B is reduced. A reasonable choice for r is 10. Finally, we define a minimal block size $N_B^{(low)}$ giving a lower bound on the working block size $N_B^{(w)}$ to avoid that the CPU's computations result in a level-2 BLAS like algorithm. For this reason setting, $N_B^{(low)}$ to 32 is a good choice because that coincides with good block sizes for the compact storage QR decomposition [72]. In general the parameters r and $N_B^{(low)}$ can be seen as additional tuning parameters.

Remark 4.21:

The restriction to a minimal block size $N_B^{(low)}$ in combination with the data distribution condition (4.100) for the working block size $N_B^{(w)}$ yields that the initial block size N_B, used for the data distribution, should fulfill

$$N_B = 2^k \cdot N_B^{(low)}. \tag{4.101}$$

Remark 4.22:

In general, we can use the same strategy if the CPU waits for the GPUs to upload the currently computed column block. In this case, we are able to increase the working block size N_B following similar rules as in the decreasing case. Due to the fact that the GPUs have a much higher performance than the CPUs, we neglect this case here because the number of required flops to perform a Householder QR decomposition is at least quadratic in the number of processed columns.

Since the change of the working block size $N_B^{(w)}$ affects the size of the matrices V_j and T_j, we either need to take care about the indices, where the block size is decreased, or we recover the compact storage representation with a fixed block size N_B. We choose the later option in order to obtain an output, which is, on the one hand, compatible to the GEQRT routine of LAPACK and, on the other hand, increases the size of the matrix-matrix products when Q is applied. Therefore, like in the derivation of the compact storage QR decomposition of Bischof and van Loan [40], we combine two neighboring $T_j \in \mathbb{C}^{N_B^{(w)} \times N_B^{(w)}}$ and $T_{j+1} \in \mathbb{C}^{N_B^{(w)} \times N_B^{(w)}}$ to $T_j^{(*)} \in \mathbb{C}^{2N_B^{(w)} \times 2N_B^{(w)}}$ by

$$\left(I - V_j^{(*)} T_j^{(*)} V_j^{(*)H} \right) = \left(I - \begin{bmatrix} V_j & V_{j+1} \end{bmatrix} \begin{bmatrix} T_j & -T_j V_j^H V_{j+1} T_{j+1} \\ & T_{j+1} \end{bmatrix} \begin{bmatrix} V_j & V_{j+1} \end{bmatrix}^H \right). \tag{4.102}$$

This is done recursively until $T_j^{(*)}$ is of size $N_B \times N_B$. Since this can be done independently for all blocks of size N_B, we are able to parallelize the work using several approaches using OpenMP:

1. We loop over all $N_B^{(w)} = N_B^{(low)}, 2N_B^{(low)}, \ldots, N_B$ and combine all blocks of size $N_B^{(w)} \times N_B^{(w)}$ with the corresponding neighbors in parallel. Therefore, the standard OpenMP loop parallelization is used on the innermost loop level.

2. We loop over all columns of the block size N_B and perform the recursive scheme (4.102) for each iterate independently. Again, the standard OpenMP loop parallelization can be used here but this time on the outermost loop level.

3. We can use the recursive dependencies in formula (4.102) and formulate it using task dependencies in OpenMP 4, here, as well. Thus, we create a task for each computation of (4.102), which depends on the computations one recursion level below.

Depending on the availability of OpenMP 4, the last approach is the preferred one. Otherwise, the first approach is a good alternative providing more independent tasks than the second approach.

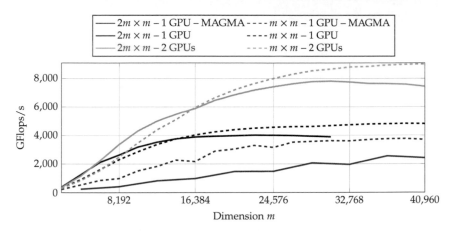

Figure 4.18.: Performance Comparison of the MAGMA QR decomposition and our GPU implementation on an accelerated 20-core IBM POWER 8 System with 2x NVidia P100.

Finally, Figure 4.18 shows the performance improvements of our implementation compared to the one in MAGMA. In both cases decomposing either an $m \times m$ or and $2m \times m$ matrix we see a large performance increase. Especially, our variant scales perfectly across both GPUs in the IBM POWER 8 system. Comparing the achieved performance with the theoretical peak performance of the GPUs, we see that our implementation achieves up to 85.2% in the $m \times m$ case and 73.7% in the $2m \times m$ case. Regarding the performance of MAGMA on a single GPU, since the multi-GPU implementation is not competitive as seen in Figure 4.11, we obtain a speed up of 2.43 for the square case with $m = 40\,960$ and 2.38 for the $2m \times m$ case with $m = 30\,720$. The performance loss for large inputs in the $2m \times m$ case is caused by CPU slow down due to the power capping of the IBM POWER 8 CPUs.

Application of Q. Beside the fast computation of the QR decomposition, we need to compute QC or $Q^H C$, too, where C is an arbitrary matrix. In the matrix disc function, the matrix $\begin{bmatrix} 0 & I \end{bmatrix}^H$ is used as C but some variants of the generalized matrix sign function require the QR decomposition as well and the application of Q. Since the matrix C has $O(m)$ columns in our case, we distribute the matrix C in a column block cyclic way across all GPUs, like the matrix A in the GPU accelerated QR decomposition. Afterwards, the Householder vectors V_j and the Householder factors T_j are uploaded to all GPUs and the updates can be computed in a distributed parallel manor, as already shown in Figure 4.15. If the memory for V_j and T_j is allocated twice on all GPUs, double buffering can be used to upload the Householder data for the block $j \pm 1$, while the GPUs still compute the update. Thereby, with the ability of modern GPUs to copy data between CPU and GPUs during computations is used again. Since V_j is unit lower trapezoidal and T_j is upper triangular, they can be transferred in a single operation as discussed for the QR decomposition.

NUMERICAL EXPERIMENTS FOR THE EIGENVALUE PROBLEM

Contents

In the previous chapters we presented the Divide-Shift-and-Conquer and the Divide-Scale-and-Conquer algorithm to approximate the generalized eigenvalue problem and how they can be implemented on modern hardware. In this chapter we present numerical results on several current hardware platforms. First, we introduce the hardware and software setup and show the influence of tuning parameters, like the block size, on the single operations as well as their performance. Finally, we show a qualitative and a quantitative comparison between our eigenvalue solvers and the QZ algorithm, i.e., its state-of-the-art implementations.

The runtime measurements are done with respect to the wall-time, i.e., the time the user waits for the result. In some cases the CPU-time is mentioned, which is the accumulated time of all CPU cores in work. The CPU-time is used to show the utilization of the whole system. If a CPU utilizes all its n cores for a runtime of t seconds, a CPU-time of $n \cdot t$ is spent.

5.1. Hardware and Software Setup

As test hardware we use a set of compute servers, whose micro architectures were developed between 2014 and 2018. In order to not stick to architecture specific properties, we use x86-64 based systems and the IBM POWER 8 architecture, which is also used in current HPC environments.

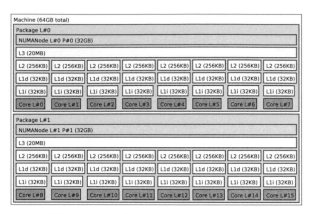

Figure 5.1.: CPU structure and Memory hierarchy of the Haswell system.

5.1.1. Compute Servers

We use three different systems for evaluating the performance of our algorithms.

Intel® Haswell based Compute Server – "Haswell".   The first system is based on the Intel® Haswell architecture. The system consists of two eight-core Intel® Xeon® E5-2640v3 running at 2.6 GHz. Each socket has a shared level-3 cache of 20 MB and each core utilizes a level-2 cache of 256 kB. Furthermore, the system is equipped with 64 GB DDR4 memory. For acceleration, there are two Nvidia® K20m GPUs available, each equipped with 5 GB memory and connected with PCIexpress generation 2. The system runs on Linux, CentOS 7.6, using the Intel® Compiler Suite version 18.1 and the Intel® MKL 2018.1 library as BLAS and LAPACK implementation. The GPUs are accessed using Nvidia® CUDA 8.0. Figure 5.1 shows the CPU core and memory hierarchy of the system as obtained by hwloc [48].

Intel® Skylake based Compute Server – "Skylake".   As successor of the Haswell micro architecture on the x86-64 platform we use an Intel® Skylake system equipped with two eight-core Intel® Xeon® Silver 4110 CPUs running at 2.1 GHz. Each socket has 11.2 MB level-3 cache shared over all CPU cores. Each individual CPU core has 1 MB level-2 cache. The overall system has 192 GB DDR4 memory. The system has a single Nvidia® P100 accelerator connected via PCIexpress generation 3. The software setup is identical to the Haswell based system with the exception that Nvidia® CUDA 9.2 is used here. The CPU core and memory hierarchy layout is similar to the Haswell system as shown in Figure 5.2.

IBM POWER 8 based Compute Server – "POWER 8".   In contrast to both systems presented before, the last system is a two socket system equipped with ten-core IBM POWER 8 CPUs running at 4 GHz in little-endian mode. Each CPU core has 8 MB of level-3 cache and 512 kB of level-2 cache. There is no shared cache across the CPU cores, but the level-3 caches are kept coherent by the hardware. Furthermore, the system has 256 GB DDR4 memory. Since the POWER 8 CPUs have a direct high speed interconnection for Nvidia® GPUs, the system uses two Nvidia® P100 SXM2 accelerators. The interconnection, called NVLink, allows much higher data transfer rates between CPU and GPU with low latency. Furthermore, direct access from the CPU to the GPUs memory

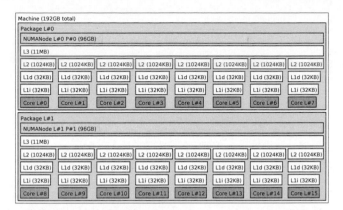

Figure 5.2.: CPU structure and Memory hierarchy of the Skylake system.

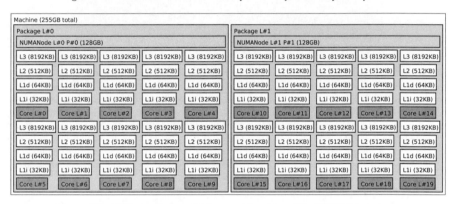

Figure 5.3.: CPU structure and Memory hierarchy of the POWER 8 system.

and vice versa is possible. The system runs on Linux, CentOS 7.6, using the IBM XLC C compiler 16.1.0, the IBM XLF Fortran compiler 16.1.0, and Nvidia® CUDA 9.2. The IBM ESSL 5.6 is used for BLAS and LAPACK calls. Since the LAPACK part of the ESSL is not feature complete, missing routines are added from the reference LAPACK implementation version 3.8.0. Figure 5.3 shows the architecture of the POWER 8 system.

5.1.2. Implementation

The whole code is implemented in C and Fortran. Algorithms which access the GPUs are written in C. Computational routines that only require the CPU are written entirely in Fortran. Thereby, the array operations introduced in Fortran 90 are widely used. Explicit parallelization is done using the features of OpenMP 4 and 4.5 described in Chapter 4. All basic linear algebra operations are performed with the help of the parallel capabilities of the BLAS and the LAPACK implementation. In the case that BLAS and LAPACK calls are executed inside an OpenMP parallelized part of the code, the used BLAS and LAPACK libraries automatically switch back to single thread mode for

Name	Dimension	Description	Reference
BCSST 07	420	Structural Engineering	[70]
BCSST 19	817	Structural Engineering	[70]
BCSST 09	1083	Structural Engineering	[70]
BCSST 12	1473	Structural Engineering	[70]
Filter 2D	1668	Tunable Optical Filter	[136, 93]
BCSST 26	1992	Seismic analysis	[70]
BCSST 13	2003	Fluid Flow	[70]
BCSST 23	3134	Structural Engineering	[70]
BCSST 21	3600	Structural Engineering	[70]
Steel 5	5177	Heat Transfer Problem	[135, 34]
Flow Meter	9669	Convective Thermal Flow Problem	[163, 130]
BCSST 25	15439	Structural Engineering	[70]
Chip Cooling	20082	Convective Thermal Flow Problem	[163, 130]
Steel 20	20209	Heat Transfer Problem	[135, 34]
Gyro	34722	Micro-Mechanical Gyroscope	[134, 38]

Table 5.1.: Benchmark Matrices from the Oberwolfach and the SuiteSparse Collection

the time they run inside the OpenMP parallel region.

Since many algorithms depend on hardware specific parameters, like the block size, switching points, etc., the implementation needs to be adjusted to the hardware and software setup. Typically, this is done at compile time like in the ATLAS project [172] with the help of auto-tuning. This results in long compilation times and makes fast adjustments impossible. Hence, we go for the approach used in the last version of the AMD Math Kernel Library released in 2014, where the automatic GPU offloading is controlled by evaluating small scripts at runtime. Therefore, we use the LUA programming language [95]. It is an imperative scripting language whose interpreter can be easily embedded into applications without introducing a large overhead. All tuning parameters are implemented as functions inside a single LUA script. In order to avoid further overhead and failures of the applications due to errors in the tuning codes, the whole tuning file is executed once at the application start. Thereby, a syntax check is performed and the code is translated into the internal representation for LUA's stack based virtual machine.

All computations are performed in IEEE-754.2008 binary64, i.e. double precision, arithmetic.

5.1.3. Reference Results and Data Sources

The reference results for the generalized eigenvalue problem are obtained using the QZ algorithm available in LAPACK with and without the level-3 generalized Hessenberg reduction. These routines are GGES, based on the work of Moler and Stewart [129], and its extension GGES3 by Kågström et al. [104]. Furthermore, we use the multi-shift QZ algorithm with aggressive early deflation from Kågström and Kressner [103] together with the level-3 generalized Hessenberg reduction. This procedure is referenced as KKQZ. Distributed parallel versions like the implementation of Adlerborn, Kågström, and Kressner [8] are not used, since their distributed architecture lies outside our focus.

The random matrices are generated using subsequent calls to the random number generator of LAPACK with a uniform-$(-1, 1)$ value distribution. These matrices are used for the performance tests of the Gauss-Jordan Elimination, the generalized matrix sign function, the QR decompositions, and the inverse free iteration. Most tuning parameters, like the optimal block sizes, depend only on the size of the input rather than the specific values. The eigenvalue solvers are tested with the matrices listed in Table 5.1 obtained from the SuiteSparse Collection [58] and the Oberwolfach Benchmark Collection [116]. Thereby, sparsity, definiteness, or symmetry properties are neglected and not taken into account for any of the computations, since our algorithms do not take care about this information.

The Gyro example [134, 38] is a bit different from the remaining ones. The matrices M and K describe a second order differential equation

$$M\ddot{x}(t) + \left(\alpha M + \beta K\right)\dot{x}(t) + Kx(t) = B_{\text{control}}u(t), \tag{5.1}$$

where $\alpha = 10^{-6}$ and $\beta = 0$. The mass matrix M and the stiffness matrix K are of order $17\,361$. The matrix B_{control} is used to control the system and not of interest in our application. In order to obtain a generalized eigenvalue problem from Equation (5.1) we transform it into a first order differential equation neglecting the control B_{control}:

$$\begin{bmatrix} I \\ & M \end{bmatrix} \begin{bmatrix} \dot{x} \\ \ddot{x} \end{bmatrix} = \begin{bmatrix} & I \\ -K & -\left(\alpha M + \beta K\right) \end{bmatrix} \begin{bmatrix} x \\ \dot{x} \end{bmatrix}. \tag{5.2}$$

By setting the matrix pair (A, B) to

$$(A, B) = \left(\begin{bmatrix} & I \\ -K & -\left(\alpha M + \beta K\right) \end{bmatrix}, \begin{bmatrix} I \\ & M \end{bmatrix} \right) \tag{5.3}$$

we obtain our generalized eigenvalue problem of order $34\,722$.

5.2. Performance of the Building Blocks

Before we present the results for the generalized eigenvalue solvers, we have to tune the Gauss-Jordan Elimination and the QR decompositions together with the two most time-consuming algorithms, Newton's method for the generalized matrix sign function and the inverse free iteration. All experiments focus on medium and large-scale problems beginning with a dimension of $m = 1\,024$ or $m = 2\,000$. For smaller problems, effects caused by environmental effects of the hardware and the operating system are too large to give reliable results. Furthermore, we focus on large problems. GPUs are used as long as the matrices fit into their memories. Out of core algorithms lie outside our focus.

5.2.1. Gauss-Jordan Elimination

The Gauss-Jordan Elimination is our alternative to the classical LU decomposition based approach for solving and inverting linear systems. Although our derivations and optimizations in Chapter 4 focus on the GPU acceleration, we regard the CPU version here as well. Since both, the LU decomposition and the Gauss-Jordan Elimination, are built on top of general matrix-matrix multiplies, we expect that both will result in a proper CPU utilization. However, the Gauss-Jordan Elimination needs 12.5% more flops than the LU decomposition, when solving a linear system of dimension m with m right-hand sides, see Section 4.2.4. We use a set of benchmarks to determine the optimal block sizes in the Gauss-Jordan Elimination. Furthermore, we search for switching points between Gauss-Jordan Elimination and the LU decomposition. Since only the dimension m is of interest for the performance of the algorithms, and we do not want to tune with respect to a special pivoting caused by artificial data, each experiment in this section is done with ten random matrices. The random matrices are generated with uniform-$(-1, 1)$ distribution and an initial seed of $(1, 1, 1, 1)$ using the LARNV random number generator of LAPACK.

First, we determine the optimal block size for the matrix inversion on our three systems. The block size is chosen as a multiple of 64 in the range between 128 and $1\,536$ and assumed to be optimal if it either yields a locally maximum flop rate or the flop rate stagnates if the block size is

	Haswell		Skylake		POWER 8	
Dimension m	GFlops/s	Opt. N_B	GFlops/s	Opt. N_B	GFlops/s	Opt. N_B
1 024	175	192	93	192	103	192
2 048	249	384	205	192	185	320
3 072	316	256	290	256	259	448
4 096	357	256	311	128	212	704
6 144	356	256	365	128	349	1 536
10 240	377	192	417	128	387	1 536
14 336	391	192	430	128	408	1 536
18 432	418	192	441	128	416	1 536
22 528	453	192	451	128	424	1 536

Table 5.2.: Optimal block sizes for the Gauss-Jordan Elimination on the CPU. (Computing the inverse)

increased further. In Table 5.2, we present the optimal block sizes for selected problem dimensions on the three target systems. We see that for both systems based on Intel® CPUs, the optimal performance is obtained by using a small block size of 128 to 192 for large problems. In the case of the POWER 8 system, the flops stagnate for large problems if 1 536 is reached. For small problems, both Intel® architectures need a slightly larger block size. A possible reason is that in these cases almost all required data fits into the CPU caches. Since it is not possible to sample all possible matrix dimensions, the runtime selection of the parameter N_B chooses the block size from the samples which is the nearest neighbor to the actual problem dimension. This behavior is implemented as LUA function depending on the problem dimension m, which is called at the beginning of the Gauss-Jordan-Elimination.

We compare the performance of the Gauss-Jordan Elimination based matrix inversion using these optimal block sizes with the LU decomposition based one implemented in LAPACK. Figure 5.4 shows that on both Intel® architectures, the LU based inversion from Intel® MKL's LAPACK implementation performs inferior in all cases. For $m = 30 720$, the Gauss-Jordan approach is already 1.44 times faster on the Haswell system. On the Skylake system, we obtain a ratio of 1.32. Regarding the POWER 8 system with IBM ESSL's BLAS and LAPACK implementation, the LU decomposition and the subsequent inversion is more competitive. There we see that for small problems it is not worth to use the Gauss-Jordan Elimination approach but for increasing problem sizes, beginning at $m \approx 6\,000$ our approach gets faster as well. In the end, for $m = 30 720$, we are 7% faster than the vendor's implementation.

The unsatisfactory performance of the LU based inversion on the Intel® systems compared to the POWER 8 system calls for further investigations. Figure 5.4d presents an additional benchmark, where BLAS and LAPACK on the Skylake system is provided by the OpenBLAS 0.3.5 library [137] instead of the Intel® MKL. We see that using OpenBLAS makes the Gauss-Jordan Elimination 14% slower but the ratio between the LU decomposition based inversion and our approach, a factor of 1.33, is approximately the same. This leads to the assumption that the inversion procedure for LU decomposed matrices, namely GETRI, is not well suited for current Intel® architectures.

If the generalized matrix sign function requires the solution of a linear system $AX = B$, i.e. we use the naive implementation, the situation on the CPU slightly changes, since solving a linear system of order m with m right hand sides takes $\left(2 + \frac{2}{3}\right) m^3$ flops using the LU decomposition and $3m^3$ flops using the Gauss-Jordan Elimination. Thus, it is only worth to use the Gauss-Jordan Elimination if we achieve a 12.5% higher performance. Again we determine the optimal block sizes first, as shown in Table 5.3, and compare the performance of the Gauss-Jordan Elimination with the LU decomposition with subsequent forward and backward substitution in Figure 5.5. The graphs show that both strategies lead to nearly the same performance on the Intel® based systems

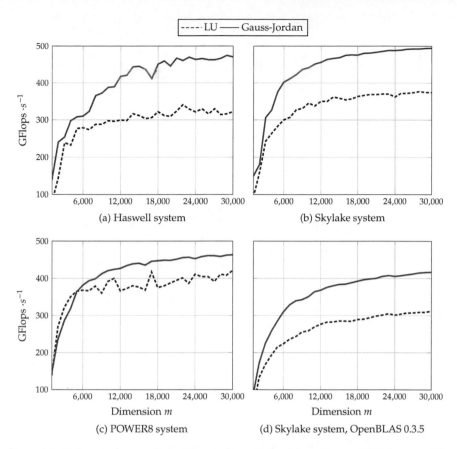

Figure 5.4.: Performance comparison between Gauss-Jordan Elimination and LU decomposition based matrix inversion.

and hence the Gauss-Jordan approach runs approximately 12.5% longer. Thus, we only use the LU decomposition on the Intel® systems for solving the appearing linear systems. In contrast, the POWER 8 system with the ESSL library can benefit from the Gauss-Jordan Elimination approach even for solving linear systems. For problems larger than $m = 8\,000$, we obtain an average runtime gain of 11.1%. Regarding the performance, we increase the flop rate by 26.7% in average. This shows that even on CPUs we have to check whether the Gauss-Jordan Elimination or the LU decomposition results in the shortest runtime for our problems. The switching point, depending on the input's dimension, is implemented as a LUA function returning whether the Gauss-Jordan Elimination approach or the LU decomposition should be for the solution of linear systems.

Finally, the automatic selection between Gauss-Jordan Elimination and LU decomposition on the CPU is performed as follows. On both Intel® systems, the Gauss-Jordan Elimination is used to compute the inverse of a matrix and the LU decomposition is used for the solution of the

Dimension m	Haswell		Skylake		POWER 8	
	GFlops/s	Opt. N_B	GFlops/s	Opt. N_B	GFlops/s	Opt. N_B
1 024	303	512	278	256	214	512
2 048	303	512	278	256	214	512
3 072	303	512	278	256	214	512
4 096	303	512	278	256	214	512
6 144	404	384	425	384	337	1 536
10 240	445	384	452	384	382	1 536
14 336	466	576	466	576	404	1 536
18 432	476	576	476	576	418	1 536
22 528	495	576	483	576	423	1 536

Table 5.3.: Optimal block sizes for the Gauss-Jordan Elimination on the CPU. (Solving a linear system with m right-hand sides.)

Dimension m	Haswell 2 × Nvidia® K20m		Skylake 1 × Nvidia® P100		POWER 8 2 × Nvidia® P100	
	GFlops/s	Opt. N_B	GFlops/s	Opt. N_B	GFlops/s	Opt. N_B
2 048	260	256	337	256	295	64
6 144	1 007	384	1 770	128	1 609	64
10 240	1 341	512	2 776	384	3 326	64
14 336	1 485	640	3 160	640	4 821	128
18 432	1 563	640	3 432	768	5 853	128
22 528	1 630	768	3 632	896	6 835	256
26 624	1 684	1 280	3 778	1 408	7 496	256
30 720	1 743	1 408	3 869	1 536	7 708	512
40 960	—	—	4 017	1 536	8 274	576
51 200	—	—	—	—	8 609	896
61 440	—	—	—	—	8 741	1 024

Table 5.4.: Optimal block sizes and performance for the Gauss-Jordan Elimination with GPU acceleration. (Computing the inverse.)

Dimension m	Haswell 2 × Nvidia® K20m		Skylake 1 × Nvidia® P100		POWER 8 2 × Nvidia® P100	
	GFlops/s	Opt. N_B	GFlops/s	Opt. N_B	GFlops/s	Opt. N_B
2 048	260	256	337	256	445	64
6 144	1 007	384	1 770	128	2 454	64
10 240	1 341	512	2 776	384	4 702	128
14 336	1 485	640	3 160	640	6 463	128
18 432	1 563	640	3 432	768	7 325	256
22 528	1 630	768	3 632	896	7 943	256
26 624	—	—	3 778	1 408	8 399	448
30 720	—	—	3 869	1 536	8 576	512
40 960	—	—	—	—	8 908	896
43 008	—	—	—	—	8 959	896

Table 5.5.: Optimal block sizes for the Gauss-Jordan Elimination on the GPUs. (Solving a linear system with m right hand sides.)

linear systems. On the POWER 8 system the matrix inversion uses the LU decomposition for $m < 5\,000$ and for larger problems the Gauss-Jordan Elimination approach. In case of solving a linear system, the POWER 8 system uses the LU decomposition up to $m = 8\,000$ and for larger problems the Gauss-Jordan Elimination.

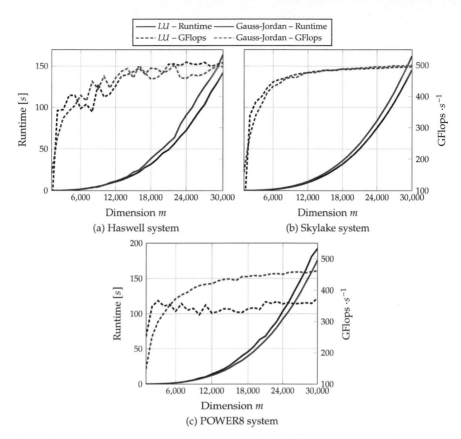

Figure 5.5.: Performance comparison between Gauss-Jordan Elimination and LU decomposition based linear system solve.

After analyzing the pure CPU performance, we switch over to the GPU accelerated implementations. On the Haswell and the POWER 8 system we use two accelerator cards, while on the Skylake based system we only have one available. Since the GPUs have a higher performance for matrix-matrix products we extend the dimension of the benchmark inputs as long as they fit in the memory of the GPUs. This results in a maximum problem size for the inversion procedure of 30 720, 40 960, and 61 440 on the Haswell, Skylake, and POWER 8 systems. In case of solving a linear system with m right-hand sides, the maximum problem size is limited to 22 528, 30 720, and 43 008, respectively. Again, the first benchmark determines the optimal block sizes for the algorithm. On the accelerator based systems we obtain Table 5.4 for inverting a matrix and Table 5.5 for the solution of a linear system with m right-hand sides. Regarding the inversion procedure, we achieve 74%, 85%, and 82% of the theoretical peak performance of the GPUs on the Haswell, the Skylake, and the POWER 8 systems, respectively. In case of solving a linear system with m right-hand sides we reach up to 68.4%, 82.3%, and 84.5% of the peak performance on these systems.

Algorithm	Haswell	Skylake	POWER 8
Naive	$m \leq 500$	$m \leq 1\,000$	$m \leq 2\,000$
One-Sided TRTRSM	—	—	—
One-Sided TRMM	—	—	—
Two-Sided Bidiag.	$500 < m \leq 2\,500$	$1\,000 < m \leq 3\,000$	$2\,000 < m \leq 3\,000$
Two-Sided Band.	$m > 2\,500$	$m > 3\,000$	$m > 3\,000$

Table 5.6.: Algorithm selection for the computation of the generalized matrix sign function using Newton's method.

Furthermore, the tables show a difference between the systems. While the Intel® based systems require a relatively large block size to obtain the best performance on the GPUs, the POWER 8 system needs only small to moderate block sizes to achieve the highest performance. This is caused by the different interconnects between the CPU and the GPUs. On the POWER 8 system, NVLink is used and thus we have a higher bandwidth available. The improved architecture of the NVLink interconnection can transfer small chunks between CPU and GPU memories without spending too much time in the auxiliary parts of the GPU-CPU communication.

5.2.2. Generalized Matrix Sign Function

After tuning the Gauss-Jordan Elimination and determining the switching points between Gauss-Jordan Elimination and LU decomposition, we consider the generalized matrix sign function. Different variants are developed in Chapter 4. Here, they are benchmarked to determine the fastest algorithm depending on the size of the input matrices. Thereby, we use the results of the previous benchmarks to select between Gauss-Jordan Elimination and LU decomposition based solvers and their optimal block sizes. On the CPU we test five approaches, the naive one, the one-sided transformation with triangular solves (TRTRSM), the one-sided transformation with triangular matrix-matrix products (TRMM), the two-sided transformation with bidiagonal reduction, and the two-sided transformation with band reduction. Regarding the capabilities of the GPU accelerated Gauss-Jordan Elimination, we only use the naive implementation, the one-sided transformation with triangular matrix-matrix multiplies, and the two-sided transformation with initial band reduction on the CPU. The approach using a one-sided transformation together with backward substitution for triangular matrices, as shown in Section 4.2.2, requires multi-GPU aware operations on triangular matrices. These operations are either not implemented in GPU accelerated libraries or they are too communication-intensive and thus slow down the computation.

The benchmarks work as follows: for each algorithm, we fix the number of iteration steps to 25 without checking for convergence because we are interested in the best-performing algorithm here and not in eventually faster convergence by the initial transformation. This number coincides with the typical average number of iterations taken by the generalized matrix sign function iteration, compare [31, 18]. We use random matrices of dimension $1\,000$ to $30\,000$ for the CPU tests and $1\,000$ to $30\,000$, $40\,000$, or $60\,000$ depending on the memory of the GPUs for the accelerated tests. Based on the results of these benchmarks, we implement the selection between the algorithms in the startup LUA function.

From the synthetic benchmark with a fixed number of iteration steps, we obtain Table 5.6 for selecting the proper algorithm inside the Divide-Shift-and-Conquer scheme. We see that for small problems we can still use the naive implementation since the overhead of the initial transformations used by the non-naive algorithms does not disappear. Increasing the problem to a moderate size, the bidiagonal reduction variant becomes the fastest version. Although the performance of these methods gets unsatisfactory for increasing problem sizes, it is favorable until we have enough data for the parallelization inside the band reduction and the ratio between the bandwidth and the

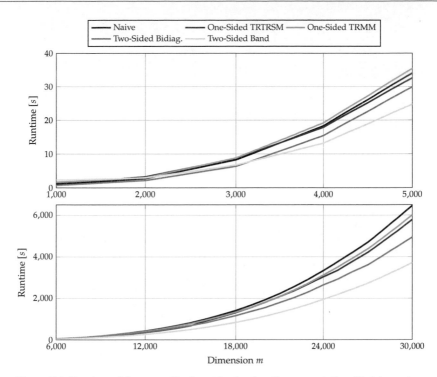

Figure 5.6.: Runtime of the generalized matrix sign function computation, Skylake system

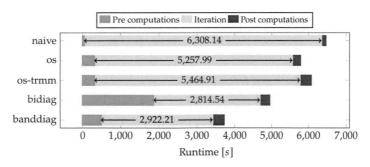

Figure 5.7.: Runtime distribution in the computation of the generalized matrix sign function computation, $m = 30\,000$ Skylake system

problem size gets larger. For large-scale problems, the band reduction is the best choice. Figure 5.6 exemplary shows the runtime for a CPU only implementation on the Skylake system. There, we see a drastic runtime reduction by our band reduction approach. For the $m = 30\,000$ case, Figure 5.7 shows in detail, where the runtime is spent in the computation. The band reduction approach

Matrix	Dim. m.	# Iter	Runtime [s]	Difference to Naive	Runtime [s] GGES3
BCSST09	1 083	18	**0.41**	-11.56%	9.03
BCSST12	1 473	20	**0.78**	-11.70%	33.26
BCSST24	3 562	26	10.14	-26.89%	273.66
Flow Meter	9 996	20	**120.80**	-34.42 %	7 600.23
BCSST25	15 439	26	**558.46**	-41.03%	35 898.86
Chip Cooling	20 082	15	**783.60**	-34.78%	76 814.84
Gyro	34 722	18	**4490.41**	-44.24%	352 406.85

Table 5.7.: Runtime of the generalized matrix sign function for real world problems on the Skylake system, CPU-only. (Bold means the fastest algorithm was selected.)

needs nearly twice the pre-computation time as the one-sided approaches, as expected by the required number of flops. The bidiagonal approach takes 381% longer in the preparation phase than the band reduction while saving only 3.7% during the iteration. Overall the band reduction approach gives a performance increase of a factor of 1.74 compared to the naive implementation and a factor of 1.33 compared to the bidiagonal approach. The results on the Haswell and the POWER 8 systems are similar.

Afterwards, we confirm the selection between the different generalized matrix sign function implementations using some of our real world data, compare Table 5.1, and a convergence criterion of $\tau_S = 10^{-7} \approx 10 \cdot \sqrt{\varepsilon}$. Once the criterion is fulfilled we perform two additional steps. Table 5.7 shows the runtime of the selected algorithm from Table 5.16. Only in one case the selection mechanism does not select the fastest algorithm but the difference in runtime is below 3%, which may be caused by operating system interactions. Even more, we see that compared to the naive solution we win between 10% and 45% execution time. This makes our selection strategy valid for the future use inside the Divide-Shift-and-Conquer scheme.

Furthermore, Table 5.7 shows that computation of a single generalized matrix sign function takes drastically less runtime than the QZ algorithm of the same size. A runtime difference of a factor of 22 to 98 is possible. This raises the hope that a recursive application of the generalized matrix sign function, as proposed in the Divide-Shift-and-Conquer algorithm, is able to decrease the runtime of the overall computation of the generalized Schur decomposition.

The GPU benchmarks are performed in the same manor. Thereby, an additional parameter appears in the selection. We not only compare the GPU enabled approaches, but the fastest CPU code for the given problem size as well, since we have to decide whether it is worth to switch to the GPUs at all. Figure 5.8 compares the three accelerated implementations on our three systems. The selection of the optimal algorithm gets adjusted to the values shown in Table 5.8. Although the triangular matrix-matrix multiply based approach is the fastest for moderate size problems on the GPU, we switch to the band reduction based scheme as soon as the problem gets too large for the GPUs memory. In this case only one $m \times m$ matrix needs to be stored on the accelerator instead of two. If the problem size is larger than the memory of the accelerator devices, we use the CPU implementation as fallback, see Table 5.6.

5.2.3. QR Decompositions

The second important algorithm used in our Divide-Scale-and-Conquer algorithm is the inverse free iteration, whose key operation is the QR decomposition. Furthermore, the QR decomposition is already used in the one-sided flop reducing strategies for the generalized matrix sign function, where it has only a small impact on the runtime. Both applications of the QR decomposition lead

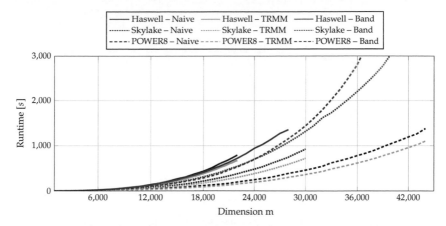

Figure 5.8.: Runtime of the GPU accelerated generalized matrix sign function, 25 iterations.

Algorithm	Haswell	Skylake	POWER 8
CPU Selection	$m \leq 1\,000, m > 30\,000$	$m \leq 1\,000, m > 44\,000$	$m < 1\,400, \ m > 62\,000$
GPU			
Naive	—	—	—
One-Sided TRMM	$1\,000 \leq m \leq 22\,000$	$1\,000 \leq m \leq 30\,000$	$1\,400 \leq m < 44\,000$
Two-Sided Band.	$22\,000 < m \leq 30\,000$	$30\,000 < m \leq 44\,000$	$44\,000 \leq m < 62\,000$

Table 5.8.: Algorithm selection for the computation of the generalized matrix sign function using Newton's method with GPU acceleration.

to two scenarios we have to regard. First, problems of size $m \times m$ for the initial transformations and second of size $2m \times m$ appearing in the inverse free iteration. Since some algorithms use the tile matrix storage idea rather than the standard column major storage scheme, we set the tile size to 256×256. This value is already verified by the PLASMA project [5, 51, 151] to be a proper choice. Beside the general purpose QR decomposition, a specialized one for the use inside the inverse free iteration was developed in Section 4.3.2. This algorithm is treated in the next section, since it is directly integrated into the inverse free iteration and is not used outside of this.

First, we compare the level-3 enabled storage efficient QR decomposition [40], namely $WY - QR$, with the performance of Tile-QR [51, 151], which can be seen as a successor on multi-core architectures. We measure the runtime and the performance to decompose a matrix and apply the unitary factor to m right-hand sides from the left. This is the typical usage of the QR decomposition in our applications. Figure 5.9 shows the performance of both algorithms on the Skylake system for problems of size $m \times m$ and $2m \times m$. We see that the Tile-QR decomposition yields a better runtime than the classical QR implementations for medium and large-scale problems. Only for small problems it is worth to use the classical implementation, since it does not involve the overhead of converting the matrices back and forth to the tile storage scheme. Similar results are obtained on the Haswell and the POWER 8 systems and thus we use the Tile-QR for medium and large-scale problems on all platforms.

The structure-preserving QR decomposition from Section 4.3.2 requires to have the bottom part of the matrix in upper triangular form or at least block upper triangular form before the

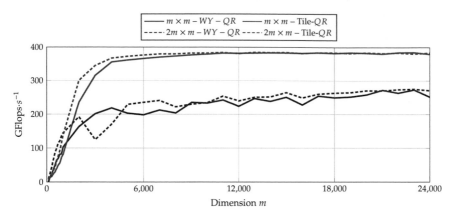

Figure 5.9.: Performance of the $WY - QR$ and the Tile-QR decomposition on the Skylake system.

Algorithm	Haswell	Skylake	POWER 8
QR Decomposition	$m < 2\,000$	$m < 2\,000$	$m < 6\,000$
Block Upper Triangular	$m \geq 2\,000$	$m \geq 2\,000$	$m \geq 6\,000$

Table 5.9.: Switching points for the upper triangularization in the inverse free iteration.

inverse free iteration begins. An algorithm for block upper triangularization based on generalized Householder transformations was shown in Subsection 4.3.3. Table 5.9 shows when it is worth to switch from a storage efficient $WY - QR$ decomposition to the block upper triangularization in the initialization of the inverse free iteration. The block upper triangularization is rich in matrix-matrix multiplies and thus it achieves a better performance on large problems. Table 5.10 shows the speed up of the block triangularization compared to using the $WY - QR$ from LAPACK.

If GPUs come into play, the situation changes since OpenMP does not provide a task based parallelization on accelerator devices. For this reason we cannot use the tile-based parallelization approaches on the GPUs. Instead, we use the hybrid QR decomposition approach shown in Section 4.3.4. Its two main tuning parameters are the block size on the GPUs and the selection of the CPU's factorization algorithm. First, we tune the factorization of a single panel of the matrix, which is the CPU part of the hybrid CPU-GPU algorithm. Thereby, we fix the number of columns of the panel to $n = 32, 64, \ldots, 512$, which can be seen as typical for the application inside our hybrid QR decomposition. We vary the number of rows to determine the switching point between the storage efficient $WY - QR$ decomposition and the communication avoiding QR decomposition with Householder reconstruction. Table 5.11 shows the switching points for all three used architectures.

Using these switching rules for the CPU part of the QR decomposition, we end up with Table 5.12 for the GPU block sizes. Obviously, the QR decomposition achieves its best performance with smaller block sizes than the Gauss-Jordan Elimination procedure, compare Tables 5.4 and 5.5. Furthermore, we see that with an increasing problem size, the optimal block size increases slower than for the Gauss-Jordan Elimination. We see a dependence of the optimal block size on the hardware. The Haswell system has the slowest interconnect between CPU and GPU and yields the largest block sizes. Since the NVLink architecture of the POWER 8 system allows small transfers between CPU and GPU without noteworthy overheads, we can already exchange small datasets.

Dimension m	Haswell	Skylake	POWER 8
1 000	0.55	0.37	0.35
2 000	1.08	1.19	0.59
4 000	2.77	2.26	0.98
6 000	3.84	2.33	1.06
10 000	3.61	2.20	1.35
14 000	4.23	2.07	1.42
18 000	4.03	2.12	1.38
20 000	3.70	2.19	1.68

Table 5.10.: Speed up of the block upper triangularization compared to the LAPACK $WY - QR$ decomposition.

Columns n	Haswell	Skylake	POWER 8
64	$m > 0$	$m > 0$	$m > 16\,000$
128	$m < 50\,000$	$m > 0$	$m > 91\,000$
192	$m < 10\,000$	$m > 0$	$m > \infty$
256	$m < 23\,000$	$m > 223\,000$	never
320	$m < 2\,000$	never	never
384	never	never	never
448	never	never	never
512	never	never	never

Table 5.11.: Selection rules based on the number of rows to use the communication avoiding QR decomposition with Householder reconstruction depending on the number of columns n.

In this way, selecting a smaller block size yields a faster computation on the CPU and the GPU does not have to wait for the data.

Table 5.13 shows the optimal block sizes on the GPU for the $2m \times m$ case. We see that this case requires smaller block sizes and does not lead to the same high performance as the square case. The reason lies in the inefficient data access properties during the factorization of the panels on the host. In the $2m \times m$ case, the panels, which are factorized on the CPU, get taller and more skinny for an increasing problem dimension. It is known for the QR decomposition that the tall and skinny case is the killer application from a performance point of view [59, 17]. Although we already use the communication avoiding QR decomposition, this only works well for a few columns. This performance decreasing behavior affects the runtime of the panel factorization on the CPU and hence the GPUs have to wait for data and stay idle during this time. Finally, this leads to a slow down of the overall computation.

5.2.4. Inverse Free Iteration

After evaluating the different variants of the QR decomposition, we benchmark the performance of the inverse free iteration. We use the selection strategies for the QR decomposition inside the different implementations. As in the case of the generalized matrix sign function, we fix the number of iteration steps to 25, first, and find the optimal algorithm with respect to the dimension of the input. This is done with and without GPU acceleration. Afterwards, we use some of the real world problems to confirm our selection.

Table 5.14 presents the runtime of 25 iteration steps of the inverse free iteration for $m \geq 2\,000$ using the CPUs only. The naive implementation uses the Tile-QR approach to compute and apply the QR decomposition since we have seen in the previous section, that this presents the fastest solver for medium and large-scale problems. The trapezoidal QR decomposition based approach delivers the best runtime for all problems beginning at $m = 2\,000$. Remarkable is the high performance

Dimension	Haswell 2 × Nvidia® K20m		Skylake 1 × Nvidia® P100		POWER 8 2 × Nvidia® P100	
m	GFlops/s	Opt. N_B	GFlops/s	Opt. N_B	GFlops/s	Opt. N_B
2 048	199	512	275	192	303	128
6 144	922	384	826	64	1 384	64
10 240	1 361	384	2 146	192	2 861	64
14 336	1 718	384	3 039	192	4 378	64
18 432	1 872	384	3 612	192	5 853	128
22 528	1 991	384	3 905	192	7 055	128
26 624	2 049	384	4 033	192	7 887	128
30 720	2 065	384	4 101	192	8 361	128
40 960	—	—	4 274	384	8 895	128
43 008	—	—	4 333	384	8 944	128
51 200	—	—	—	—	8 958	128
61 440	—	—	—	—	9 028	512

Table 5.12.: Optimal block sizes and performance for the QR decomposition of $m \times m$ matrices on the GPU.

Dimension	Haswell 2 × Nvidia® K20m		Skylake 1 × Nvidia® P100		POWER 8 2 × Nvidia® P100	
m	GFlops/s	Opt. N_B	GFlops/s	Opt. N_B	GFlops/s	Opt. N_B
2 048	229	64	314	64	380	64
6 144	982	128	1 478	64	1 745	64
10 240	1 328	128	2 553	128	3 473	64
14 336	1 497	128	3 159	128	4 902	64
18 432	1 564	256	3 416	128	5 999	128
22 528	1 658	384	3 557	128	6 823	128
26 624	—	—	3 605	128	7 309	128
30 720	—	—	3 549	128	7 509	128
40 960	—	—	—	—	7 312	128

Table 5.13.: Optimal block sizes and performance for the QR decomposition of $2m \times m$ matrices on the GPU.

loss of the Tile-QR decomposition on the POWER 8 system. Possible reasons for this behavior are the missing optimized routines inside the ESSL library and a not yet feature complete OpenMP 4 support in the IBM XLF 16.1 compiler. However, the trapezoidal QR approach introduces an overhead caused by the initial transformation and conversion between standard column major storage and the tile storage scheme. This overhead increases for smaller problems. Figure 5.10 shows the behavior for problems of size $m < 2\,000$. There we observe that the critical dimension to use the trapezoidal algorithm rather than the naive implementation lies between 1 000 and 1 600 depending on the architecture. Summing up, Table 5.15 shows when which of the presented implementations of the inverse free iteration is used depending on the dimension of the input. The determined switching points between are confirmed with the help of the real world data, as in the case of the generalized matrix sign function. Table 5.16 shows that our automatic selection scheme still works reliably with enabled convergence criterion as well. Except for the smallest example, our selection mechanism always selects the best suited algorithm. Comparing Figure 5.10, we see that the dimension of the small problems lies in a region, where all algorithms yield a similar performance and thus the decision can be seen to be right.

As in the generalized matrix sign function case, Table 5.16 also shows the runtime of the QZ algorithm. We see that the computation of a single disc function is faster than the QZ algorithm by

Dimension	Haswell		Skylake		POWER 8	
m	Tile QR	Trap. QR	Tile QR	Trap. QR	Tile QR	Trap. QR
2 000	7.54	6.60	8.38	7.26	9.01	8.19
4 000	52.57	41.07	55.27	44.51	54.67	44.38
6 000	171.17	134.44	181.61	142.42	182.72	140.71
8 000	405.16	308.77	425.31	328.87	439.24	325.13
10 000	763.59	597.90	827.07	634.96	897.60	626.96
12 000	1 364.58	1 027.51	1 426.70	1 092.78	1 678.33	1 071.18
14 000	2 079.32	1 632.86	2 246.68	1 728.40	3 096.84	1 697.62
16 000	3 197.79	2 408.70	3 389.71	2 580.92	5 486.50	2 528.40
18 000	4 418.97	3 382.91	4 776.33	3 660.96	9 781.69	3 606.20
20 000	6 080.22	4 656.26	6 554.19	4 977.87	16 607.10	4 965.36

Table 5.14.: CPU runtime (in seconds) of the inverse free iteration.

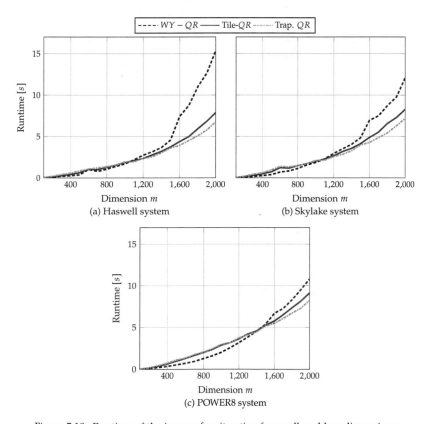

(a) Haswell system

(b) Skylake system

(c) POWER8 system

Figure 5.10.: Runtime of the inverse free iteration for small problem dimensions.

Algorithm	Haswell	Skylake	POWER 8
Naive with WY-QR	$m \leq 1\,000$	$m \leq 1\,100$	$m \leq 1\,500$
Naive with Tile-QR	—	—	—
Trap. QR	$m > 1\,000$	$m \geq 1\,100$	$m \geq 1\,500$
GPU QR	$3500 < m \leq 22\,000$	$1\,000 < m \leq 30\,000$	$800 < m \leq 42\,000$

Table 5.15.: Algorithm selection for the computation of the matrix disc function function using the inverse free iteration, 25 iterations.

Matrix	Dim. m.	# Iter	Runtime [s]	Difference to Naive	Runtime [s] GGES3
BCSST09	1\,083	4	0.54	+5.9%	9.03
BCSST12	1\,473	4	**0.80**	-13.14%	33.26
BCSST24	3\,562	6	**9.27**	-59.96%	273.66
Flow Meter	9\,996	9	**224.44**	-40.24%	7\,600.23
BCSST25	15\,439	11	**1\,092.23**	-39.34%	35\,898.86
Chip Cooling	20\,082	6	**1\,403.63**	-32.72%	76\,814.84
Gyro	34\,722	5	6186.32	-28.27%	352\,406.85

Table 5.16.: Runtime of the inverse free iteration for real world problems on the Skylake architecture, CPU-only. (Bold means the fastest algorithm was selected.)

a factor of 16.7 to 57. Since the matrix disc function requires more operations than the generalized matrix sign function, we expect that the Divide-Scale-and-Conquer approach will not be as fast as the Divide-Shift-and-Conquer algorithm.

After evaluating the CPU performance, we treat the GPU enabled algorithm now. In contrast to the generalized matrix sign function we only have to check one implementation here, which relies on the previously tuned QR decomposition. Figure 5.11 presents the runtimes of the GPU enabled implementation and the speedup compared to the trapezoidal CPU implementation. For increasing problem dimensions, we obtain a speedup of 2.8, 5.5, and 13.0 on the Haswell, the Skylake and the POWER 8 systems, respectively. Comparing the performance of the GPU algorithm to the CPU algorithms extends Table 5.15 by the GPU algorithm.

5.3. Computing the Generalized Schur Decomposition

During the previous sections, we have determined the optimal runtime parameters for the most time-consuming parts of the divide-and-conquer based algorithms. Now, we apply the algorithms derived in Chapters 3 and 4 to real world problems to show their impact on practical applications.

The approximation quality of the generalized Schur decomposition is measured using the following quantities. Assuming that our algorithms compute the generalized Schur decomposition of (A, B) as

$$(A, B) \approx \left(Q A_S Z^H, Q B_S Z^H \right), \tag{5.4}$$

we define the *reconstruction error* η as

$$\eta = \max \left\{ \frac{\left\| A - Q A_S Z^H \right\|_F}{\|A\|_F}, \frac{\left\| B - Q B_S Z^H \right\|_F}{\|B\|_F} \right\}. \tag{5.5}$$

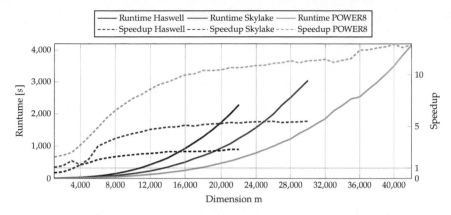

Figure 5.11.: Runtime and Speedup of the GPU accelerated inverse free iteration, 25 iterations.

In order to compare the approximation of the eigenvalues to the results of the QZ algorithm, we define $\Lambda^{\approx}$ as the vector of the finite eigenvalues obtained from the approximation (5.4) and Λ^{QZ} as the vector of the finite eigenvalues computed using the QZ algorithm GEES from LAPACK. Then we set the *global normalized eigenvalue deviation* δ_G as

$$\delta_G = \frac{\left\|\Lambda^{\approx} - \Lambda^{QZ}\right\|_2}{\left\|\Lambda^{QZ}\right\|_2} \tag{5.6}$$

and the *local normalized eigenvalue deviation* δ_L as

$$\delta_L = \max_{k=1,\dots,m} \frac{\left|\Lambda_k^{\approx} - \Lambda_k^{QZ}\right|}{\left|\Lambda_k^{QZ}\right|}. \tag{5.7}$$

For both deviations we match the eigenvalues of the sets Λ^{QZ} and $\Lambda^{\approx}$ before, such that corresponding eigenvalues have the same index k. Since the exact eigenvalues of the problems are not known and the QZ algorithm is an iterative procedure, we regard the global and local eigenvalue deviation as measure how good we can reconstruct the results of the QZ algorithm.

The runtime comparison is done employing the execution time and the CPU time to verify whether the multi-core capabilities of the system are properly used. Since the building blocks are already optimized, the benchmarks only vary the subspace extraction algorithm. The three variants GEQPX, GEQP3, and HQRRP, which are used in our experiments, are explained in Section 4.1.5. These algorithms differ in performance as well as in accuracy, especially the randomized algorithm HQRRP [125] is known to perform well on current hardware, but is less accurate. On all systems we perform the test with and without the GPU accelerators to validate that our approach does not depend on the high performance of the GPUs to be competitive to the different implementations of the QZ algorithm.

Performance of the QZ algorithm. The QZ algorithm is used to compute the reference eigenvalues for the global and local eigenvalue deviation as well as the reference runtime for the computation of the generalized Schur decomposition. We take the already mentioned implementations of the QZ algorithm into account.

Tables 5.17, 5.18, and 5.19 present the runtimes and the CPU times to compute the generalized Schur decomposition using the three algorithms and the real world problems from Section 5.1.3. We see that for the GGES and the GGES3 implementation, the runtime and the CPU time nearly coincide for large-scale problems on both Intel® systems. This means that the implementation does not benefit from the multiple CPU cores on these systems. On the POWER 8 system, we have a ratio between 7.5 and 8.5 between CPU time and the runtime for those two implementations. That means that still less than half of the available CPU cores on this system are used at all for the full time. Switching to the KKQZ approach, the situation slightly changes. On both Intel® systems we obtain a better ratio between CPU and runtime. Still, it does not utilize all available CPU cores. Comparing the speedup between GGES (or GGES3) and the KKQZ implementations we see a factor of at most 2.4 for medium- to large-scale problems on the Intel® systems. Only for the extremely large Gyro problem, this behavior does not hold true. Regarding the size of the data, the CPU cache is updated too often and thus data, which is still in use, is thrown out of the cache and fetched again from the memory. Regarding the POWER 8 system again, we see a huge speed up for the KKQZ approach compared to the approaches available in LAPACK. Reasons for this behavior lie in the higher memory bandwidth, the higher clock frequency, and the different cache architecture. However, we still need more than 6.5 hours for a problem of size $m = 20\,209$ or 29.6 hours for a problem of size $m = 34\,722$ using the fastest algorithm on the best suited architecture. Using the fastest algorithm implemented in LAPACK, these problems take between 27.5 hours to 68.8 hours or 81.1 hours to 121.2 hours to be solved on our systems.

Table 5.20 shows the reconstruction error of the three QZ implementations. We see that we obtain similar results for all problems with a small increase for increasing problem dimensions. All generalized Schur decompositions have a reconstruction error in the order of $O\left(10^{-13}\right)$ or $O\left(10^{-14}\right)$.

General Setup for the Divide and Conquer Schemes. Independent of the way to compute the deflating subspaces, the following parameters are set. The trivial problem size is set to $m_{triv} = 80$ on the Haswell system, to $m_{triv} = 170$ on the Skylake system, and to $m_{triv} = 120$ on the POWER 8 system. Since the POWER 8 system has two cache levels CPU-core exclusive, a trivial problem size of $m_{triv} = 500$ is possible, too, and used later in a single experiment. The derivation of these values was shown in Section 4.1.4. The tolerance factor for the maximum depth of the recursion tree τ_{rec} is set to 12 in all test cases. It serves only to detect undividable spectra. The algorithm to compute the deflating subspaces is limited to 50 iteration steps. A relative change stopping criterion of 10^{-9} is used with three additional steps once it is fulfilled. The algorithms to compute the generalized matrix sign function and the inverse free iteration are selected with respect to the results from the previous section.

5.3.1. Divide-Shift-and-Conquer

Before we evaluate the runtime for the different hardware setups, we do a qualitative comparison between the results obtained by the Divide-Shift-and-Conquer algorithm and the results obtained by the QZ algorithm GGES. Thereby, we extract the deflating subspaces using the different QR decompositions, the column pivoted QR decomposition GEQP3, the randomized sampling QR decomposition HQRRP, and the rank revealing QR decomposition GEQPX (see Section 4.1.5). The trivial problem size for these experiments is set to $m_{triv} = 80$, since this is the smallest trivial size of the used hardware in the comparison.

Tables 5.21 and 5.22 show the reconstruction error, the global, and the local eigenvalue deviation for the GEQP3 and the HQRRP approach. We see that both variants to extract the deflating subspaces result in comparable reconstruction errors and eigenvalue deviations. The approximations by

Problem	m		GGES	GGES3	KKQZ
BCSST 07	420	*Runtime*	0.50	0.50	0.66
		CPU time	3.79	3.77	9.81
BCSST 19	817	*Runtime*	2.23	2.22	2.91
		CPU time	5.88	5.83	28.47
BCSST 09	1 083	*Runtime*	8.07	8.18	7.40
		CPU time	12.11	12.21	58.58
BCSST 12	1 473	*Runtime*	31.58	31.55	18.41
		CPU time	36.19	36.24	121.33
Filter 2D	1 668	*Runtime*	37.36	37.45	24.68
		CPU time	42.44	42.49	149.88
BCSST 26	1 922	*Runtime*	41.32	41.18	34.29
		CPU time	47.14	46.95	150.52
BCSST 13	2 003	*Runtime*	58.94	58.36	36.30
		CPU time	65.10	64.52	114.64
BCSST 23	3 134	*Runtime*	203.96	203.02	150.51
		CPU time	218.91	217.91	526.77
BCSST 21	3 600	*Runtime*	307.44	307.49	214.71
		CPU time	326.39	326.97	614.33
Steel 5	5 177	*Runtime*	1 672.03	1 674.93	740.64
		CPU time	1 716.83	1 719.50	2 953.81
Flow Meter	9 669	*Runtime*	10 346.36	10 312.70	4 599.00
		CPU time	10 510.63	10 473.68	15 729.28
BCSST 25	15 439	*Runtime*	30 244.58	30 341.34	17 524.05
		CPU time	30 876.28	30 950.82	41 001.79
Chip Cooling	20 082	*Runtime*	62 888.24	62 908.66	37 285.27
		CPU time	64 196.24	64 200.79	89 371.27
Steel 20	20 209	*Runtime*	98 827.96	98 980.95	42 087.87
		CPU time	100 165.53	100 364.78	120 859.10
Gyro	34 722	*Runtime*	328 795.58	292 089.77	286 784.14
		CPU time	335 700.15	299 341.51	557 712.98

Table 5.17.: Runtime and CPU time of the QZ algorithms on the Haswell system.

the Divide-Shift-and-Conquer algorithms are not as accurate as the ones achieved by the QZ algorithm, compare Table 5.20, but still sufficient for most applications. Especially, it is smaller than the single precision machine epsilon ($\epsilon_s = 6 \cdot 10^{-8}$). This means that our approximation is more accurate than the generalized Schur decomposition computed in single precision arithmetic. We see in Chapter 6 that this is important for the direct solution of Sylvester-type matrix equations. Both subspace extraction variants approximate the eigenvalue set in a comparable quality. But the local eigenvalue deviation tells us that at least in one of algorithms, the Divide-Shift-and-Conquer or the QZ algorithm, some eigenvalues are perturbed.

Regarding the results obtained using the rank revealing QR decomposition GEQPX for subspace extraction in Table 5.23 the situation changes. In almost all cases, the reconstruction error is not sufficient. The eigenvalue deviation show that many eigenvalues are not approximated such that we can obtain useful information from them. The reason is the internal rank detection used by the rank revealing QR decomposition. This procedure is similar to the Incremental Condition Estimation [41] procedure, see Section 3.1.1. Thereby, cut-off tolerances need to be selected, which strongly depend on the input data. Furthermore, the rank revealing QR procedure only relies on this condition estimation and ignores the rank of the deflating subspace determined by the properties of the generalized matrix sign function, compare Corollary 3.10. Hence, when using the rank-revealing QR decomposition the rank decision is computed by it as well. Due to additional

Problem	m		GGES	GGES3	KKQZ
BCSST 07	420	*Runtime*	0.39	0.40	0.62
		CPU time	3.67	3.70	9.66
BCSST 19	817	*Runtime*	2.28	2.23	2.98
		CPU time	5.79	5.80	30.83
BCSST 09	1 083	*Runtime*	8.92	9.03	7.46
		CPU time	13.43	12.81	61.54
BCSST 12	1 473	*Runtime*	32.85	33.26	19.10
		CPU time	37.18	37.55	128.62
Filter 2D	1 668	*Runtime*	34.67	33.57	23.86
		CPU time	39.33	39.10	158.64
BCSST 26	1 922	*Runtime*	39.06	37.76	32.77
		CPU time	44.55	44.40	154.73
BCSST 13	2 003	*Runtime*	61.21	60.76	35.62
		CPU time	66.98	66.75	114.58
BCSST 23	3 134	*Runtime*	188.85	187.34	135.69
		CPU time	199.93	198.37	499.00
BCSST 21	3 600	*Runtime*	287.78	286.58	190.38
		CPU time	301.83	300.65	557.34
Steel 5	5 177	*Runtime*	1 820.94	1 780.56	732.44
		CPU time	1 855.31	1 816.61	2 798.89
Flow Meter	9 669	*Runtime*	12 429.67	11 907.14	4 913.48
		CPU time	12 577.83	12 054.61	14 596.85
BCSST 25	15 439	*Runtime*	36 340.86	35 898.86	19 462.10
		CPU time	36 910.94	36 469.15	38 910.71
Chip Cooling	20 082	*Runtime*	79 459.74	76 814.84	44 212.78
		CPU time	80 647.65	78 013.49	82 967.32
Steel 20	20 209	*Runtime*	121 422.87	119 258.14	48 105.21
		CPU time	122 545.77	120 388.67	111 161.92
Gyro	34 722	*Runtime*	362 756.17	352 406.85	255 551.01
		CPU time	368 902.83	358 598.55	458 538.50

Table 5.18.: Runtime and CPU time of the QZ algorithms on the Skylake system.

parameters, like tolerances in this procedure, it happens that computed rank not coincide with the one computed from the generalized matrix sign function. For example full rank or a zero matrix are not detected properly. This leads to a possibly wrong splitting, which is not detectable by the relative decoupling residual. Due to this problem in the rank decision and its poor performance, we discard the rank revealing QR decomposition for the deflating subspace extraction.

Since the QZ algorithm only approximates the eigenvalues of the matrix pair (A, B) using an iterative procedure, we take a closer look at the results of the Gyro example. The eigenvalues which cause the maximum local eigenvalue deviation is

$$\lambda_{QZ} = -2.079614498465631 \cdot 10^3 \pm 6.464696728361261 \cdot 10^4 i$$

as computed by the QZ algorithm and

$$\lambda_{DSC} = -2.091865566678822 \cdot 10^3 \pm 6.464790853135486 \cdot 10^4 i$$

as computed with the Divide-Shift-and-Conquer algorithm. The quality of an eigenvalue can be checked using the minimal singular σ_{min} value of $A - \lambda B$, which gives information about the perturbation that would cause this eigenvalue to be computed in exact arithmetic [80]. In our case we obtain

$$\sigma_{min}\left(A - \lambda_{QZ}B\right) = 1.531065727744096 \cdot 10^{-9}$$

Problem	m		GGES	GGES3	KKQZ
BCSST 07	420	Runtime	0.56	0.55	1.35
		CPU time	4.30	4.85	17.03
BCSST 19	817	Runtime	3.03	2.69	5.62
		CPU time	23.68	23.01	70.21
BCSST 09	1 083	Runtime	11.22	10.14	12.39
		CPU time	87.65	82.91	156.28
BCSST 12	1 473	Runtime	26.54	23.23	25.40
		CPU time	207.26	189.18	320.61
Filter 2D	1 668	Runtime	35.45	32.93	33.95
		CPU time	272.39	268.10	415.35
BCSST 26	1 922	Runtime	38.85	31.29	35.86
		CPU time	303.77	260.94	415.68
BCSST 13	2 003	Runtime	41.02	36.77	29.98
		CPU time	320.65	303.95	329.93
BCSST 23	3 134	Runtime	201.34	139.64	124.49
		CPU time	1 572.42	1 141.45	1 370.24
BCSST 21	3 600	Runtime	282.21	230.11	158.36
		CPU time	2 201.85	1 869.15	1 650.22
Steel 5	5 177	Runtime	2 356.72	1 546.22	581.10
		CPU time	18 356.42	12 248.17	6 703.82
Flow Meter	9 669	Runtime	24 470.31	15 442.78	3 007.18
		CPU time	190 588.06	121 375.85	32 642.06
BCSST 25	15 439	Runtime	67 047.28	53 476.30	9 796.06
		CPU time	586 375.61	420 805.73	95 129.16
Chip Cooling	20 082	Runtime	207 772.25	193 775.92	22 322.27
		CPU time	1 737 057.04	1 521 796.44	216 729.21
Steel 20	20 209	Runtime	261 813.77	247 785.85	23 595.78
		CPU time	2 160 818.01	1 940 695.82	243 616.22
Gyro	34 722	Runtime	423 964.40	436 369.78	106 720.60
		CPU time	3 807 409.04	3 500 402.58	1 058 734.59

Table 5.19.: Runtime and CPU time of the QZ algorithms on the POWER 8 system.

Problem	GGES	GGES3	KKQZ	Problem	GGES	GGES3	KKQZ
BCSST 07	1.49e-14	1.49e-14	9.82e-15	BCSST 19	1.70e-14	1.70e-14	2.02e-14
BCSST 09	3.23e-14	3.23e-14	1.97e-14	BCSST 12	8.62e-14	8.62e-14	2.07e-14
Filter 2D	4.48e-14	4.48e-14	2.79e-14	BCSST 26	4.15e-14	4.15e-14	4.62e-14
BCSST 13	3.85e-14	3.85e-14	2.81e-14	BCSST 23	6.78e-14	6.78e-14	4.05e-14
BCSST 21	2.89e-14	2.89e-14	3.01e-14	Steel 5	1.85e-13	2.05e-13	1.23e-13
Flow Meter	1.86e-13	1.86e-13	1.95e-13	BCSST 25	2.49e-13	2.49e-13	1.81e-13
Chip Cooling	2.50e-13	2.50e-13	2.66e-13	Steel 20	4.85e-13	4.95e-13	4.18e-13
Gyro	5.13e-13	5.12e-13	3.71e-13				

Table 5.20.: Reconstruction error η of the QZ algorithms.

Problem	η	δ_G	δ_L	Problem	η	δ_G	δ_L
BCSST 07	7.29e-14	7.63e-14	1.41e-09	BCSST 19	3.28e-13	8.56e-14	9.88e-05
BCSST 09	7.28e-11	5.95e-12	5.82e-10	BCSST 12	1.48e-12	1.03e-11	1.15e-08
Filter 2D	1.48e-12	6.47e-15	1.36e-10	BCSST 26	4.47e-12	1.16e-13	1.44e-06
BCSST 13	3.22e-13	1.41e-13	1.25e-07	BCSST 23	7.57e-12	6.04e-10	5.77e-05
BCSST 21	1.75e-11	4.17e-15	5.05e-11	Steel 5	9.21e-13	1.85e-14	4.45e-10
Flow Meter	8.97e-10	5.75e-09	1.43e-05	BCSST 25	5.88e-11	5.25e-10	1.78e-02
Chip Cooling	1.02e-09	2.31e-07	6.52e-05	Steel 20	3.06e-11	3.77e-14	6.53e-09
Gyro	1.98e-09	1.99-07	1.91e-04				

Table 5.21.: Divide-Shift-and-Conquer – reconstruction error, global, and local eigenvalue deviation. Subspace extraction with GEQP3, $m_{triv} = 80$.

Problem	η	δ_G	δ_L	Problem	η	δ_G	δ_L
BCSST 07	6.42e-14	7.63e-14	1.54e-09	BCSST 19	4.15e-13	8.48e-14	8.42e-05
BCSST 09	1.23e-10	1.04e-11	1.22e-09	BCSST 12	2.23e-12	7.14e-12	6.00e-09
Filter 2D	1.67e-12	6.38e-15	1.38e-10	BCSST 26	4.60e-12	1.10e-13	5.30e-06
BCSST 13	2.14e-13	1.40e-13	1.21e-07	BCSST 23	6.82e-12	6.04e-10	2.32e-05
BCSST 21	2.77e-11	4.06e-15	5.89e-11	Steel 5	8.83e-13	1.85e-14	1.59e-10
Flow Meter	8.96e-10	6.74e-09	2.90e-05	BCSST 25	4.01e-11	4.50e-10	9.45e-02
Chip Cooling	2.18e-09	3.78e-07	2.27e-04	Steel 20	3.14e-11	3.77e-14	7.42e-10
Gyro	3.03e-09	1.99e-07	1.89e-04				

Table 5.22.: Divide-Shift-and-Conquer – reconstruction error, global, and local eigenvalue deviation. Subspace extraction with HQRRP, $m_{triv} = 80$.

for the QZ algorithm and

$$\sigma_{min}\left(A - \lambda_{DSC}B\right) = 1.356097108198963 \cdot 10^{-15}$$

for the Divide-Shift-and-Conquer scheme. This shows that the QZ computes these eigenvalues with a perturbation, which is six orders of magnitude larger than using our scheme. In contrast to this, for the BCSST25 example, our approach is four orders of magnitude worse than the QZ algorithm for the eigenvalue causing the maximum local eigenvalue deviation.

The trivial problem size has an impact on the accuracy since a larger trivial problems size means less spectral division steps. This decreases the number of non-unitary transformation done on the matrix pencil. On the POWER 8 system, two trivial problem sizes are possible. A smaller one, $m_{triv} = 120$, like on the Intel® systems, and a larger one, $m_{triv} = 500$ (compare Section 4.1.4). Hence, we check the influence of a larger trivial problem size on the approximation quality of the Divide-Shift-and-Conquer algorithm. Table 5.24 shows the reconstruction error and the eigenvalue deviations using the column pivoted QR decomposition for subspace extraction with a trivial problem size $m_{triv} = 500$. We see that both reasonable trivial problem sizes yield comparable results to the ones for $m_{triv} = 80$ in Tables 5.21 and 5.22 except for two cases. In this way, we use the runtime to decide which trivial size should be on the POWER 8 system.

Runtime Comparison   Besides the accuracy, the runtime plays an important role. Thus, we test the runtime behavior of the Divide-Shift-and-Conquer algorithm next. Thereby, we first perform all operations on the CPU. Second, we switch to the GPUs, wherever the benchmarks in Section 5.2 show that this leads to an acceleration. In both cases we check how the deflating subspace extraction with GEQP3 or HQRRP influences the runtime as well.

Table 5.25 shows the CPU only runtime for the selected example problems from Section 5.1.3 on the Haswell system. It shows that using the column pivoted QR decomposition GEQP3 leads to

Problem	η	δ_G	δ_L	Problem	η	δ_G	δ_L
BCSST 07	5.97e-05	5.93e-06	4.86e-02	BCSST 19	1.81e-02	5.74e-02	4.92e+07
BCSST 09	8.22e-02	4.54e-01	1.07e+04	BCSST 12	2.59e-03	2.55e-01	4.18e+02
Filter 2D	3.31e-06	6.94e-09	2.15e-06	BCSST 26	2.18e-04	2.02e-07	1.42e+03
BCSST 13	4.98e-11	2.52e-13	1.21e-07	BCSST 23	2.36e-07	2.98e-01	7.46e+05
BCSST 21	7.87e-06	9.32e-08	1.79e-05	Steel 5	1.09e-12	1.86e-14	2.52e-10
Flow Meter	1.43e-06	1.80e-06	1.44e-04	BCSST 25	9.30e-05	2.54e-02	1.50e+09
Chip Cooling	3.51e-02	1.50e+00	4.29e+02	Steel 20	1.42e-10	9.02e-14	8.94e-08
Gyro	3.12e-07	1.99e-07	4.73e-01				

Table 5.23.: Divide-Shift-and-Conquer – reconstruction error, global, and local eigenvalue deviation. Subspace extraction with GEQPX, $m_{triv} = 80$

Problem	η	δ_G	δ_L	Problem	η	δ_G	δ_L
BCSST 07	1.48e-14	6.74e-15	1.49e-10	BCSST 19	1.08e-12	3.62e-12	3.24e-04
BCSST 09	8.81e-12	5.59e-13	6.32e-10	BCSST 12	7.68e-12	3.24e-11	1.10e-07
Filter 2D	4.59e-13	6.07e-15	4.47e-11	BCSST 26	1.38e-11	2.62e-13	7.93e-06
BCSST 13	7.35e-13	1.43e-13	1.44e-07	BCSST 23	4.75e-11	6.04e-10	9.45e-05
BCSST 21	1.55e-11	3.81e-15	7.40e-11	Steel 5	9.55e-13	1.85e-14	1.38e-10
Flow Meter	1.40e-11	1.39e-11	5.53e-08	BCSST 25	1.60e-09	5.68e-09	7.13e-01
Chip Cooling	6.03e-06	3.97e-07	1.39e-04	Steel 20	2.29e-11	3.67e-14	9.37e-09
Gyro	1.73e-09	1.99e-07	1.89e-04				

Table 5.24.: Divide-Shift-and-Conquer – reconstruction error, global, and local eigenvalue deviation. Subspace extraction with GEQP3, $m_{triv} = 500$.

a significant runtime increase for large-scale problems compared to the random sampling based decomposition HQRRP. For problems of dimension larger than 1 000, the Divide-Shift-and-Conquer approach is faster than both approaches from LAPACK. Only the multi-shift QZ algorithm, KKQZ, is faster in a few cases. Due to the limited amount of memory on this system and the memory requirements of the Divide-Shift-and-Conquer scheme, the Gyro example exceeds the capabilities of the system. Table 5.26 presents the results on the Skylake system, which has a comparable peak performance but a higher memory bandwidth. There, the Divide-Shift-and-Conquer algorithm beat even the KKQZ approach in almost all cases of a dimension larger than 1 000. A direct comparison of the subspace extraction methods on the Haswell and the Skylake systems shows that the higher memory bandwidth is particularly beneficial for the standard column pivoted QR decomposition. Speedups of up to 17.5 compared to the QZ algorithms make our algorithms competitive to them. Furthermore, we see that our algorithm with the GEQP3 subspace extraction benefits from the improved hardware of the Skylake system, while the QZ algorithms show the opposite behavior, compare Tables 5.17 and 5.18. The behavior on the POWER 8 system is shown in Table 5.27. The KKQZ benefits from the high memory bandwidth as well as from the high clock frequency of the CPU. Nevertheless, using the HQRRP algorithm for the subspace extraction, the Divide-Shift-and-Conquer algorithm is constantly faster than the KKQZ algorithm. Table 5.28 shows the runtime and the speedup with an increased trivial problem size of $m_{triv} = 500$. This change does not lead to a notable effect on the runtimes, except for the Gyro example where between 500 and 900 seconds can be saved with a larger trivial problem size.

If we restrict the comparison in Tables 5.25 to 5.28 to publicly available implementations of the QZ algorithm, i.e. the GGES and GGES3 implementation from LAPACK, we see that huge speedups on all platforms are obtained for moderate and large-scale problems. Only in a few cases, the LAPACK implementation is faster than ours. On the POWER 8 system, the speedup compared to both LAPACK codes can even be larger than the number of physical CPU cores denoting a super

Problem	Runtime w. GEQP3	Speedup w.r.t.			Runtime w. HQRRP	Speedup w.r.t.		
		GGES	GGES3	KKQZ		GGES	GGES3	KKQZ
BCSST 07	0.73	0.69	0.69	0.90	0.59	0.85	0.85	1.11
BCSST 19	3.25	0.69	0.68	0.89	3.35	0.66	0.66	0.87
BCSST 09	1.96	4.12	4.17	3.78	2.22	3.63	3.68	3.33
BCSST 12	4.99	6.33	6.32	3.69	5.65	5.59	5.59	3.26
Filter 2D	5.61	6.66	6.67	4.40	6.54	5.71	5.72	3.77
BCSST 26	33.56	1.23	1.23	1.02	36.81	1.12	1.12	0.93
BCSST 13	11.29	5.22	5.17	3.22	15.48	3.81	3.77	2.34
BCSST 23	178.98	1.14	1.13	0.84	134.84	1.51	1.51	1.12
BCSST 21	104.76	2.93	2.94	2.05	84.43	3.64	3.64	2.54
Steel 5	247.33	6.76	6.77	2.99	168.72	9.91	9.93	4.39
Flow Meter	1 394.64	7.42	7.39	3.30	864.06	11.97	11.94	5.32
BCSST 25	24 239.16	1.25	1.25	0.72	10 666.87	2.84	2.84	1.64
Chip Cooling	9 912.39	6.34	6.35	3.76	6 407.57	9.81	9.82	5.82
Steel 20	10 387.97	9.51	9.53	4.05	6 781.82	14.57	14.60	6.21
Gyro	out of memory				out of memory			

Table 5.25.: Divide-Shift-and-Conquer – runtime and speedup on the Haswell system without GPU support, $m_{triv} = 80$.

linear speedup. The Gyro example with order 34 722 is the largest tested example and can be computed 8.45 to 13 times faster than with LAPACK. This reduced the runtime from more than four and a half day to at most half a day.

During the development of the Divide-Shift-and-Conquer algorithm, we derived GPU aware versions as well. Tables 5.29, 5.30, and 5.31 show the runtimes for those using all available GPUs in the systems. Thereby, the subspace extraction is still performed on the host either with GEQP3 or HQRRP. Especially the two systems equipped with the P100 accelerators show enormous speedups compared to the QZ algorithms. On the POWER 8 system, we even obtain speedups larger than 100 compared to the LAPACK variants. For the Steel 20 example the Divide-Shift-and-Conquer algorithm with HQRRP subspace extraction only needs 28.4 minutes, while the fastest QZ algorithm still takes 6.6 hours. In the case of the Gyro example, we are able to reduce the runtime from 1.24 days with KKQZ to 122.8 minutes. The Haswell system, which is equipped with two older K20 accelerators benefits from the GPU acceleration, but since the peak performance ratio between GPU and CPU is not as high as for the P100 cards, the overall performance increase is not as high as on the newer architectures. Regarding the small-scale problems, we see that the runtime of the GPU accelerated code is similar to the one that runs only on CPUs. The reason behind this is that our auto-selection, based on the experiments in Section 5.2, will choose the CPU based algorithms. The problems are so small in these cases such that the overhead, introduce by moving the data between CPU and GPUs, is not compensated by the obtained acceleration of the computations. In this way GPUs are only used if the auto-selection detects that it is worth using GPUs. In general, we can say that the approximation by the Divide-Shift-and-Conquer algorithm can solve problems within one to two hours which previously needed six hours up to more than one day. Thereby an accuracy, which is suitable for many applications, is achieved.

Problem	Runtime w. GEQP3	Speedup			Runtime w. HQRRP	Speedup		
		GGES	GGES3	KKQZ		GGES	GGES3	KKQZ
BCSST 07	0.81	0.48	0.50	0.76	0.73	0.53	0.55	0.85
BCSST 19	3.51	0.65	0.64	0.85	4.39	0.52	0.51	0.68
BCSST 09	2.90	3.08	3.11	2.57	2.76	3.23	3.27	2.70
BCSST 12	5.22	6.29	6.37	3.66	6.74	4.87	4.93	2.83
Filter 2D	6.35	5.46	5.29	3.76	7.24	4.79	4.64	3.30
BCSST 26	31.13	1.25	1.21	1.05	38.91	1.00	0.97	0.84
BCSST 13	12.77	4.80	4.76	2.79	17.18	3.56	3.54	2.07
BCSST 23	119.64	1.58	1.57	1.13	132.69	1.42	1.41	1.02
BCSST 21	76.22	3.78	3.76	2.50	83.60	3.44	3.43	2.28
Steel 5	167.80	10.85	10.61	4.37	166.75	10.92	10.68	4.39
Flow Meter	957.65	12.98	12.43	5.13	849.13	14.64	14.02	5.79
BCSST 25	17 265.24	2.10	2.08	1.13	9 328.70	3.90	3.85	2.09
Steel 20	7 970.40	15.23	14.96	6.04	6 935.32	17.51	17.20	6.94
Chip Cooling	7 696.36	10.32	9.98	5.74	6 538.23	12.15	11.75	6.76
Gyro	41 685.99	8.70	8.45	6.13	32 032.21	11.32	11.00	7.98

Table 5.26.: Divide-Shift-and-Conquer – runtime and speedup on the Skylake system without GPU support, $m_{triv} = 170$.

Problem	Runtime w. GEQP3	Speedup			Runtime w. HQRRP	Speedup		
		GGES	GGES3	KKQZ		GGES	GGES3	KKQZ
BCSST 07	0.54	1.05	1.02	2.49	0.55	1.02	1.00	2.44
BCSST 19	3.54	0.86	0.76	1.59	3.53	0.86	0.76	1.59
BCSST 09	2.04	5.49	4.97	6.07	2.27	4.95	4.48	5.47
BCSST 12	4.88	5.44	4.76	5.20	5.51	4.82	4.21	4.61
Filter 2D	5.94	5.97	5.54	5.71	6.68	5.31	4.93	5.09
BCSST 26	31.59	1.23	0.99	1.14	34.72	1.12	0.90	1.03
BCSST 13	12.07	3.40	3.05	2.48	14.38	2.85	2.56	2.08
BCSST 23	108.01	1.86	1.29	1.15	120.59	1.67	1.16	1.03
BCSST 21	67.84	4.16	3.39	2.33	76.64	3.68	3.00	2.07
Steel 5	153.26	15.38	10.09	3.79	158.19	14.90	9.77	3.67
Flow Meter	944.52	25.91	16.35	3.18	795.34	30.77	19.42	3.78
BCSST 25	15 401.43	4.35	3.47	0.64	8 599.89	7.80	6.22	1.14
Chip Cooling	7 415.60	28.02	26.13	3.01	6 545.74	31.74	29.60	3.41
Steel 20	7 658.14	34.19	32.36	3.08	6 812.90	38.43	36.37	3.46
Gyro	44 645.44	9.50	9.77	2.49	33 985.11	12.48	12.84	3.14

Table 5.27.: Divide-Shift-and-Conquer – runtime and speedup on the POWER 8 system without GPU support, $m_{triv} = 120$.

Problem	Runtime w. GEQP3	Speedup			Runtime w. HQRRP	Speedup		
		GGES	GGES3	KKQZ		GGES	GGES3	KKQZ
BCSST 07	0.54	1.04	1.02	2.49	0.54	1.05	1.03	2.51
BCSST 19	3.70	0.82	0.73	1.52	3.70	0.82	0.73	1.52
BCSST 09	3.10	3.62	3.27	4.00	3.32	3.38	3.05	3.73
BCSST 12	5.37	4.94	4.32	4.73	5.98	4.44	3.89	4.25
Filter 2D	6.46	5.49	5.10	5.26	7.15	4.96	4.60	4.75
BCSST 26	34.75	1.12	0.90	1.03	36.68	1.06	0.85	0.98
BCSST 13	12.43	3.30	2.96	2.41	14.76	2.78	2.49	2.03
BCSST 23	108.02	1.86	1.29	1.15	121.35	1.66	1.15	1.03
BCSST 21	68.36	4.13	3.37	2.32	77.01	3.66	2.99	2.06
Steel 5	155.43	15.16	9.95	3.74	159.27	14.80	9.71	3.65
Flow Meter	909.70	26.90	16.98	3.31	790.58	30.95	19.53	3.80
BCSST 25	15 253.88	4.40	3.51	0.64	8 611.62	7.79	6.21	1.14
Chip Cooling	7 381.88	28.15	26.25	3.02	6 324.96	32.85	30.64	3.53
Steel 20	7 660.12	34.18	32.35	3.08	6 734.98	38.87	36.79	3.50
Gyro	43 719.85	9.70	9.98	2.44	33 467.10	12.67	13.04	3.19

Table 5.28.: Divide-Shift-and-Conquer – runtime and speedup on the POWER 8 system without GPU support, $m_{triv} = 500$.

Problem	Runtime w. GEQP3	Speedup w.r.t.			Runtime w. HQRRP	Speedup w.r.t.		
		GGES	GGES3	KKQZ		GGES	GGES3	KKQZ
BCSST 07	0.59	0.86	0.86	1.12	0.62	0.82	0.82	1.07
BCSST 19	3.30	0.68	0.67	0.88	3.24	0.69	0.68	0.90
BCSST 09	2.55	3.17	3.21	2.91	2.81	2.87	2.91	2.63
BCSST 12	5.36	5.89	5.88	3.43	6.49	4.87	4.86	2.84
Filter 2D	6.88	5.43	5.44	3.59	7.03	5.31	5.33	3.51
BCSST 26	34.93	1.18	1.18	0.98	40.10	1.03	1.03	0.86
BCSST 13	13.86	4.25	4.21	2.62	16.84	3.50	3.47	2.16
BCSST 23	175.34	1.16	1.16	0.86	117.36	1.74	1.73	1.28
BCSST 21	93.85	3.28	3.28	2.29	72.86	4.22	4.22	2.95
Steel 5	184.46	9.06	9.08	4.02	110.83	15.09	15.11	6.68
Flow Meter	1 124.36	9.20	9.17	4.09	527.60	19.61	19.55	8.72
BCSST 25	20 764.96	1.46	1.46	0.84	5 814.89	5.20	5.22	3.01
Chip Cooling	6 755.69	9.31	9.31	5.52	3 522.38	17.85	17.86	10.59
Steel 20	6 689.55	14.77	14.80	6.29	3 535.59	27.95	28.00	11.90
Gyro		out of memory				out of memory		

Table 5.29.: Divide-Shift-and-Conquer – runtime and speedup on the Haswell system with GPU support, $m_{triv} = 80$. (2× Nvidia® K20)

Problem	Runtime w. GEQP3	Speedup w.r.t.			Runtime w. HQRRP	Speedup w.r.t.		
		GGES	GGES3	KKQZ		GGES	GGES3	KKQZ
BCSST 07	0.57	0.68	0.71	1.08	0.73	0.53	0.55	0.84
BCSST 19	4.34	0.52	0.51	0.69	3.52	0.65	0.64	0.85
BCSST 09	2.93	3.05	3.09	2.55	3.04	2.93	2.97	2.45
BCSST 12	5.37	6.11	6.19	3.55	7.06	4.65	4.71	2.70
Filter 2D	6.65	5.22	5.05	3.59	7.24	4.79	4.64	3.30
BCSST 26	29.82	1.31	1.27	1.10	39.02	1.00	0.97	0.84
BCSST 13	13.79	4.44	4.41	2.58	18.73	3.27	3.24	1.90
BCSST 23	102.88	1.84	1.82	1.32	109.76	1.72	1.71	1.24
BCSST 21	62.58	4.60	4.58	3.04	69.05	4.17	4.15	2.76
Steel 5	101.67	17.91	17.51	7.20	101.48	17.94	17.55	7.22
Flow Meter	492.39	25.24	24.18	9.98	396.51	31.35	30.03	12.39
BCSST 25	11 710.18	3.10	3.07	1.66	4 183.46	8.69	8.58	4.65
Chip Cooling	3 395.23	23.40	22.62	13.02	2 280.04	34.85	33.69	19.39
Steel 20	3 332.21	36.44	35.79	14.44	2 240.97	54.18	53.22	21.47
Gyro	17 651.45	20.55	19.96	14.48	10 644.50	34.08	33.11	24.01

Table 5.30.: Divide-Shift-and-Conquer – runtime and speedup on the Skylake system with GPU support, $m_{triv} = 170$. (1× Nvidia® P100)

Problem	Runtime w. GEQP3	Speedup w.r.t.			Runtime w. HQRRP	Speedup w.r.t.		
		GGES	GGES3	KKQZ		GGES	GGES3	KKQZ
BCSST 07	0.55	1.03	1.01	2.46	0.56	1.02	0.99	2.42
BCSST 19	3.73	0.81	0.72	1.51	3.75	0.81	0.72	1.50
BCSST 09	2.37	4.73	4.27	5.22	2.59	4.33	3.91	4.78
BCSST 12	4.88	5.43	4.76	5.20	5.55	4.78	4.18	4.57
Filter 2D	5.64	6.28	5.83	6.02	6.43	5.52	5.12	5.28
BCSST 26	27.70	1.40	1.13	1.29	32.20	1.21	0.97	1.11
BCSST 13	12.99	3.16	2.83	2.31	15.38	2.67	2.39	1.95
BCSST 23	68.85	2.92	2.03	1.81	82.58	2.44	1.69	1.51
BCSST 21	50.22	5.62	4.58	3.15	51.22	5.51	4.49	3.09
Steel 5	69.05	34.13	22.39	8.42	72.24	32.62	21.40	8.04
Flow Meter	395.11	61.93	39.08	7.61	287.57	85.09	53.70	10.46
BCSST 25	10 666.37	6.29	5.01	0.92	3 547.01	18.90	15.08	2.76
Chip Cooling	2 776.56	74.83	69.79	8.04	1 677.90	123.83	115.49	13.30
Steel 20	2 690.21	97.32	92.11	8.77	1 705.48	153.51	145.29	13.84
Gyro	15 035.40	28.20	29.03	7.10	7 366.48	57.56	59.24	14.49

Table 5.31.: Divide-Shift-and-Conquer – runtime and speedup on the POWER 8 system with GPU support, $m_{triv} = 120$. (2× Nvidia® P100)

Problem	η	δ_G	δ_L	Problem	η	δ_G	δ_L
BCSST 07	4.96e-15	4.43e-14	1.89e-10	BCSST 19	5.40e-15	3.09e-14	2.49e-06
BCSST 09	4.69e-15	2.96e-13	1.09e-12	BCSST 12	5.52e-15	7.18e-13	2.88e-09
Filter 2D	5.30e-15	5.87e-15	1.60e-11	BCSST 26	8.16e-15	1.05e-13	1.13e-09
BCSST 13	7.87e-15	1.40e-13	1.22e-07	BCSST 23	8.27e-15	6.04e-10	6.01e-06
BCSST 21	6.98e-15	2.85e-15	7.41e-12	Steel 5	6.53e-15	1.84e-14	5.31e-12
Flow Meter	8.24e-15	2.47e-12	1.92e-08	BCSST 25	1.69e-10	6.93e-12	1.80e-04
Chip Cooling	8.80e-15	4.01e-11	6.76e-09	Steel 20	8.28e-15	3.55e-14	1.05e-10
Gyro	1.66e-14	1.88e-07	5.10e-05				

Table 5.32.: Divide-Scale-and-Conquer – reconstruction error, global and local eigenvalue deviation. Subspace extraction with GEQP3, $m_{triv} = 80$.

Problem	η	δ_G	δ_L	Problem	η	δ_G	δ_L
BCSST 07	5.04e-15	4.45e-14	1.19e-10	BCSST 19	5.86e-15	3.09e-14	2.26e-06
BCSST 09	4.70e-15	3.05e-13	1.09e-12	BCSST 12	5.48e-15	7.18e-13	2.90e-09
Filter 2D	5.25e-15	5.80e-15	1.62e-11	BCSST 26	8.17e-15	1.08e-13	1.13e-09
BCSST 13	8.02e-15	1.40e-13	1.22e-07	BCSST 23	8.28e-15	6.04e-10	7.76e-06
BCSST 21	6.97e-15	2.84e-15	7.39e-12	Steel 5	6.47e-15	1.84e-14	5.03e-12
Flow Meter	9.08e-15	2.47e-12	1.92e-08	BCSST 25	5.20e-11	6.45e-12	1.80e-04
Chip Cooling	8.70e-15	4.00e-11	7.92e-09	Steel 20	8.23e-15	3.55e-14	1.03e-10
Gyro	1.77e-14	1.88e-07	5.10e-05				

Table 5.33.: Divide-Scale-and-Conquer – reconstruction error, global and local eigenvalue deviation. Subspace extraction with HQRRP, $m_{triv} = 80$.

5.3.2. Divide-Scale-and-Conquer

First, we focus on the accuracy of the Divide-Scale-and-Conquer algorithm before we compare its runtimes to the QZ algorithms. Tables 5.32 and 5.33 show the reconstruction error as well as the global and local deviation. Thereby, we regard the subspace extraction using the column pivoted QR decomposition GEQP3 and the randomized sampling QR decomposition HQRRP. The previous experiments in Section 5.3.1 with the Divide-Shift-and-Conquer algorithm showed that the rank revealing QR decomposition GEQPX leads to inaccurate results and thus we do not use it here. The results for the subspace extraction with GEQP3 and HQRRP are comparable and only small differences appear. From this point of view, one can select the subspace extraction algorithm based on the performance. In the case of the Divide-Shift-and-Conquer algorithm in Section 5.3.1 this was the randomized sampling QR decomposition HQRRP.

A direct comparison to the results between the accuracy of the Divide-Shift-and-Conquer algorithm in Tables 5.21 and 5.22 and the Divide-Scale-and-Conquer algorithm in Tables 5.32 and 5.33 shows that the Divide-Scale-and-Conquer algorithm leads to smaller errors in almost all test cases. The main reason for this is that, in contrast to the generalized matrix sign function, the inverse free iteration heavily relies on unitary operations and thus does not amplify rounding errors. Especially the reconstruction error is close to machine precision. Even the results obtained with the QZ algorithm in Table 5.20 are not better in almost all cases. In this way the Divide-Scale-and-Conquer algorithm gives a better approximation of the generalized Schur decomposition than the QZ algorithms. Only in the case of the BCSST25 example we get a result which only differs in two orders of magnitude from the one of the QZ algorithms. From the accuracy point of view the Divide-Scale-and-Conquer algorithm is competitive to the QZ algorithm. Regarding the global and the local deviation we see that we get close to the QZ algorithm in most cases and obtain a better reconstruction than the Divide-Shift-and-Conquer algorithm.

Problem	Runtime w. GEQP3	Speedup w.r.t			Runtime w. HQRRP	Speedup w.r.t.		
		GGES	GGES3	KKQZ		GGES	GGES3	KKQZ
BCSST 07	1.30	0.39	0.39	0.50	1.28	0.39	0.39	0.51
BCSST 19	7.22	0.31	0.31	0.40	7.50	0.30	0.30	0.39
BCSST 09	5.11	1.58	1.60	1.45	5.79	1.39	1.41	1.28
BCSST 12	10.32	3.06	3.06	1.78	10.56	2.99	2.99	1.74
Filter 2D	13.76	2.72	2.72	1.79	15.16	2.46	2.47	1.63
BCSST 26	51.96	0.80	0.79	0.66	55.38	0.75	0.74	0.62
BCSST 13	25.19	2.34	2.32	1.44	26.50	2.22	2.20	1.37
BCSST 23	193.40	1.05	1.05	0.78	186.03	1.10	1.09	0.81
BCSST 21	128.70	2.39	2.39	1.67	126.12	2.44	2.44	1.70
Steel 5	207.57	8.06	8.07	3.57	173.17	9.66	9.67	4.28
Flow Meter	2 813.86	3.68	3.66	1.63	22 66.55	4.56	4.55	2.03
BCSST 25	22 564.41	1.34	1.34	0.78	17 459.65	1.73	1.74	1.00
Chip Cooling	24 364.47	2.58	2.58	1.53	20 515.72	3.07	3.07	1.82
Steel 20	19 473.63	5.07	5.08	2.16	14 386.68	6.87	6.88	2.93
Gyro	out of memory				out of memory			

Table 5.34.: Divide-Scale-and-Conquer – runtime and speedup on the Haswell system without GPU support, $m_{triv} = 80$.

Regarding the Gyro example in detail, the eigenvalue causing the maximum local eigenvalue deviation is

$$\lambda_{QZ} = -2.079614498465631 \cdot 10^3 \pm 6.464696728361261 \cdot 10^4 i$$

computed by the QZ algorithm and

$$\lambda_{DSC} = -2.091863262445210 \cdot 10^3 \pm 6.464790885446857 \cdot 10^4 i$$

computed by the Divide-Scale-and-Conquer algorithm. Computing the minimal singular value of $A - \lambda B$ for these eigenvalue approximations we get

$$\sigma_{min} \left(A - \lambda_{QZ} B \right) = 1.531065727744096 \cdot 10^{-9}$$

for the QZ algorithm and

$$\sigma_{min} \left(A - \lambda_{DSC} B \right) = 6.342629318670011 \cdot 10^{-16}$$

for the Divide-Scale-and-Conquer scheme. Especially the result for our approach is close to machine precision. This shows that the large local eigenvalue deviation is caused by an insufficient approximation of the QZ algorithm. Regarding the BCSST25 example, we see that our approach gives a result, which is five orders of magnitude better than the one computed by the QZ algorithm. This is a contrast to the approximation computed by the Divide-Shift-and-Conquer algorithm, where our approach results in the worse approximation. This shows again that the Divide-Scale-and-Conquer algorithm is superior to the Divide-Shift-and-Conquer approach from an accuracy point of view.

Runtime Comparison The performance comparison between the Divide-Scale-and-Conquer algorithm and the QZ implementations differs from the Divide-Shift-and-Conquer algorithm since the inverse free iteration is much more expensive than the computation of the generalized matrix sign function. Regarding the CPU only cases in Tables 5.34, 5.35, and 5.36, the Divide-Scale-and-Conquer algorithm is in many cases slower than the Divide-Shift-and-Conquer algorithm,

Problem	Runtime w. GEQP3	Speedup w.r.t.			Runtime w. HQRRP	Speedup w.r.t.		
		GGES	GGES3	KKQZ		GGES	GGES3	KKQZ
BCSST 07	1.37	0.28	0.30	0.45	1.40	0.28	0.29	0.44
BCSST 19	8.03	0.28	0.28	0.37	7.95	0.29	0.28	0.37
BCSST 09	6.44	1.38	1.40	1.16	6.00	1.49	1.50	1.24
BCSST 12	10.96	3.00	3.03	1.74	11.29	2.91	2.95	1.69
Filter 2D	16.06	2.16	2.09	1.49	17.31	2.00	1.94	1.38
BCSST 26	56.94	0.69	0.66	0.58	61.57	0.63	0.61	0.53
BCSST 13	27.38	2.24	2.22	1.30	29.69	2.06	2.05	1.20
BCSST 23	191.32	0.99	0.98	0.71	197.65	0.96	0.95	0.69
BCSST 21	128.47	2.24	2.23	1.48	132.55	2.17	2.16	1.44
Steel 5	178.65	10.19	9.97	4.10	181.18	10.05	9.83	4.04
Flow Meter	2 467.85	5.04	4.82	1.99	2 319.80	5.36	5.13	2.12
BCSST 25	20 221.72	1.80	1.78	0.96	18 511.44	1.96	1.94	1.05
Chip Cooling	23 807.30	3.34	3.23	1.86	22 637.78	3.51	3.39	1.95
Steel 20	16 927.59	7.17	7.05	2.84	15 699.44	7.73	7.60	3.06
Gyro	295 411.30	1.29	1.19	0.87	278 214.65	1.30	1.27	0.92

Table 5.35.: Divide-Scale-and-Conquer – runtime and speedup on the Skylake system without GPU support, $m_{triv} = 170$.

Problem	Runtime w. GEQP3	Speedup w.r.t.			Runtime w. HQRRP	Speedup w.r.t.		
		GGES	GGES3	KKQZ		GGES	GGES3	KKQZ
BCSST 07	1.35	0.42	0.41	1.00	1.35	0.42	0.41	1.00
BCSST 19	7.71	0.39	0.35	0.73	7.79	0.39	0.35	0.72
BCSST 09	5.60	2.00	1.81	2.21	5.75	1.95	1.76	2.15
BCSST 12	12.06	2.20	1.93	2.11	12.47	2.13	1.86	2.04
Filter 2D	17.43	2.03	1.89	1.95	18.18	1.95	1.81	1.87
BCSST 26	61.42	0.63	0.51	0.58	64.73	0.60	0.48	0.55
BCSST 13	25.79	1.59	1.43	1.16	27.21	1.51	1.35	1.10
BCSST 23	186.66	1.08	0.75	0.67	194.63	1.03	0.72	0.64
BCSST 21	126.92	2.22	1.81	1.25	131.56	2.15	1.75	1.20
Steel 5	173.49	13.58	8.91	3.35	175.39	13.44	8.82	3.31
Flow Meter	2 411.68	10.15	6.40	1.25	2 223.43	11.01	6.95	1.35
BCSST 25	19 998.41	3.35	2.67	0.49	17 422.02	3.85	3.07	0.56
Chip Cooling	22 169.54	9.37	8.74	1.01	21 139.64	9.83	9.17	1.06
Steel 20	15 820.74	16.55	15.66	1.49	14 743.18	17.76	16.81	1.60
Gyro	292 242.55	1.45	1.49	0.365	275 246.03	1.54	1.58	0.39

Table 5.36.: Divide-Scale-and-Conquer – runtime and speedup on the POWER 8 system without GPU support, $m_{triv} = 120$.

Problem	Runtime w. GEQP3	Speedup w.r.t. GGES	GGES3	KKQZ	Runtime w. HQRRP	Speedup w.r.t. GGES	GGES3	KKQZ
BCSST 07	1.27	0.40	0.40	0.52	1.24	0.41	0.41	0.53
BCSST 19	8.26	0.27	0.27	0.35	7.51	0.30	0.30	0.39
BCSST 09	5.25	1.54	1.56	1.41	5.69	1.42	1.44	1.30
BCSST 12	10.56	2.99	2.99	1.74	10.15	3.11	3.11	1.81
Filter 2D	14.76	2.53	2.54	1.67	14.64	2.55	2.56	1.69
BCSST 26	53.03	0.78	0.78	0.65	55.96	0.74	0.74	0.61
BCSST 13	25.81	2.28	2.26	1.41	27.63	2.13	2.11	1.31
BCSST 23	194.03	1.05	1.05	0.78	185.82	1.10	1.09	0.81
BCSST 21	124.10	2.48	2.48	1.73	124.21	2.48	2.48	1.73
Steel 5	167.38	9.99	10.01	4.42	139.65	11.97	11.99	5.30
Flow Meter	1 761.08	5.88	5.86	2.61	1 219.22	8.49	8.46	3.77
BCSST 25	12 856.04	2.35	2.36	1.36	7 798.21	3.88	3.89	2.25
Chip Cooling	12 651.94	4.97	4.97	2.95	9 009.33	6.98	6.98	4.14
Steel 20	9 912.35	9.97	9.99	4.25	6 559.55	15.07	15.09	6.42
Gyro	out of memory				out of memory			

Table 5.37.: Divide-Scale-and-Conquer – runtime and speedup on the Haswell system with GPU support, $m_{triv} = 80$. (2× Nvidia® K20)

compare Tables 5.25, 5.26, and 5.27. This leads to more cases where the KKQZ is faster. In a few cases, when the spectral division performed by inverse free iteration leads to better distributed subproblems, the Divide-Scale-and-Conquer algorithm is faster. This is for example the case in the Steel 5 example on the Haswell system. The opposite is true for the Gyro example, where an adverse eigenvalue distribution leads to many spectral division step with a nearly full-sized problem. The result of this unfavorable distribution are long runtimes. On the POWER 8 system we see that the Divide-Scale-and-Conquer algorithm can hardly compete with the KKQZ algorithm. The reason therefore is that the KKQZ benefits much from the high memory bandwidth and the high clock rate. If we only regard the publicly available QZ implementations from LAPACK, the Divide-Scale-and-Conquer algorithm is still competitive from a performance point of view.

Moving on to the GPU accelerated versions, the Divide-Scale-and-Conquer algorithm delivers a better performance than the CPU based implementations of the QZ algorithm. Tables 5.37, 5.38, and 5.39 show the results obtained by using the available accelerators. We see that the GPU acceleration helps to gain runtime advantages compared to the "CPU only case" in many cases. Due to the limited GPU memory, the Gyro example slows down on the Skylake system. There, the first spectral division steps are performed on the CPU until the problem dimension is reduced such that it fits into the memory of the accelerator device. Since the POWER 8 system is equipped with two GPUs, the Gyro example can use them from the beginning and thus the high computational performance of the GPU accelerator is used along the whole algorithm. With the help of GPU accelerators, the Divide-Scale-and-Conquer algorithm gets competitive again such that it can be used as a replacement for the QZ algorithms. A comparison to the GPU accelerated Divide-Shift-and-Conquer algorithm shows that the Divide-Scale-and-Conquer algorithm is between 30% to 40% slower in most cases. Again, the Gyro example is impeded by the eigenvalue distribution, which causes a larger difference to the Divide-Shift-and-Conquer approach.

Nevertheless, it results in much smaller reconstruction errors as already shown in Tables 5.32 and 5.33.

Problem	Runtime w. GEQP3	Speedup w.r.t. GGES	GGES3	KKQZ	Runtime w. HQRRP	Speedup w.r.t. GGES	GGES3	KKQZ
BCSST 07	1.24	0.31	0.33	0.50	1.32	0.29	0.31	0.47
BCSST 19	7.83	0.29	0.29	0.38	8.27	0.28	0.27	0.36
BCSST 09	6.83	1.31	1.32	1.09	5.93	1.50	1.52	1.26
BCSST 12	11.17	2.94	2.98	1.71	12.20	2.69	2.73	1.56
Filter 2D	15.69	2.21	2.14	1.52	18.81	1.84	1.78	1.27
BCSST 26	55.22	0.71	0.68	0.59	59.83	0.65	0.63	0.55
BCSST 13	27.46	2.23	2.21	1.30	29.40	2.08	2.07	1.21
BCSST 23	138.25	1.37	1.36	0.98	147.22	1.28	1.27	0.92
BCSST 21	91.87	3.13	3.12	2.07	98.22	2.93	2.92	1.94
Steel 5	120.61	15.10	14.76	6.07	124.50	14.63	14.30	5.88
Flow Meter	1 005.12	12.37	11.85	4.89	861.63	14.43	13.82	5.70
BCSST 25	6 886.63	5.28	5.21	2.83	5 156.62	7.05	6.96	3.77
Chip Cooling	6 701.39	11.86	11.46	6.60	5 592.97	14.21	13.73	7.91
Steel 20	5 447.11	22.29	21.89	8.83	4 323.68	28.08	27.58	11.13
Gyro	206 154.63	1.76	1.71	1.24	186 636.90	1.94	1.89	1.37

Table 5.38.: Divide-Scale-and-Conquer – runtime and speedup on the Skylake system with GPU support, $m_{triv} = 170$. (1× Nvidia® P100)

Problem	Runtime w. GEQP3	Speedup w.r.t. GGES	GGES3	KKQZ	Runtime w. HQRRP	Speedup w.r.t. GGES	GGES3	KKQZ
BCSST 07	1.37	0.41	0.40	0.99	1.38	0.41	0.40	0.98
BCSST 19	7.87	0.39	0.34	0.71	7.86	0.39	0.34	0.72
BCSST 09	4.90	2.29	2.07	2.53	5.02	2.23	2.02	2.47
BCSST 12	8.22	3.23	2.83	3.09	8.62	3.08	2.69	2.95
Filter 2D	11.12	3.19	2.96	3.05	11.90	2.98	2.77	2.85
BCSST 26	36.89	1.05	0.85	0.97	40.08	0.97	0.78	0.89
BCSST 13	19.31	2.12	1.90	1.55	20.75	1.98	1.77	1.44
BCSST 23	91.01	2.21	1.53	1.37	98.68	2.04	1.42	1.26
BCSST 21	63.25	4.46	3.64	2.50	67.94	4.15	3.39	2.33
Steel 5	81.78	28.82	18.91	7.11	83.48	28.23	18.52	6.96
Flow Meter	686.72	35.63	22.49	4.38	534.86	45.75	28.87	5.62
BCSST 25	5 068.30	13.23	10.55	1.93	3 229.88	20.76	16.56	3.03
Chip Cooling	4 516.33	46.00	42.91	4.94	3 456.87	60.10	56.06	6.46
Steel 20	3 814.06	68.64	64.97	6.19	2 792.30	93.76	88.74	8.45
Gyro	65 306.78	6.49	6.68	1.63	38 701.65	10.95	11.28	2.76

Table 5.39.: Divide-Scale-and-Conquer – runtime and speedup on the POWER 8 system with GPU support, $m_{triv} = 120$. (2× Nvidia® P100)

Summary. The numerical experiments showed that we are able to compute the generalized Schur decomposition with the help of level-3 BLAS enabled spectral divide-and-conquer schemes. The algorithms are designed in a way such that the most time-consuming building block can be replaced with GPU accelerated versions whenever it is worth to do so. Although the algorithms require a much higher number of floating-point operations than the QZ algorithms, they lead to a better utilization of the CPUs without invoking the GPUs at all. The performance of the algorithms can be increased further, if we only eigenvalues in parts of the spectrum are desired. Then there is no need to continue the recursion until we obtained the generalized Schur decomposition for the whole matrix pair. If the desired eigenvalues are included in a deflating subspace computed during the Divide-Shift-and-Conquer or the Divide-Scale-and-Conquer procedure, the remaining parts of the matrix pair can be neglected. The spectral bounds computed during recursion can be used to check if the desired eigenvalue are included in a computed deflating subspace.

APPLICATION TO SYLVESTER-TYPE MATRIX EQUATIONS

Contents

The direct solution of Sylvester-type matrix equations is one application benefiting from the generalized Schur decomposition of matrix pairs. Therefore, we consider the **generalized Sylvester equation**:

$$AXB + CXD = Y, \qquad \text{(GSYLV)}$$

where (A, C) is a matrix pair of order m, (D, B) is a matrix pair of order n, and the right-hand side Y and the solution X are matrices of size $m \times n$. Depending on the choices of A, B, C, and D, a set of related equations appear:

Sylvester equation The Sylvester equation

$$AX + XD = Y \qquad \text{(SYLV)}$$

is used, among other, in the diagonalization of Schur decompositions of (quasi) upper triangular matrices, e.g. [80]. Furthermore, it has applications in system and control theory, e.g. [12], as well.

Discrete Time Sylvester equation Another variant of the generalized Sylvester equation is the discrete-time Sylvester equation:

$$AXB \pm X = Y. \qquad \text{(DSYLV)}$$

It has its application for example in the computation of the cross gramian for discrete-time model order reduction [98].

Generalized Coupled Sylvester Equation The extension of the standard Sylvester equation for the block diagonalization to the generalized Schur decomposition is the generalized coupled Sylvester equation, compare Section 2.3. In contrast to the generalized Sylvester equation it consists of two equations requiring to be solved simultaneously for $X^{(1)}$ and $X^{(2)}$:

$$\begin{aligned} AX^{(1)} + X^{(2)}D &= Y^{(1)} \\ CX^{(1)} - X^{(2)}B &= Y^{(2)} \end{aligned} \cdot \qquad \text{(GCSYLV)}$$

By setting $CX = X^{(2)}$, $XB = X^{(1)}$, $Y^{(1)} = Y$, and $Y^{(2)} = 0$, the generalized Sylvester equation (GSYLV) can be rewritten as generalized coupled Sylvester equation.

Lyapunov equation The Lyapunov equation

$$AX + XA^H = Y \qquad \text{(LYAP)}$$

and the generalized Lyapunov equation

$$AXC^H + CXA^H = Y \qquad \text{(GLYAP)}$$

with symmetric matrices X and Y are important tools in system and control theory for linear ODEs, e.g. [12].

Stein equation The standard Stein equation is given by

$$AXA^H - X = Y, \qquad \text{(STEIN)}$$

and the generalized Stein equation is given by

$$AXA^H - CXC^H = Y. \qquad \text{(GSTEIN)}$$

Again, the right-hand side Y and the solution X are symmetric. The Stein equation is the discrete-time counter part of the Lyapunov equation for linear difference equations, e.g. [12].

In all of these equations the matrix pairs (A, C) or (B, D) can be replaced by their hermitian transposes $\left(A^H, C^H\right)$ or $\left(B^H, D^H\right)$, respectively. The equations can be classified either by the symmetry of the solution or the fact that the solution is multiplied from one or two sides with a matrix. Table 6.1 illustrates this classification for the equations mentioned before.

In this chapter we recall direct solution strategies for the generalized Sylvester equation and develop a high performance algorithm well tailored to current computer architectures and their compilers. The algorithm, as well as the theory, can be transferred to the aforementioned equations using the definitions provided in Table 6.1. The treatment of special properties, like symmetry, is explained where ever they are of importance.

6.1. Mathematical Basics

This section covers the basic theoretical aspects of the solvability of Sylvester-type matrix equations. In order to transform the generalized Sylvester equation (GSYLV) into a standard linear system we need the following definition:

Name	$\tilde{A}$	$\tilde{B}$	$\tilde{C}$	$\tilde{D}$	One-Sided	Sym-metric	References
	Generalized Sylvester Eq. $\tilde{A}X\tilde{B} + \tilde{C}X\tilde{D} = Y$						
LYAP	A	I	I	A^H	×	×	[19]
STEIN	A	A^H	I	I	×	×	[19]
GLYAP	A	C^H	C	A^H		×	[19]
GSTEIN	A	A^H	C	C^H		×	[19]
SYLV	A	I	I	D	×		[19]
DSYLV	A	B	I	I			[19]
GCSYLV	A	B	C	D	×		[80]
GSYLV	A	B	C	D			[78]

Table 6.1.: Classification of Sylvester-type matrix equations

Definition 6.1 ([170, 86]):
*Let A be a matrix of size $m \times n$ and B be a matrix of size $p \times q$. Then the **Kronecker product** $C := A \otimes B$ is a matrix of size $mp \times nq$ and given by*

$$C := \begin{bmatrix} a_{11}B & \cdots & a_{1n}B \\ \vdots & \ddots & \vdots \\ a_{m1}B & \cdots & a_{mn}B. \end{bmatrix}. \tag{6.1}$$

The vec ***operator*** *stacks the columns of a matrix underneath each other. It transforms a matrix A of size $m \times n$ into a vector of length mn:*

$$\text{vec}(A) = \begin{bmatrix} a_1 \\ a_2 \\ \vdots \\ a_n \end{bmatrix}, \tag{6.2}$$

where a_i is the i-th column of A.

Using this definition, the next theorem presents a connection between the structure used in Equation (GSYLV), the Kronecker product (6.1), and the vectorization operator (6.2):

Theorem 6.2 ([56, 78]):
Let A be a matrix of order m, B be a matrix of order n, and X and Y be matrices of size $m \times n$. Then the product

$$AXB = Y$$

can be rewritten as

$$\left(B^\mathsf{T} \otimes A\right) \text{vec}(X) = \text{vec}(Y). \tag{6.3}$$

Utilizing this we can rewrite the generalized Sylvester equation (GSYLV) as

$$\left(B^\mathsf{T} \otimes A + D^\mathsf{T} \otimes C\right) \text{vec}(X) = \text{vec}(Y). \tag{6.4}$$

Now, this is a standard linear system of order mn and standard techniques to determine the solvability and the uniqueness can be applied, and we obtain the next theorem:

Theorem 6.3 ([56]):

The generalized Sylvester equation (GSYLV) has a unique solution if and only if the following conditions hold:

1. *the matrix pairs (A, C) and (B, D) are regular and*

2. $\Lambda(A, C) \cap \Lambda(-D, B) = \emptyset.$

In order to reformulate Theorem 6.3 in terms of eigenvalues of the matrices or the matrix pairs we need the following Lemma:

Lemma 6.4 ([154]):

Let A be a matrix of order m and B be a matrix of order n. Furthermore, let λ_i, $i = 1, \ldots, m$, be the eigenvalues of A and μ_j, $j = 1, \ldots, n$, the eigenvalues of B. Then the eigenvalues ν_k of $A \otimes B$ are given by $\nu_k = \lambda_i \mu_j$, $i = 1, \ldots, m$, $j = 1, \ldots, n$, and $k = (j - 1)m + i$.

Combining Theorem 6.3 and Lemma 6.4 leads us to the following corollary for an eigenvalue based solvability condition:

Corollary 6.5:

Theorem 6.3 gives a general condition about the solvability of the generalized Sylvester equation. In terms of eigenvalues, condition 2 of Theorem 6.3 translates into the following condition for the particular equations:

1. *Generalized and generalized coupled Sylvester equation (GSYLV): $\lambda_i + \mu_j \neq 0$ for all eigenvalues $\lambda_i \in \Lambda(A, C)$ and $\mu_j \in \Lambda(D, B)$.*

2. *Sylvester equation (SYLV): $\lambda_i + \mu_j \neq 0$ for all $\lambda_i \in \Lambda(A)$ and $\mu_j \in \Lambda(D)$.*

3. *Discrete time Sylvester equation (DSYLV): $\lambda_i \mu_j \neq \mp 1$ for all $\lambda_i \in \Lambda(A)$ and $\mu_j \in \Lambda(B)$.*

4. *Lyapunov equation (LYAP): $\lambda_i + \lambda_j \neq 0$ for all $\lambda_i, \lambda_j \in \Lambda(A)$*

5. *Generalized Lyapunov equation (GLYAP): $\lambda_i + \lambda_j \neq 0$ for all $\lambda_i, \lambda_j \in \Lambda(A, C)$*

6. *Stein equation (STEIN): $\lambda_i \lambda_j \neq 1$ for all $\lambda_i, \lambda_j \in \Lambda(A)$*

7. *Generalized Stein equation (GSTEIN): $\lambda_i \lambda_j \neq 1$ for all $\lambda_i, \lambda_j \in \Lambda(A, C)$*

Regarding the differential or difference equation context, where most of these matrix equations have an application, we see that for those arising from continuous-time problems, the condition looks like $\lambda_i + \mu_j \neq 0$. For those equations that have an origin in discrete-time applications the condition looks like $\lambda_i \lambda_j \neq 1$. Further conditions regarding the deflating subspaces of the involved matrix pairs and how the eigenvalues of the right-hand side Y transfer into the solution X can be found in the literature [12, 19, 107, 56].

6.2. Classic Direct Solution Techniques

Using the connection of the generalized Sylvester equation (GSYLV) to the corresponding linear system (6.4) seems to be a straight forward approach to solve the equation, but leads to the following problem: Assume both matrix pairs are of order $1\,000$, then the linear system in Equation (6.4) is of order $1\,000\,000$ and would take 7.2 TB memory if all entries are stored in double precision. Regarding the computational complexity we end up with $\approx 6.6 \cdot 10^{17}$ operations for an

LU decomposition of the linear system. Having a computer with a performance of 500 GFlop-s/s, the factorization phase would take 15.4 days. Since the *LU* decomposition of the Kronecker system (6.4) takes $2/3(mn)^3$ flops, the runtime will increase dramatically for an increasing problem dimension and makes this approach even more unusable.

The basic idea for a reduction of the numerical complexity was developed by Bartels and Stewart [19] for the solution of the standard Sylvester equation (SYLV) and the Lyapunov equation (LYAP). Therefore, we first need to increase the structure in the matrix pairs (A, C) and (D, B) transforming them into (quasi) upper triangular matrices. In order to obtain a numerically stable algorithm, this transformation needs to be done using unitary transformations. In this way, we can use the computation of the generalized Schur decomposition from Chapters 2 to 5. Having computed the generalized Schur decompositions (QA_SZ^H, QC_SZ^H) of (A, C) and (UD_SV^H, UB_SV^H) of (D, B), we plug them into the generalized Sylvester equation (GSYLV). This reads

$$QA_SZ^HXUB_SV^H + QC_SZ^HXUD_SV^H = Y. \tag{6.5}$$

By setting $\tilde{X} = Z^HXU$ and $\tilde{Y} = Q^HYV$ we get

$$A_S\tilde{X}B_S + C_S\tilde{X}D_S = \tilde{Y} \tag{6.6}$$

with the structure

where the triangular matrices can be quasi upper triangular as well. Since this idea is the foundation of most of the direct solution techniques developed during the last decades, we set $A := A_S$, $B := B_S$, $C := C_S$, $D := D_S$, $X := \tilde{X}$, and $Y = \tilde{Y}$ in the following description of the direct solution techniques.

6.2.1. Bartels-Stewart Algorithm

The first algorithm using the triangular structure (6.6) was the Bartels-Stewart Algorithm [19] to solve the Sylvester (SYLV) and the Lyapunov (LYAP) equation. Starting with

$$A = \begin{bmatrix} a_{11} & \cdots & a_{1m} \\ 0 & \ddots & \vdots \\ 0 & 0 & a_{mm} \end{bmatrix} \quad \text{and} \quad D = \begin{bmatrix} d_{11} & \cdots & d_{1n} \\ 0 & \ddots & \vdots \\ 0 & 0 & d_{nn} \end{bmatrix} \tag{6.8}$$

and X and Y

$$X = \begin{bmatrix} x_{11} & \cdots & x_{1n} \\ \vdots & \vdots & \vdots \\ x_{m1} & \cdots & x_{mn} \end{bmatrix} \quad \text{and} \quad Y = \begin{bmatrix} y_{11} & \cdots & y_{1n} \\ \vdots & \vdots & \vdots \\ y_{m1} & \cdots & y_{mn} \end{bmatrix}, \tag{6.9}$$

each value in the right-hand side of the Sylvester equation (SYLV) is given as

$$y_{kl} = \sum_{j=k}^{m} a_{kj}x_{jl} + \sum_{j=1}^{l} x_{kj}d_{jl}. \tag{6.10}$$

From this we can determine x_{kl} for $k = m, \ldots, 1$ and $l = 1, \cdots, n$ by

$$a_{kk}x_{kl} + x_{kl}d_{ll} = y_{kl} - \left(\sum_{j=k+1}^{m} a_{kj}x_{jl} + \sum_{j=1}^{l-1} x_{kj}d_{jl} \right) = \hat{y}_{kl}. \tag{6.11}$$

In case of the Lyapunov equation (LYAP), where we have $d_{kj} = \overline{a_{jk}}$ and $m = n$, we obtain

$$a_{kk}x_{kl} + x_{kl}\overline{a_{ll}} = y_{kl} - \left(\sum_{j=k+1}^{m} a_{kj}x_{jl} + \sum_{j=l+1}^{m} x_{kj}\overline{a_{lj}} \right) = \hat{y}_{k,l} \tag{6.12}$$

for $l = m, \cdots, 1$ and $k = l, \ldots, 1$. The remaining entries x_{kl}, $l = m, \cdots, 1$ and $k = m, \ldots, l + 1$, are restored by symmetry. The increased structure reduces the number of required operations.

This idea assumes that A and D are upper triangular. If we stay in real arithmetic, i.e. A and D are in real Schur form, we can have 2×2 blocks on their diagonals. In this case, the linear system in Equation (6.11) or (6.12) becomes either

$$\left(1 \otimes \begin{bmatrix} a_{k,k} & a_{k,k+1} \\ a_{k+1,k} & a_{k+1,k+1} \end{bmatrix} + d_{ll} \otimes \begin{bmatrix} 1 \\ & 1 \end{bmatrix} \right) \begin{bmatrix} x_{k,l} \\ x_{k+1,l} \end{bmatrix} = \begin{bmatrix} \hat{y}_{k,l} \\ \hat{y}_{k+1,l} \end{bmatrix}, \tag{6.13a}$$

or

$$\left(\begin{bmatrix} 1 \\ & 1 \end{bmatrix} \otimes a_{k,k} + \begin{bmatrix} d_{l,l} & d_{l+1,l} \\ d_{l,l+1} & d_{l+1,l+1} \end{bmatrix} \otimes 1 \right) \begin{bmatrix} x_{k,l} \\ x_{k,l+1} \end{bmatrix} = \begin{bmatrix} \hat{y}_{k,l} \\ \hat{y}_{k,l+1} \end{bmatrix}, \tag{6.13b}$$

or

$$\left(\begin{bmatrix} 1 \\ & 1 \end{bmatrix} \otimes \begin{bmatrix} a_{k,k} & a_{k,k+1} \\ a_{k+1,k} & a_{k+1,k+1} \end{bmatrix} + \begin{bmatrix} d_{l,l} & d_{l+1,l} \\ d_{l,l+1} & d_{l+1,l+1} \end{bmatrix} \otimes \begin{bmatrix} 1 \\ & 1 \end{bmatrix} \right) \begin{bmatrix} x_{k,l} \\ x_{k+1,l} \\ x_{k,l+1} \\ x_{k+1,l+1} \end{bmatrix} = \begin{bmatrix} \hat{y}_{k,l} \\ \hat{y}_{k+1,l} \\ \hat{y}_{k,l+1} \\ \hat{y}_{k+1,l+1} \end{bmatrix}, \tag{6.13c}$$

depending on the 2×2 block structure on the diagonals of A and D. The linear systems in (6.13) are at most of order four and thus easily solvable. For enhancing the numerical stability these systems are solved using a completely pivoted LU decomposition [19, 140]. The procedure to solve the triangular Sylvester equations needs $m^2n + mn^2$ flops. Employing the symmetric structure in the case of the Lyapunov equation reduces this to m^3 flops.

Kågström and Westin [107] extended the idea straight forward to the generalized coupled Sylvester equation (GCSYLV). The implementation of TGSYL in LAPACK follows their work directly [105]. Although this implementation is designed to be aware of level-3 BLAS and LAPACK operations, the block size parameters are too small values, i.e. one is used. Consequently, it behaves like a level-2 BLAS algorithm and thus yields no acceptable performance on current computer architectures.

In the case of the Lyapunov and the Stein equation with stable matrices or matrix pairs, i.e. all their eigenvalues lie in the open left half-plane, an extension of the Bartels-Stewart idea was presented, by Hammarling [85], to compute directly a Cholesky factor $ZZ^H = X$ of the solution. This approach was extended by Penzl [140] in 1997 to the corresponding generalized equations. A block approach of Hammarlings method was presented by Kressner [117] in 2008, which only covers the case $B = C = I$.

6.2.2. One- and Two-Solve Schemes

The Bartels-Stewart approach as well as Kågström-Westin approach solve for only one or a few elements of the solution at once. This limits the performance and the parallel scalability since no or only few level-3 enabled operations can be used. One improvement is the one-solve scheme, e.g. shown by Sorensen and Zhou [153]. Again, we regard the Sylvester equation where the coefficient matrices are transformed to (quasi) upper triangular form. Then we partition the solution X as well as the right-hand side Y into columns

$$X = \begin{bmatrix} X_1 & X_2 & \cdots & X_n \end{bmatrix} \quad \text{and} \quad Y = \begin{bmatrix} Y_1 & Y_2 & \cdots & Y_n \end{bmatrix}. \tag{6.14}$$

Now, we reformulate the Sylvester equation (SYLV) to

$$A \begin{bmatrix} X_1 & X_2 & \cdots & X_n \end{bmatrix} + \begin{bmatrix} X_1 & X_2 & \cdots & X_n \end{bmatrix} D = \begin{bmatrix} Y_1 & Y_2 & \cdots & Y_n \end{bmatrix}. \tag{6.15}$$

Assuming that both A and D are upper triangular, i.e. we have only real eigenvalues in A and D, or we use complex arithmetic, we are able to solve for X_k, $k = 1, \ldots, n$, by subsequently using

$$\left(A + d_{k,k} I \right) X_k = Y_k - \sum_{j=1}^{k-1} d_{j,k} X_j. \tag{6.16}$$

In the case of real arithmetic and a 2×2 block on the diagonal of D we have to solve for k and $k+1$ at once using the following subproblem:

$$A \begin{bmatrix} X_k & X_{k+1} \end{bmatrix} + \begin{bmatrix} X_k & X_{k+1} \end{bmatrix} \begin{bmatrix} d_{k,k} & d_{k,k+1} \\ d_{k+1,k} & d_{k+1,k+1} \end{bmatrix}$$
$$= \begin{bmatrix} Y_k - \sum_{j=1}^{k-1} d_{j,k} X_j & Y_{k+1} - \sum_{j=1}^{k-1} d_{j,k+1} X_j \end{bmatrix} = \begin{bmatrix} \tilde{Y}_k & \tilde{Y}_{k+1} \end{bmatrix}. \tag{6.17}$$

Naively, using the Kronecker representation (6.4), this would lead to the assembly and the solution of a linear system of dimension $2m \times 2m$

$$\begin{bmatrix} A + d_{k,k} I & d_{k+1,k} I \\ d_{k,k+1} I & A + d_{k+1,k+1} I \end{bmatrix} \begin{bmatrix} X_k \\ X_{k+1} \end{bmatrix} = \begin{bmatrix} \tilde{Y}_k \\ \tilde{Y}_{k+1} \end{bmatrix}. \tag{6.18}$$

This does not take into account the structure of A and D. Furthermore, it increases the memory requirement dramatically, as well as the number of required flops. The setup of this huge matrix is avoided by splitting (6.17) into two equations [153]:

$$AX_k + d_{k,k} X_k + d_{k+1,k} X_{k+1} = \tilde{Y}_k$$
$$AX_{k+1} + d_{k,k+1} X_k + d_{k+1,k+1} X_{k+1} = \tilde{Y}_{k+1}. \tag{6.19}$$

From the first one we get

$$X_{k+1} = \frac{1}{d_{k+1,k}} \left(\tilde{Y}_k - \left(A - d_{k,k} I \right) X_k \right). \tag{6.20}$$

Together with the second equation we can solve for X_k

$$\left(A^2 + \left(d_{k,k} + d_{k+1,k+1} \right) A + \left(d_{k,k} d_{k+1,k+1} - d_{k+1,k} d_{k,k+1} I \right) \right) X_k$$
$$= A\tilde{Y}_k + d_{k+1,k+1} \tilde{Y}_k - d_{k+1,k} \tilde{Y}_{k+1}. \tag{6.21}$$

Afterwards, X_{k+1} is restored using Equation (6.20). Since A is (quasi) upper triangular its square has the same structure. The independence from D allows us to compute A^2 once for all 2×2 blocks appearing on the diagonal of D. Beside storing this additional matrix, its increased condition number may lead to stability issues. The overall scheme solves the Sylvester equation column-wise, processing one column of the solution after each other. The name *one-solve scheme* originates from this behavior.

An improvement to this scheme is obtained by resolving Equation (6.19) for X_k and computing X_{k+1} directly similar to Equation (6.21) [153]. Since the appearing linear system has the same coefficient matrix we are able to solve for X_k and X_{k+1} at once using

$$\left(A^2 + \left(d_{k,k} + d_{k+1,k+1} \right) A + \left(d_{k,k} d_{k+1,k+1} - d_{k+1,k} d_{k,k+1} I \right) \right) \begin{bmatrix} X_k & X_{k+1} \end{bmatrix}$$
$$= \begin{bmatrix} A\tilde{Y}_k + d_{k+1,k+1} \tilde{Y}_k - d_{k+1,k} \tilde{Y}_{k+1} & A\tilde{Y}_{k+1} + d_{k,k} \tilde{Y}_{k+1} - d_{k,k+1} \tilde{Y}_k \end{bmatrix}. \tag{6.22}$$

We still require the setup of A^2 and hence the problem of the increasing condition number obtained form squaring does not vanish by this approach. However, we solve for two columns of the solution in a single step, which leads to the name *two-solve scheme*. The main advantage compared to the one-solve approach is that block operations from BLAS and LAPACK can be used to update parts of the right-hand side and to compute the solution. Sorensen and Zhou [153] presented extensions to this scheme for solving the Lyapunov equation (LYAP) and came up with an eigenvalue decomposition approach.

The one-solve scheme, as well as the two-solve scheme, use the triangular structure of the coefficient matrix to apply forward and backward substitution. Therefore, it is necessary to compute two (generalized) Schur decompositions in the beginning to obtain the triangular structure. To overcome this computationally expensive task Golub, Nash, and Van Loan presented already an improvement in 1979 [79]. They suggest reducing A or, in case of generalized equations, (A, C) to upper Hessenberg form to save one of the time-consuming Schur decompositions. Then the coefficient matrix in (6.16) stays upper Hessenberg and thus it can be solved using a specialized LU decomposition within $2m^2$ flops. Even in the two-solve scheme this approach can be used since the square of an upper Hessenberg matrix has only two subdiagonals. In this way, the linear system appearing in Equation (6.22) can be solved within $4m^2$ flops.

The idea of the one-solve scheme was transferred to the generalized Sylvester equation (GSYLV) by Gardiner et al. [78]

$$\left(b_{k,k}A + d_{k,k}C\right) X_k = Y_k - \sum_{j=1}^{k-1} \left(b_{j,k}AX_k + d_{j,k}CX_k\right). \tag{6.23}$$

In the case of a quasi upper triangular matrix D or B this yields the large linear system

$$\begin{bmatrix} b_{k,k}A + d_{k,k}C & b_{k+1,k}A + d_{k+1,k}C \\ b_{k,k+1}A + d_{k,k+1}C & b_{k+1,k+1}A + d_{k+1,k+1}C \end{bmatrix} \begin{bmatrix} X_k \\ X_{k+1} \end{bmatrix} = \begin{bmatrix} \tilde{Y}_k \\ \tilde{Y}_{k+1} \end{bmatrix}, \tag{6.24}$$

where $\tilde{Y}_k$ and $\tilde{Y}_{k+1}$ are given by

$$\tilde{Y}_k = Y_k - \sum_{j=1}^{k-1} \left(b_{j,k}A + d_{j,k}C\right) X_j \ \text{ and } \ \tilde{Y}_{k+1} = Y_{k+1} - \sum_{j=1}^{k-1} \left(b_{j,k+1}A + d_{j,k+1}C\right) X_j. \tag{6.25}$$

Gardiner and Laub showed, that changing the natural order $(1, 2, \ldots, 2m)$ of the rows and columns of the linear systems to $(1, m+1, 2, m+2, \ldots, m, 2m)$ will make it block upper triangular with 2×2 and 4×4 blocks on the diagonal. This makes the system easily solvable again although it needs to be set up or the permutation needs to be implemented implicitly.

The Hessenberg-Schur scheme of Golub, Nash, and Van Loan can be used in Equation (6.24) as well. The permutation introduced by Gardiner et al. [78] makes the linear system from Equation (6.24) upper triangular with three non-zero subdiagonals in this case. This reduces the costs for the solution of one linear system to $16m^2$ flops.

Although the presented algorithms are well understood and partly transformed into software, they have several issues from a high performance computing point of view. First, they mostly involve only level-2 BLAS operations that prevent a good scalability on current multi-core CPUs. Second, in order to avoid the complex arithmetic, either a $2m \times 2m$ system needs to be assembled or we have to use the square of a matrix. Third, although the Hessenberg-Schur approach reduces the number of flops it does not pay off to use it for the symmetric equations, since there we use the triangular structure of the coefficient matrices twice.

6.2.3. Recursive Blocking

All previously shown techniques mostly involve matrix-vector or scalar operations. First approaches to obtain a block algorithm were presented by Kågström and Poroma [105] for the integration in LAPACK. Unfortunately this solver only handles the generalized coupled Sylvester equation (GCSYLV) and the LAPACK implementation only uses a block size of two as reported by the environment information function ILAENV of LAPACK. In fact this causes the algorithm to behave like a level-2 BLAS implementation. Furthermore, the other matrix equations introduced at the beginning of this chapter can only be solved with the cost of the solution of further subsequent linear systems to recover the generalized Sylvester equation.

In order to overcome this issue, Jonsson and Kågström presented a recursive blocking approach [100, 101], which focuses on the usage of large matrix-matrix products. Their main idea is the partitioning of the matrices X and Y into 1×2, 2×2, or 2×1 block matrices depending on the ratio between m and n. Thereby they distinguish between three cases. Exemplary, for the generalized Sylvester equation (GSYLV) this gives the following:

Case 1: $1 \le n \le \frac{m}{2}$ We only split the matrix pair (A, C) into a 2×2 block structure. This provides

$$\begin{bmatrix} A_{11} & A_{12} \\ & A_{22} \end{bmatrix} \begin{bmatrix} X_1 \\ X_2 \end{bmatrix} B + \begin{bmatrix} C_{11} & C_{12} \\ & C_{22} \end{bmatrix} \begin{bmatrix} X_1 \\ X_2 \end{bmatrix} D = \begin{bmatrix} Y_1 \\ Y_2 \end{bmatrix}, \tag{6.26}$$

where X_1 and Y_1 are of dimension $\lceil \frac{m}{2} \rceil \times n$ and the dimensions of the other matrices are chosen accordingly. Then, we have to solve the following two equations:

$$A_{22}X_2B + C_{22}X_2D = Y_2,$$
$$A_{11}X_1B + C_{11}X_1D = Y_1 - A_{12}X_2B - C_{12}X_2D.$$

Case 2: $1 \le m \le \frac{n}{2}$ This case is similar to the first one, but we partition B and D into a 2×2 block structure. This yields

$$A \begin{bmatrix} X_1 & X_2 \end{bmatrix} \begin{bmatrix} B_{11} & B_{12} \\ & B_{22} \end{bmatrix} + C \begin{bmatrix} X_1 & X_2 \end{bmatrix} \begin{bmatrix} D_{11} & D_{12} \\ & D_{22} \end{bmatrix} = \begin{bmatrix} Y_1 & Y_2 \end{bmatrix}, \tag{6.27}$$

where X_1 and Y_1 are of dimension $m \times \lceil \frac{n}{2} \rceil$. Here, we have to solve the following two equations:

$$AX_1B_{11} + CX_1D_{11} = Y_1,$$
$$AX_2B_{22} + CX_2D_{22} = Y_2 - AX_1B_{12} - CX_1D_{12}.$$

Case 3: $\frac{n}{2} < m < 2n$ If the dimensions of the solution and the right-hand side are close to square matrices, we partition all matrices into a 2×2 block structure. Then, we obtain

$$\begin{bmatrix} A_{11} & A_{12} \\ & A_{22} \end{bmatrix} \begin{bmatrix} X_{11} & X_{12} \\ X_{21} & X_{22} \end{bmatrix} \begin{bmatrix} B_{11} & B_{12} \\ & B_{22} \end{bmatrix} +$$

$$\begin{bmatrix} C_{11} & C_{12} \\ & C_{22} \end{bmatrix} \begin{bmatrix} X_{11} & X_{12} \\ X_{21} & X_{22} \end{bmatrix} \begin{bmatrix} D_{11} & D_{12} \\ & D_{22} \end{bmatrix} = \begin{bmatrix} Y_{11} & Y_{12} \\ Y_{21} & Y_{22} \end{bmatrix}, \tag{6.28}$$

where X_1 and Y_1 are of order $\lceil \frac{m}{2} \rceil$, and the dimension of the other matrices are chosen accordingly. This partitioning yields the following four smaller equations:

$$A_{22}X_{21}B_{11} + C_{22}X_{21}D_{11} = Y_{21}$$
$$A_{22}X_{22}B_{22} + C_{22}X_{22}D_{22} = Y_{22} - A_{22}X_{21}B_{12} - C_{22}X_{21}D_{12},$$
$$A_{11}X_{11}B_{11} + C_{11}X_{11}D_{11} = Y_{11} - A_{12}X_{21}B_{11} - C_{12}X_{21}D_{11},$$
$$A_{11}X_{12}B_{22} + C_{11}X_{12}D_{22} = Y_{12} - A_{11}X_{11}B_{12} - A_{12}X_{21}B_{12} - A_{12}X_{22}B_{22}$$
$$- C_{11}X_{11}D_{12} - C_{12}X_{21}D_{12} - C_{12}X_{22}D_{22}.$$

Thereby, the last equation can be rewritten into

$$A_{11}X_{12}B_{22} + C_{11}X_{12}D_{22} = Y_{12} - (A_{11}X_{11} + A_{12}X_{21})B_{12} - A_{12}X_{22}B_{22}$$
$$- (C_{11}X_{11} + C_{12}X_{21})D_{12} - C_{12}X_{22}D_{22}$$

or

$$A_{11}X_{12}B_{22} + C_{11}X_{12}D_{22} = Y_{12} - A_{11}X_{11}B_{12} - A_{12}(X_{21}B_{12} + X_{22}B_{22})$$
$$- C_{11}X_{11}D_{12} - C_{12}(X_{21}D_{12} + X_{22}D_{22})$$

to reduce the number of flops.

This partition is applied recursively to the emerging subproblems until they are small enough, i.e. the Kronecker product equivalent problem (6.4) is trivial-to-solve. It is easy to see that the overall scheme is rich in level-3 matrix-matrix multiplies. Since the (sub-)matrices in the block partitioning are either full or triangular, the BLAS routines should be selected in correspondence to their structure. For operations that involve (quasi) triangular matrices, the three-operand TRMM3 operation from Subsection 4.2.2 can be used to avoid unnecessary copy operations. Furthermore, matrix-matrix multiplies of the form AXB are evaluated either as

$$(AX)B \quad \text{or } A(XB),$$

depending on the dimensions of A and B, to further reduce the number of required flops. The overall scheme is implemented in the RECSY library of Jonsson and Kågström [100, 101].

Symmetric Matrix Equations In the case of symmetric matrix equations, we observe that the recursion only needs the symmetric solver again to compute X_{11} and X_{22} in the first case. The X_{12} block is computed using a solver for the non-symmetric matrix equation and the symmetry is recovered by setting $X_{21} = X_{12}^H$. This reduces the number of required operations, since we have to solve for one equation less. Furthermore, filling X_{21} from its symmetric counterpart guarantees that the final solution will be symmetric, as well. The number of required operations can be reduced further if matrix-matrix multiplies that incorporate X_{11} or X_{22} use the symmetric matrix-matrix multiply SYR2K of BLAS. Overall, the solution of the symmetric matrix equations requires symmetric, as well as non-symmetric solvers, when recursive blocking is applied.

Parallelization Since the whole procedure is rich in level-3 BLAS operations, the parallelization of the BLAS library can be used inside each operation. Regarding the third case, further parallel execution paths are possible. The main ingredient for this is that once X_{21} is computed, the blocks X_{22} and X_{11} can be obtained independently. The implementation of Jonsson and Kågström [100, 101] solves for them in parallel. Figure 6.1 shows that this does not lead to a higher performance on current computers. This is caused by the behavior of the BLAS library. Once the program enters an OpenMP parallel execution block the BLAS library detects this and switches to single thread operations. Since the computations of X_{11} and X_{22} are encapsulated in OpenMP parallel sections, in the implementation of Jonsson and Kågström, only a single thread is used to compute them. This prohibits a parallelization outside the BLAS library. Some variants, like Intel®'s MKL, support nested parallelization but this breaks portability and requires further tuning for optimal performance. Furthermore, the fast decrease of the dimension in the matrix-matrix products leads to an algorithm which cannot benefit from the parallel capabilities of the BLAS library any longer. very fast.

An extension to the recursive blocking, focusing on distributed memory systems, is contained in the SCASY library, developed by Granat and Kågström [83, 82]. This approach is not regarded here, due to its bad performance on current multi-core systems, as shown by Köhler and Saak in [112].

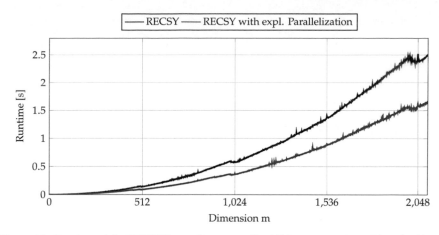

Figure 6.1.: Runtime of the RECSY library for a generalized Sylvester equation with and without explicit parallelization.

6.2.4. Matrix Sign Function Solvers

The previous sections show different approaches which can be used to solve the Sylvester-type matrix equation if the involved coefficient matrices are in (quasi) upper triangular form or at least in upper Hessenberg form. Regarding the origin of the matrix sign function, as we discussed in detail in Chapter 3, Roberts shows in his fundamental paper [144] a connection between the matrix sign function and the solution of the Sylvester equation (SYLV). Although it is not a standard direct solver we recall this approach briefly here, since it has a close connection to the matrix sign function. Unfortunately, this idea is restricted to the case where all eigenvalues of A and B lie in open left half-plane.

First, we need the connection between the Sylvester equation (SYLV) and the block diagonalization of a block upper triangular matrix:

Lemma 6.6 (e.g. [80]):

Let A be a matrix of order m, B be a matrix of order n, and C and X be matrices of size $m \times n$. If X solves

$$AX + XB = C \tag{6.29}$$

it holds

$$\begin{bmatrix} I_m & -X \\ & I_n \end{bmatrix} \begin{bmatrix} A & -C \\ & -B \end{bmatrix} \begin{bmatrix} I_m & X \\ & I_n \end{bmatrix} = \begin{bmatrix} A & \\ & -B \end{bmatrix}. \tag{6.30}$$

This Lemma together with the properties of the matrix sign function, compare Theorem 3.5, yields:

$$\operatorname{sign}\left(\begin{bmatrix} A & -C \\ & -B \end{bmatrix}\right) = \operatorname{sign}\left(\begin{bmatrix} I_m & X \\ & I_n \end{bmatrix} \begin{bmatrix} A & \\ & -B \end{bmatrix} \begin{bmatrix} I_m & -X \\ & I_n \end{bmatrix}\right)$$
$$\overset{\text{Thm. 3.5/3.}}{=} \begin{bmatrix} I_m & X \\ & I_n \end{bmatrix} \operatorname{sign}\left(\begin{bmatrix} A & \\ & -B \end{bmatrix}\right) \begin{bmatrix} I_m & -X \\ & I_n \end{bmatrix}.$$

If the matrices A and B only have eigenvalues in the open left half-plane we obtain

$$\text{sign}\left(\begin{bmatrix} A & -C \\ & -B \end{bmatrix}\right) = \begin{bmatrix} I_m & X \\ & I_n \end{bmatrix} \begin{bmatrix} -I & \\ & I \end{bmatrix} \begin{bmatrix} I_m & -X \\ & I_n \end{bmatrix}$$

$$= \begin{bmatrix} -I & -2X \\ & I \end{bmatrix},$$

which gives us the solution X of the standard Sylvester equation (SYLV). Using the Newton approach from Chapter 3 we are able to compute the solution X without requiring A and B to be (quasi) upper triangular. Since setting up the matrix

$$\mathcal{A}_0 = \begin{bmatrix} A & -C \\ & -B \end{bmatrix} \tag{6.31}$$

would require too much memory we rewrite the iteration, compare (3.25),

$$\mathcal{A}_{k+1} = \frac{1}{2}\left(\gamma_k \mathcal{A}_k + \gamma_k^{-1} \mathcal{A}^{-1}\right) \tag{6.32}$$

as

$$A_0 = A, \quad B_0 = -B, \text{ and } C_0 = -C \tag{6.33}$$

$$A_{k+1} = \frac{1}{2}\left(\gamma_k A_k + \gamma_k^{-1} A_k^{-1}\right) \tag{6.34}$$

$$B_{k+1} = \frac{1}{2}\left(\gamma_k B_k + \gamma_k^{-1} B_k^{-1}\right) \tag{6.35}$$

and

$$C_{k+1} = \frac{1}{2}\left(\gamma_k C_k - \gamma_k^{-1} A_k^{-1} C_k B_k^{-1}\right). \tag{6.36}$$

In this way, we are able to use the matrix sign function to compute the solution of the standard Sylvester equation without computing the Schur decomposition before. The accelerated computation of A^{-1} can be used here as well, compare Section 4.2.4

Moving forward to the generalized case requires a property of the generalized matrix sign function, which is not necessary for the computation of our eigenvalue problem in Chapters 3 and 4. The following lemma generalizes the two-sided transformation of the matrix sign function, see Theorem 3.5, to matrix pairs:

Lemma 6.7:

Let (A, B) be a matrix pair of order m. Furthermore, B, L and R are nonsingular matrices of order m. Then the following equivalence holds true:

$$\text{sign}\,(LAR, LBR) = L\,\text{sign}\,(A, B)\,R. \tag{6.37}$$

Proof. From Definition 3.8 of the generalized matrix sign function we obtain

$$\text{sign}\,(LAR, LBR) = LBR\,\text{sign}\left((LBR)^{-1} LAR\right)$$

$$= LBR\,\text{sign}\left(R^{-1} B^{-1} AR\right).$$

Using Theorem 3.5 this yields

$$= LBRR^{-1}\,\text{sign}\left(B^{-1}A\right) R.$$

Finally, this gives

$$\text{sign}\,(LAR, LBR) = LB\,\text{sign}\,(A, B)\,R = L\,\text{sign}\,(A, B)\,R. \qquad \square$$

From the eigenvalue reordering in Section 2.3, we already know a connection between the generalized coupled Sylvester equation and the generalized eigenvalue problem [107]. If L and R solve

$$\begin{aligned} AL + RD &= Y \\ CL - RB &= 0 \end{aligned} \quad , \tag{6.38}$$

we can diagonalize the following block matrix pair:

$$\left(\begin{bmatrix} A & -Y \\ & -D \end{bmatrix}, \begin{bmatrix} C & \\ & B \end{bmatrix} \right) = \begin{bmatrix} I & -L \\ & I \end{bmatrix} \left(\begin{bmatrix} A & \\ & -D \end{bmatrix}, \begin{bmatrix} C & \\ & B \end{bmatrix} \right) \begin{bmatrix} I & R \\ & I \end{bmatrix} \tag{6.39}$$

Then $X = C^{-1}L = RB^{-1}$ solves the generalized Sylvester equation (GSYLV). Now, we regard

$$\begin{aligned} \mathrm{sign}\left(\begin{bmatrix} A & -Y \\ & -D \end{bmatrix}, \begin{bmatrix} C & \\ & B \end{bmatrix} \right) &= \begin{bmatrix} I & -L \\ & I \end{bmatrix} \mathrm{sign}\left(\begin{bmatrix} A & \\ & -D \end{bmatrix}, \begin{bmatrix} C & \\ & B \end{bmatrix} \right) \begin{bmatrix} I & R \\ & I \end{bmatrix} \\ &= \begin{bmatrix} I & -L \\ & I \end{bmatrix} \begin{bmatrix} C & \\ & B \end{bmatrix} \mathrm{sign}\left(\begin{bmatrix} C^{-1} & \\ & B^{-1} \end{bmatrix} \begin{bmatrix} A & \\ & -D \end{bmatrix} \right) \begin{bmatrix} I & R \\ & I \end{bmatrix} \end{aligned}$$

If the matrix pairs (A, C) and (D, B) only have eigenvalues in the open left half-plane we get

$$\begin{aligned} \mathrm{sign}\left(\begin{bmatrix} A & -Y \\ & -D \end{bmatrix}, \begin{bmatrix} C & \\ & B \end{bmatrix} \right) &= \begin{bmatrix} I & -L \\ & I \end{bmatrix} \begin{bmatrix} C & \\ & B \end{bmatrix} \begin{bmatrix} -I & \\ & I \end{bmatrix} \begin{bmatrix} I & R \\ & I \end{bmatrix} \\ &= \begin{bmatrix} I & -L \\ & I \end{bmatrix} \begin{bmatrix} -C & \\ & B \end{bmatrix} \begin{bmatrix} I & R \\ & I \end{bmatrix} \\ &= \begin{bmatrix} -C & -CR - LB \\ & B \end{bmatrix} = \begin{bmatrix} -C & -2CXB \\ & B \end{bmatrix}. \end{aligned}$$

and recover the solution of the generalized Sylvester equation from this. For practical implementations the Newton iteration used to compute the generalized matrix sign function is rewritten in the same way as for the standard scheme, compare Equation (6.31). The optimizations to compute the inverse of a matrix or to solve a linear system with many right-hand sides, from Section 4.2.4, can be used in this, again.

Both schemes can be adopted to the (generalized) Lyapunov equation [31, 144], which benefits from the symmetric properties of the equation. High performance implementations of this method are presented, e.g., in [30, 33].

6.3. Block Algorithms and their Implementation

The first block algorithms appeared in the 1990s by Kågström and Poroma [105] and Marqués and Hernández [123]. However, they either have been implemented with bad block parameters, covering only a special case of equations, or they are not available in software at all. In this section we show how block algorithms for the Sylvester-type matrix equations can be developed and optimized with respect to current computer architectures. We use the generalized Sylvester equation (GSYLV) as model problem since it covers all problems that may appear in the implementation and its building blocks are necessary for the remaining matrix equations, as well. This section contains a set of explanatory experiments. These are performed with randomly generated generalized Schur decompositions on the Intel® Skylake systems, compare Appendix A. All examples solve a generalized Sylvester equation with $m = n$ and a block size $M_B = N_B = 64$.

The basic idea behind a level-3 enabled solver is the partitioning of X and Y into blocks of size $M_B \times N_B$. This yields a $p \times q$ block partitioning with $p = \lceil \frac{m}{M_B} \rceil$ and $q = \lceil \frac{n}{N_B} \rceil$. The last block row and block column are eventually smaller to fit the size of the matrices exactly. This yields

$$A = \begin{bmatrix} A_{11} & A_{12} & \cdots & A_{1p} \\ & A_{22} & \cdots & A_{2p} \\ & & \ddots & \vdots \\ & & & A_{pp} \end{bmatrix} \qquad C = \begin{bmatrix} C_{11} & C_{12} & \cdots & C_{1p} \\ & C_{22} & \cdots & C_{2p} \\ & & \ddots & \vdots \\ & & & C_{pp} \end{bmatrix},$$

$$B = \begin{bmatrix} B_{11} & B_{12} & \cdots & B_{1q} \\ & B_{22} & \cdots & B_{2q} \\ & & \ddots & \vdots \\ & & & B_{qq} \end{bmatrix} \qquad D = \begin{bmatrix} D_{11} & D_{12} & \cdots & D_{1q} \\ & D_{22} & \cdots & D_{2q} \\ & & \ddots & \vdots \\ & & & D_{qq} \end{bmatrix},$$

and

$$X = \begin{bmatrix} X_{11} & X_{12} & \cdots & X_{1q} \\ X_{21} & X_{22} & \cdots & X_{2q} \\ \vdots & \vdots & \cdots & \vdots \\ X_{p1} & \cdots & \cdots & X_{pq} \end{bmatrix} \qquad Y = \begin{bmatrix} Y_{11} & Y_{12} & \cdots & Y_{1q} \\ Y_{21} & Y_{22} & \cdots & Y_{2q} \\ \vdots & \vdots & \cdots & \vdots \\ Y_{p1} & \cdots & \cdots & Y_{pq} \end{bmatrix}. \qquad (6.40)$$

In case of real arithmetic, we have to take the quasi upper triangular structure of A and D into account. If the block structure would split a 2×2 block on the diagonal, the block size of the corresponding block row, or block column, is enlarged by one. Therefore, we can only assume a constant block size if all matrix pairs only have real eigenvalues or we use complex arithmetic.

For each block in the solution Y from the block partitioning (6.40) we obtain:

$$Y_{ij} = \sum_{l=1}^{j} \left(\sum_{k=i}^{p} A_{ik} X_{kl} \right) B_{lj} + \sum_{l=1}^{j} \left(\sum_{k=i}^{p} C_{ik} X_{kl} \right) D_{lj}, \qquad (6.41)$$

which gives

$$A_{ii} X_{ij} B_{jj} + C_{ii} X_{ij} D_{jj} = Y_{ij} - \sum_{l=1}^{j-1} \left(\sum_{k=i}^{p} A_{ik} X_{kl} \right) B_{lj} - \left(\sum_{k=i+1}^{p} A_{ik} X_{kl} \right) B_{jj}$$
$$- \sum_{l=1}^{j-1} \left(\sum_{k=i+1}^{p} C_{ik} X_{kl} \right) D_{lj}. - \left(\sum_{k=i+1}^{p} C_{ik} X_{kl} \right) D_{jj}. \qquad (6.42)$$

Obviously, X_{ij} is required to compute X_{kl}, with $k \le i$, $l \ge j$, $k \ne l$. From this, we see that iterating over $i = p, \ldots, 1$ and $j = 1, \ldots, p$ allows us to compute all blocks X_{ij} of the solution X by solving smaller generalized Sylvester equations. Since i and j do not depend on each other, we have two variants to construct the solution X. If we iterate over j, i.e. in direction of the block rows in X, for each iterate i, the solution is set up block row by block row. Vice versa, if we sweep over i for each iterate j the solution is obtained block column by block column. The remaining smaller generalized Sylvester equations for X_{ij} can be solved using the same scheme with a block size of one, where then only trivial-to-solve equations will appear. We see that, once the solution block X_{ij} is computed, the right-hand side block Y_{ij} is no longer required. Hence, we can overwrite Y_{ij} by X_{ij} and work in place on the right-hand side.

In order to make the evaluation of the sums in (6.42) more efficient, we compute the influence of X_{ij} in the remaining right-hand side blocks as soon as we know it. Furthermore, we use this to

---

Algorithm 19 Block Algorithm for the Generalized Sylvester Equation – Row-wise variant

---

Input: (A, C) of order m and (D, B), Y of dimension $m \times n$. A and D (quasi) upper triangular, B and C upper triangular, block sizes M_B and N_B.

Output: Y overwritten with X solving $AXB + CXD = Y$.

1: $k^* = \left\lfloor \frac{m-1}{M_B} \right\rfloor M_B + 1$
2: **for** $k = k^*, k^* - M_B, \ldots, 1$ **do**
3: $\quad k_h = \min\{M_B, m - k + 1\} + k - 1$
4: $\quad$ **for** $l = 1, 1 + N_B, \ldots, n$ **do**
5: $\quad\quad l_h = \min\{N_B, n - l + 1\} + l - 1$
6: $\quad\quad$ Solve $A_{k:k_h,k:k_h} X_{k:k_h,l:l_h} B_{l:l_h,l:l_h} + C_{k:k_h,k:k_h} X_{k:k_h,l:l_h} D_{l:l_h,l:l_h} = Y_{k:k_h,l,l_h}$ and overwrite $Y_{k:k_h,l:l_h}$
   $\quad\quad$ with $X_{k:k_h,l:l_h}$.
7: $\quad\quad Y_{k:k_h,l_h+1:n} \leftarrow Y_{k:k_h,l_h+1:n} - A_{k:k_h,k:k_h} X_{k:k_h,l:l_h} B_{l:l_h,l_h+1:n} - C_{k:k_h,k:k_h} X_{k:k_h,l:l_h} D_{l:l_h,l_h+1:n}$
8: $\quad$ **end for**
9: $\quad Y_{1:k-1,1:n} \leftarrow Y_{1:k-1,1:n} - A_{1:k-1,k:k_h} X_{k:k_h,l:l_h} B - C_{1:k-1,k:k_h} X_{k:k_h,l:l_h} D$
10: **end for**

---

enlarge the matrix-matrix multiplies appearing in (6.42) to increase the usage of level-3 operations. In case of the row-wise computation of the solution X, the influence of X_{ij} in row i is promoted to the remaining right-hand sides by overwriting:

$$Y_{i,j+1:q} \leftarrow Y_{i,j+1:q} - A_{ii} X_{ij} B_{j,j+1:q} - C_{ii} X_{ij} D_{j,j+1:q}. \tag{6.43}$$

After the i-th row is computed, we overwrite the right-hand side with

$$Y_{1:i-1,1:q} \leftarrow Y_{1:i-1,1:q} - A_{1:i-1,i} X_{i,1:q} B - C_{1:i-1,i} X_{i,1:q} D. \tag{6.44}$$

This computes the effect of $X_{i,1:q}$ in Equation (6.42) for all rows above row i. The column-wise computation of X leads to the following updates: Once the X_{ij} is known, it is promoted to the remaining right-hand sides in the current column by:

$$Y_{1:i-1,j} \leftarrow Y_{1:i-1,j} - A_{1:i-1,i} X_{ij} C_{jj} - C_{1:i-1,i} X_{ij} D_{jj}. \tag{6.45}$$

Once the j-th column of the solution is known, its effect on the remaining right-hand side is computed by

$$Y_{1:p,j+1:q} = Y_{1:p,j+1:q} - A X_{1:p,j} B_{j,j+1:q} - C X_{1:p,j} D_{j,j+1:q}. \tag{6.46}$$

Both ways to set up the solution X are shown in Algorithm 19 and Algorithm 20, respectively. Since the matrices are typically stored in Fortran column-major storage, the column-wise operation scheme is preferred. This is confirmed by Figure 6.2, too, comparing the runtime of the row- and column-wise variants. Thereby, the small subproblems occurring (in Algorithms 19 and 20) are solved using the same algorithm with a 1×1 partitioning of $X_{k:k_h,l:l_h}$ and $Y_{k:k_h,l:l_h}$. In this case, the appearing linear systems are at most size 4×4, and hence solved directly. This is done employing a completely pivoted LU decomposition to ensure accuracy.

Figure 6.2 shows characteristic peaks at $m = 512 \cdot g$, $g \in \mathbb{N}$. These are caused by the memory access and prefetching scheme of the CPU. If $m = 512 \cdot g$ and double precision arithmetic is used, each column ends at a multiple of the memory's page size, i.e. $k \cdot 4096$ bytes. If there is a continuous data access until this point, the prefetcher of the CPU will pick the next page from the memory no matter whether it is required or not. This phenomenon also appears solving a small subproblem inside Algorithm 19 or 20, if the columns of the blocks of A, B, C, D, X, or Y end close to a page boundary.

Algorithm 20 Block Algorithm for the Generalized Sylvester Equation – Column-wise variant

Input: (A, C) of order m and (D, B), Y of dimension $m \times n$. A and D (quasi) upper triangular, B and C upper triangular, block sizes M_B and N_B.

Output: Y overwritten with X solving $AXB + CXD = Y$.

1: $k^* = \left\lfloor \frac{m-1}{M_B} \right\rfloor M_B + 1$
2: **for** $l = 1, 1 + N_B, \ldots, n$ **do**
3:    $l_h = \min\{N_B, n - l + 1\} + l - 1$
4:    **for** $k = k^*, k^* - M_B, \ldots, 1$ **do**
5:      $k_h = \min\{M_B, m - k + 1\} + k - 1$
6:      Solve $A_{k:k_h,k:k_h} X_{k:k_h,l:l_h} B_{l:l_h,l:l_h} + C_{k:k_h,k:k_h} X_{k:k_h,l:l_h} D_{l:l_h,l:l_h} = Y_{k:k_h,l,l_h}$ and overwrite $Y_{k:k_h,l:l_h}$
      with $X_{k:k_h,l:l_h}$.
7:      $Y_{1:k-1,l:l_h} \leftarrow Y_{1:k-1,l:l_h} - A_{1:k-1,k:k_h} X_{k:k_h,l:l_h} B_{l:l_h,l:l_h} - C_{1:k-1,k:k_h} X_{k:k_h,l:l_h} D_{l:l_h,l:l_h}$
8:    **end for**
9:    $Y_{1:m,l_h+1:n} \leftarrow Y_{1:m,l_h+1:n} - AX_{1:m,l:l_h} B_{l:l_h,l_h+1:n} - CX_{1:m,l:l_h} D_{l:l_h,l_h+1:n}$
10: **end for**

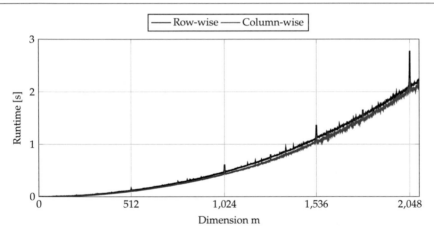

Figure 6.2.: Performance comparison for the row-wise and column-wise block algorithm with naive small solves

Before we focus on the optimization and the solution of the prefetching issue, we show a connection between the level-3 approach and the previously presented solvers. Since the block sizes M_B and N_B are freely selectable, the algorithm can be adjusted to the hardware using a set of benchmarks to obtain the optimal values, on the one hand. On the other hand, specific choices of these parameters will turn the level-3 approach into already existing algorithms:

Theorem 6.8:

Let $AXB + CXD = Y$ be a generalized Sylvester equation and (A, C) and (D, B) matrix pairs of order m and n in generalized Schur form. Then Algorithm 20 is equivalent to

1. *the Bartels-Stewart Algorithm, if $M_B = 1$, $N_B = 1$, $B = I$, and $C = I$ (compare Subsection 6.2.1, [19]),*

2. *the Gardiner et al. approach if $M_B = m$ and $N_B = 1$ (compare Subsection 6.2.2, [78]),*

3. *the Recursive Blocking approach, if M_B and N_B are set to*

> **Case 1** – $1 \leq n \leq \frac{m}{2}$: $M_B = \frac{m}{2}$ and $N_B = n$
>
> **Case 2** – $1 \leq m \leq \frac{n}{2}$: $M_B = m$ and $N_B = \frac{n}{2}$
>
> **Case 3** – $\frac{n}{2} \leq m \leq 2n$: $M_B = \frac{m}{2}$ and $N_B = \frac{n}{2}$,

and Algorithm 20 is used recursively with the same parameter selection rules (compare Subsection 6.2.3, [101]).

Proof.   1. Inserting $B = I$ and $C = I$ and setting $M_B = 1$ and $N_B = 1$, the right-hand side update in Equation (6.42) gets equal to Equation (6.11). The coefficients of the appearing linear system are equal as well.

2. If we set $M_B = m$ and $N_B = 1$, the coefficient matrix in (6.42) gets equal to the one of the Gardiner et al. approach in Equation (6.23). Then our block partitioning in Equation (6.41) with $p = \frac{m}{M_B} = 1$ turns into

$$Y_{1j} = \sum_{l=1}^{j} (AX_{1l}) B_{lj} + \sum_{l=1}^{j} (CX_{1l}) D_{lj}. \tag{6.47}$$

Using that B_{jj} and D_{jj} are scalars due to $N_B = 1$, this yields

$$\left(B_{jj}A + D_{jj}C\right) X_j = Y_j - \sum_{l=1}^{j-1} \left(AX_{1l}B_{lj} + C_{lj}\right) X_{1l},$$

which is equivalent to Equation (6.24). In the case of 2×2 blocks on the diagonal of B or D, we can rewrite (6.47) to (6.19), which gives the equivalence, here, as well.

3. The selection of the block sizes M_B and N_B directly provides the equations from Subsection 6.2.3 and recursive application of the strategy to each subproblem yields the recursive blocking scheme.   □

After we have seen the connection between the level-3 approach and other existing direct solvers, we focus on the optimization. We already saw that traversal of the solution and the right-hand side in a column-wise manor is beneficial for the runtime, compare Figure 6.2. Comparing the runtime to the recursive block approach from Subsection 6.2.3, we see that the level-3 approach is 28.6% slower. During the following subsections we derive an implementation of the level-3 solver which, on the one hand, is faster than the recursive blocking approach and on the other hand, overcomes the problem of the runtime peaks in Figure 6.2.

6.3.1. Improving the level-3 Implementation

We have already seen that the column-wise solution of the subproblems in the level-3 scheme is beneficial, therefore we use Algorithm 20 for all future ideas. In order to avoid unnecessary copies of matrices and to save memory, the solver for the small subproblem is only allowed to use the coefficient matrices read-only and to work in-place on the right-hand side, which is overwritten by the solution. The submatrices are handled, as usual in Fortran, using call by reference together with the leading dimension of the arrays. In this way, everything can be implemented in an LAPACK-like style.

The update in Step 7 of Algorithm 20 consists of two operations of the following structure

$$Y \leftarrow Y - UXV, \tag{6.48}$$

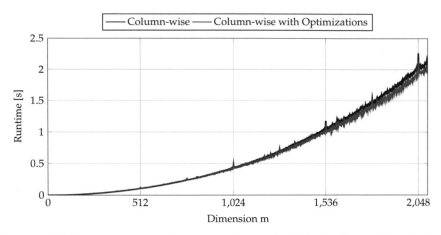

Figure 6.3.: Performance comparison between the column-wise block algorithm and its optimized version

where U is a general tall-skinny matrix and V is a (quasi) upper triangular matrix. Since the triangular matrix-matrix multiply works in-place on the non-triangular operand, we would have to back up X before applying V. Regarding the size of V, which is at most $N_B \times N_B$, we ignore the (quasi) triangular structure and use the general matrix-matrix multiply, here. On the one hand, this turns the separate copy operation into the GEMM call. On the other hand, we do not need to introduce a post processing step to make the triangular matrix-matrix multiply aware of the quasi triangular structure in V. In this way, the whole update in Step 7 can be done using four GEMM operations, with an additional workspace of at most $M_B \times N_B$ elements. Using a simplified Fortran syntax this gives:

```
CALL GEMM( 1.0,X(K:KH,L:LH),B(L:LH,L:LH), 0.0, WORK)
CALL GEMM(-1.0,A(1:K-1,K:KH),WORK,1.0D0, X(1:K-1,L:LH))

CALL GEMM( 1.0,X(K:KH,L:LH),D(L:LH,L:LH), 0.0, WORK)
CALL GEMM(-1.0,C(1:K-1,K:KH),WORK,1.0, X(1:K-1,L:LH))
```

where WORK is the auxiliary workspace.

The second update, in Step 9 of Algorithm 20, has a similar structure as the Update (6.48). However, the matrix U is (quasi) upper triangular and possibly large, the matrix V is a flat and general matrix, and X is a block column of dimension $m \times N_B$. In this way, we perform, first, the multiplication from left with U and afterwards perform the rank N_B update with V on Y. Since the matrix U is large, in contrast to the matrix V in Step 7, we have to take the (quasi) upper triangular property into account. The implementation used to obtain the results in Figure 6.2 creates a copy of the matrix X and performs the standard triangular matrix-matrix multiply from BLAS enhanced by a post processing step. This post processing performs the remaining operations caused by the quasi triangular structure. However, we avoid the explicit copy using the three operand triangular matrix-matrix multiply (TRMM3) from Subsection 4.2.2, here, and extend it to support quasi upper triangular matrices. With this routine, named QTRMM3, we obtain the following simplified implementation:

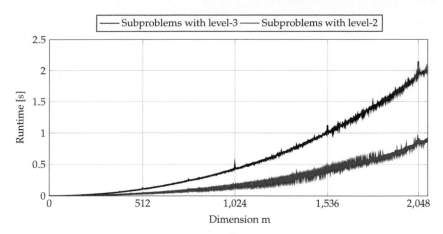

Figure 6.4.: Performance comparison between the level-3 based inner solver and the level-2 based one.

```
CALL QTRMM3(1.0,A,X(1:M,L:LH),0.0,WORK)
CALL GEMM(-1.0,WORK,B(L:LH,LH+1:N),X(1:M,LH+1:N))

CALL TRMM3(1.0,C,X(1:M,L:LH),0.0,WORK)
CALL GEMM(-1.0,WORK,D(L:LH,LH+1:N),X(1:M,LH+1:N))
```

where WORK is the auxiliary workspace of size $m \times N_B$. Hence, beside the memory required for the solution of the small Sylvester equation, the algorithm needs additional memory of size at most $m \times N_B$. The performance increase by using TRMM3 and QTRMM3 instead of the standard BLAS routines with post processing is shown in Figure 6.3. For problems of dimension $m = n = 2\,000$, the difference is already 5%. A comparison to the recursive blocking approach shows that the improved algorithm is still slower. A few flops could be saved, if we employ the quasi triangular structure for the inner updates in Step 7 of Algorithm 20. Since we are using the same BLAS routines as in the recursive blocking approach, we do not investigate more on the level-3 part of the algorithm. Instead, we focus on the inner solvers in the next subsection.

6.3.2. Efficient Solution of the small Subproblems

Until now, we use the level-3 implementation with a block size of $M_B = N_B = 1$ as solver for the small subproblems and solve the remaining linear system of dimension at most 4×4 using a completely pivoted LU decomposition. Since level-3 BLAS calls consists of highly optimized kernel routines, the implementation selects the best one depending on the size of the input data and further arguments, like transposition flags. Furthermore, the routines detect whether they are called inside a multi-threaded environment, or whether it is beneficial to use multi-threading to compute the result. In the level-3 algorithm, with sufficiently large block sizes M_B and N_B, this accelerates the algorithm, and we are able to use multiple CPU cores, automatically. Inside the solvers for the small subproblems, this selection and tuning part of the routines becomes an overhead, which slows down the overall process. Level-2 BLAS operations use such selection strategies, as well. However, they produce less overhead, since, on the one hand, fewer parameters have to be taken into account and, on the other hand, the critical size, where multi-threading is activated, lies much higher and in this way non-beneficial parallel execution for small data sets is avoided. In some BLAS implementations like OpenBLAS [137], many level-1 and level-2 BLAS routines even do not support multi-threading at all.

Due to this difference between level-1, level-2, and level-3 routines, we implement the small subproblem solvers, i.e. where $M_B = N_B = 1$, differently. The update in Step 7 of Algorithm 20

```
CALL GEMM( 1.0,X(K:KH,L:LH),B(L:LH,L:LH), 0.0, WORK)
CALL GEMM(-1.0,A(1:K-1,K:KH),WORK,1.0, X(1:K-1,L:LH))

CALL GEMM( 1.0,X(K:KH,L:LH),D(L:LH,L:LH), 0.0, WORK)
CALL GEMM(-1.0,C(1:K-1,K:KH),WORK,1.0, X(1:K-1,L:LH))
```

is replaced by the vector add routine AXPY and some auxiliary operations. In the easiest case, i.e. if we do not have 2×2 blocks on the diagonal of A or D, this results in two AXPY calls:

```
CALL AXPY(-X(K,L)*B(L,L),A(1:K-1,K),X(1:K-1,L))
CALL AXPY(-X(K,L)*D(L,L),C(1:K-1,K),X(1:K-1,L))
```

In the case where a 2×2 block on the diagonals of A or D is involved, we precompute $X_{k:k_h,l:l_h}B_{l:l_h}$ and $X_{k:k_h,l:l_h}D_{l:l_h}$ directly in Fortran to avoid the overhead of calling a BLAS routine for at most 16 flops in each product. Then the remaining GEMM calls are replaced by up to eight AXPY operations depending on whether the 2×2 block appears in A, in D, or in both.

In the larger update in Step 9, we replace the triangular matrix-matrix multiply with the triangular matrix-vector multiply (TRMV) and a post processing step to handle the quasi triangular structure correctly. The matrix-matrix multiply with parts from B and D is replaced by a rank one update using the GER routine from BLAS. If the corresponding diagonal block in D is a 2×2 block, we get, for each multiplication with A or C, two triangular matrix-vector multiplications and two rank one updates. In the $M_B = N_B = 1$ case the level-3 implementation:

```
CALL QTRMM3(1.0,A,X(1:M,L:LH),0.0,WORK)
CALL GEMM(-1.0,WORK,B(L:LH,LH+1:N),X(1:M,LH+1:N))
```

changes into (for simplicity without handling quasi triangular structures in A)

```
WORK(1:M) = X(1:M,L)
CALL TRMV(A,WORK)
CALL GER(-1.0, WORK, B(L,L+1:N), X(1:M,L+1:N)).
```

Since we only deal with small data sets here, we can create the copy of X into the auxiliary array due to the fact that it most-likely stays in the CPU´s cache when the TRMV and the GER routine are called. These replacements reduce the runtime by a factor of 2.36 for $m = n = 2\,100$, as shown in Figure 6.4. Compared to the recursive blocking approach, we are now faster by a factor of 1.91.

Reducing Function Call Overheads   The replacement of the level-3 BLAS calls with their level-1 and level-2 counterparts already leads to a notable performance boost. However, there are still many hidden overheads in the implementation. For example, each BLAS routine checks the validity of their dimensional input arguments. Furthermore, there are checks whether a loop should be unrolled or not, as in the reference implementation of AXPY [119]. Consider that we use the routine only for small problems, e.g. $m, n \leq 64$, then an AXPY operation performs at most 128 flops. Regarding a current Intel® Skylake CPU core, which performs up to 16 double precision flops per CPU cycle, this only takes eight CPU cycles to complete. The overhead of the sanity checks in the beginning, as well as the function call itself, now become a significant part of the runtime. If m and n are getting small for the subproblem, this overhead even takes more time than the actual computation. This motivates to replace all operations that perform only $O(n)$ operations directly with Fortran code. Here, it is used that Fortran, beginning with the Fortran 90/95 standard [7],

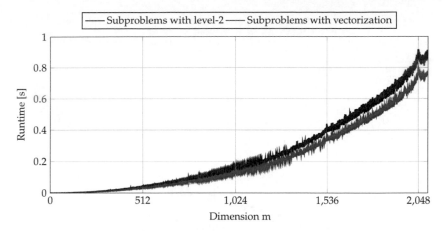

Figure 6.5.: Performance comparison between the level-2 based inner solver and the optimized one with Fortran vector operations.

supports basic vector calculations directly. In this way, the compiler generates the code to add and scale vectors without the overhead of a subroutine call. Hence, we have to replace the AXPY operations:

```
CALL AXPY(-X(K,L)*B(L,L),A(1:K-1,K),X(1:K-1,L))
CALL AXPY(-X(K,L)*D(L,L),C(1:K-1,K),X(1:K-1,L))
```

by a single statement:

```
X(1:K-1,L) = X(1:K-1,L)-X(K,L)*B(L,L)*A(1:K-1,K)
             -X(K,L)*D(L,L)*C(1:K-1,K).
```

This small change increases the performance by 14% compared to the pure level-2 case as seen in Figure 6.5. However, the peaks at $m = 512 \cdot g$, $g \in \mathbb{N}$ still exist.

Local Copies and Data Alignment The peaks visisble in Figures 6.2 to 6.5 are caused by problematic memory accesses when solving the small subproblems. A possible solution is the regularization of the memory accesses by creating local copies, with a leading dimension equal to the number of rows or the maximum block sizes M_B and N_B of the level-3 algorithm.

For each of the submatrices A_{ii}, B_{jj}, C_{ii}, D_{jj}, and Y_{ij} we create a local copy, when starting the solver for the small subproblem, and afterwards the solution X_{ij} is copied back to its final memory location. Now, the hardware prefetching scheme does no longer fetch unnecessary data from the memory. Furthermore, if M_B and N_B are chosen properly for the level-3 algorithm, like $M_B = N_N = 64$ (as we use for all experiments), the whole procedure can run inside the CPU cache without further memory interactions. In order to assist the compiler's optimization strategies, we compile the solver multiple times with different, but fixed, values of M_B and N_B.

Unfortunately, we see in Figure 6.6 that this approach slows down the computation. However, it allows further adjustments to the code, especially with respect to the optimization performed by the compiler. Thereby the fact, that modern CPUs only load vector registers fast if the data has a special alignment in memory, plays an important role. For example, to fill Intel®'s AVX registers properly, the memory address of the first byte of the dataset must be a multiple of 32 bytes. This

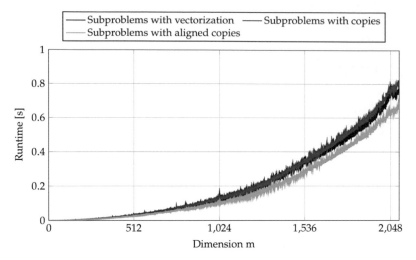

Figure 6.6.: Performance comparison between the level-2 solver with Fortran vector operations and the local copy variants.

is not valid for arbitrary memory locations. If we are able to guarantee such an alignment, the compiler can use more and better vector instructions inside the code, for example to handle the replacement of the **AXPY** operations. Unfortunately, the Fortran standard has no feature to achieve this. Nevertheless, some compiler vendors introduce specialized statements into the Fortran code to specify alignment properties of variables and arrays.

In case of the Intel® Fortran compiler [1], we can set the alignment for an array **AL** to 64 bytes by

```
!DIR$ ATTRIBUTES ALIGN: 64: AL
```

Similar techniques are supported by the IBM XLF compiler, but so far not by the GNU Fortran compiler. Beside specifying the alignment, the leading dimension of the array must be a multiple of the alignment to get a vectorized code. Furthermore, the replacement of the rank-one updates

```
CALL GER(-1.0, WORK, B(L,L+1:N), X(1:M,L+1))
```

by

```
DO I = L+1, N
    X(1:M,I) = X(1:M,I) - B(L,I)*WORK(1:M)
END DO
```

enables the compiler to use the alignment information there, as well. The overhead of calling this function is reduced, as well. Figure 6.6 shows the influence of the local copies together with the data alignment. We see that we accelerate the algorithm further by 13.7%. The peaks at dimensions close to $m = 512 \cdot g$, $g \in \mathbb{N}$ are removed as well. Compared to the recursive blocking scheme our optimized approach is now 2.48 times faster.

Remark 6.9:
If the code has to handle Sylvester-type equations with a transpose operation, like $AXB^H + CXD^H = Y$, the local copies need to contain the transposed data. If we handle the transposed case as in BLAS or LAPACK, where the transposition is done implicitly using an index switch, the compiler's automatic

vectorization would not be able to create proper code. This is caused by the need to access matrices in a row-wise fashion. Loading parts of a row to vector registers is expensive since it is non-continuous data.

Sorting Eigenvalues The last paragraph showed that using a proper data alignment in the solver for the small subproblems in Algorithm 20 increases the performance. Unfortunately, handling real problems, the Schur decomposition has 2×2 blocks on the diagonal, compare Theorem 2.11. If such a 2×2 block is split by an equally sized block partitioning, e.g. $\exists k \in \mathbb{N}$ such that $A_{kM_B+1,kM_B} \neq 0$, we have to adjust the block size for the current block by ±1. In this way, some alignment properties and data access properties get lost. Without loss of generality, we assume M_B and N_B to be greater than or equal to two and be a multiple of the CPU's vector register length. By adjusting the current block size by $\mp$ if a 2×2 block would be split, is required to keep the algorithmic approach working in real arithmetic. Then, after the block size is adjusted once, the access to all not yet processed blocks will be shifted by one element. This includes the solution of the small system, as well as the updates in Step 7 and Step 9 of Algorithm 20. The small subproblem solvers are disturbed only once, while creating the initial copy and the final copy of the result. However, if the number of rows gets adjusted by -1, the vector registers will no longer be filled completely when solving a subproblem. The adjustment of the number of columns N_B by -1, has only a smaller impact, since the implementation focuses on the column major storage and acts on columns as much as possible. The highly tuned implementations for level-3 operations in the BLAS libraries detect if the input data is aligned or not and choose the best fitting algorithm. If we provide aligned data, the BLAS library gives an increased performance for smaller input data [81].

A way to solve this problem was already presented in Section 2.3, where we presented the reordering of the eigenvalues on the diagonal of the generalized Schur form. Since we only showed how to exchange two neighboring eigenvalues, we need the following Lemma to obtain a solution for the alignment problem.

Lemma 6.10:
Let (A, C) be a real matrix pair of order $M_B + 1$ in generalized Schur form, M_B even, and $A_{M_B+1,M_B} \neq 0$. Then there exists a two-sided transformation consisting of Q and Z such that

$$Q\,(A, C)\,Z^{\mathsf{H}} = \left(\tilde{A}, \tilde{C}\right)$$

with Q and Z unitary and

$$\tilde{A}_{M_B+1,M_B} = 0$$

and $\left(\tilde{A}, \tilde{C}\right)$ still in generalized Schur form.

Proof. In Section 2.3 we recalled the techniques shown by Kågström [102] to exchange two real eigenvalues, a real eigenvalue and a complex conjugate pair, or two complex conjugate pairs on the diagonal of the generalized Schur form using unitary transformations. Since we have $A_{M_B+1,M_B} \neq 0$ and M_B is even, there must exist an odd index $1 \leq k < M_B$ such that $A_{k+1,k} = 0$. Let k^* be the largest possible index. Then we have two possibilities. First, $k = M_B - 1$ and the unitary transformation can be found as in Section 2.3. Second, if $k < M_B - 1$, then we exchange the eigenvalue at index k^* with its right neighbor which is a conjugate complex eigenvalue pair by definition. This is again achieved using unitary transformations, compare Section 2.3. Then k^* is either $M_B - 1$ or we can exchange it again with its right neighbor and so on. Since the unitary transformations can be embedded into an identity and the set of unitary matrices form a group, we obtain the transformation Q and Z as product of all performed transformations. □

The constructive proof of Lemma 6.10 shows how to create a transformation such that 2×2 blocks on the diagonal of the generalized Schur form can be moved in order to not get split by

---

Algorithm 21 Reorder Eigenvalues

Input: (A, C) of order m and an even block size M_B.

Output: Q and Z unitary and $Q(A, C) Z^\mathsf{T} = \left(\tilde{A}, \tilde{C}\right)$ such that $\tilde{A}_{jM_B+1, jM_B} = 0$ for all $j \in \mathbb{N}$.

    $Q \leftarrow I$ and $Z \leftarrow I$

    for $k = M_B, 2M_B, \ldots, m$ **do**

        Find the largest index k_s such that $k_s < k$ and $A_{k_s+1, k_s} = 0$.

        for $l = k_s, k_s + 2, \ldots, k - 1$ **do**

            Compute $\tilde{Q}$ and $\tilde{Z}$ to exchange the eigenvalue at index l with the complex conjugate eigenvalue pair at index $l + 1$ and $l + 2$ using LAPACK TGEX2.

            Update Q, Z, A, and C using $\tilde{Q}$ and $\tilde{Z}$

        end for

    end for

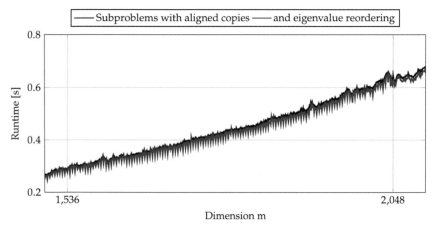

Figure 6.7.: Performance influence of sorted eigenvalues on the level-3 solver with aligned local copies in the level-2 part.

the block partitioning (6.40). Although Lemma 6.10 provides this only for a single block, we can use it to start with the first diagonal block and handle the blocks down the diagonal subsequently. With the help of the LAPACK routine TGEX2, which implements the procedure from Section 2.3, we obtain Algorithm 21 to move all 2×2 blocks away from the block boundaries.

If we apply Algorithm 21 to both matrix pairs (A, C) and (D, B) before solving the generalized Sylvester equation, our level-3 approach including all previously derived optimizations, we obtain the runtimes shown in Figure 6.7. On average this gives a speed up of 2.2%.

Remark 6.11:

The presented optimizations starting by choosing the best order to traverse the solution matrices up to the local copies and compiler assisted vectorization are applicable to all equations mentioned at the beginning of this chapter. In case of the symmetric equations, like Lyapunov and Stein Equation, we only have to solve for the upper or lower triangle of the solution and restore the remaining parts by symmetry. For example considering the generalized Lyapunov equation, small subproblems like

$$A_{ii} X_{ii} C_{ii}^\mathsf{H} + C_{ii} X_{ii} A_{ii}^\mathsf{H} = \tilde{Y}_{ii} \tag{6.49}$$

Algorithm 22 Block Algorithm for the Generalized Sylvester Equation – with instant updates

---

Input: (A, C) of order m and (D, B), Y of dimension $m \times n$. A and D (quasi) upper triangular, B and C upper triangular, block sizes M_B and N_B.

Output: Y overwritten with X solving $AXB + CXD = Y$.

1: **for** $l = 1, 1 + N_B, \ldots, n$ **do**
2:     $l_h = \min\{N_B, n - l + 1\}$
3:     $k^* = \left\lfloor \frac{m-1}{M_B} \right\rfloor M_B + 1$
4:     **for** $k = k^*, k^* - M_B, \ldots, 1$ **do**
5:        $k_h = \min\{M_B, m - k + 1\}$
6:        Solve $A_{k:k_h,k:k_h} X_{k:k_h,l:l_h} B_{l:l_h,l:l_h} + C_{k:k_h,k:k_h} X_{k:k_h,l:l_h} D_{l:l_h,l:l_h} = Y_{k:k_h,l,l_h}$ and overwrite $Y_{k:k_h,l:l_h}$
         with $X_{k:k_h,l:l_h}$.
7:        $Y_{1:k-1,l:l_h} \leftarrow Y_{1:k-1,l:l_h} - A_{1:k-1,k:k_h} X_{k:k_h,l:l_h} B_{l:l_h,l:l_h} - C_{1:k-1,k:k_h} X_{k:k_h,l:l_h} D_{l:l_h,l:l_h}$
8:        $Y_{k:k_h,l_h+1:n} \leftarrow Y_{k:k_h,l_h+1:n} - A_{k:k_h,k:m} X_{k:m,l:l_h} B_{l:l_h,l_h+1:n} - C_{k:k_h,k:m} X_{k:m,l:l_h} D_{l:l_h,l_h+1:n}$
9:     **end for**
10: **end for**

---

appear. In this case, we use the solver for the Lyapunov equation again. This is necessary to guarantee the symmetry of X_{ii} and hence the symmetry of the overall solution X. If the solver for (6.49) does not guarantee the symmetry, like the solver for the generalized Sylvester equation, small perturbations will cause an asymmetry in X_{ii}. This error propagates to the remaining parts of the solution, where X_{ii} is involved in the right-hand side. Even for small block sizes this lead to large errors and wrong solutions as observed by Köhler and Saak in [112].

6.3.3. Explicit Parallelization using OpenMP

Until now our implementation relies on the parallelization inside the BLAS and LAPACK subroutines. Especially, Algorithm 20 makes use of the parallel implementation of the matrix-matrix products. Due to the small size of the inner subproblems, these solves will be done purely sequentially, which limits the overall performance of the algorithm. Like for specialized QR decompositions in Section 4.3.2, we will derive a task based parallelization scheme for the generalized Sylvester equation, here.

Since the task parallelization does not rely on the parallelization capabilities of the BLAS library, there is no need to maximize the size of the involved matrix-matrix multiplies. This allows us to reformulate the Update (6.46) to remove the large updates in Step 9 of Algorithm 20. This yields

$$Y_{i,j+1:q} \leftarrow Y_{i,j+1:q} - A_{i,i:p} X_{i:p,j} B_{j,j+1:q} - C_{i,i:p} X_{i:p,j} D_{j,j+1:q}, \qquad (6.50)$$

which can be computed once X_{ij} is known. Now, we can update the remaining blocks in the right-hand side Y, immediately, without waiting until a block column is computed completely. Algorithm 22 shows this modification. It uses less large matrix-matrix multiplies but it is well suited for task based parallelization.

Since the task based parallelization approach needs a graph representing the dependencies between the computations, compare Section 4.3.2, we define a set of conditions to include nodes and edges in the directed acyclic graph. We see that once X_{ij} is known, we can already update $Y_{i,j+1:q}$ and $Y_{1:i-1,j}$. This leads to the following conditions:

1. The $(p, 1)$ block of the solution can be determined directly.

2. The (p, l) block, $1 < l \leq q$, can be processed once the $(p, l - 1)$ block is done.

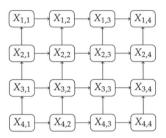

Figure 6.8.: Directed Acyclic Graph for the generalized Sylvester equation.

3. The $(k, 1)$ block, $p > k \geq 1$, can be processed once the $(k + 1, 1)$ block is computed.

4. The remaining blocks (k, l), $1 \leq k < p$ and $q \geq l > 1$, can be processed once the blocks $(k + 1, l)$ and $(k, l - 1)$ are computed.

Figure 6.8 shows the underlying graph. Using theses conditions, the graph can be implemented easily using the OpenMP task depend statement [138], like for the QR decomposition in Subsection 4.3.2. Again, we have to introduce pseudo edges as in-out dependencies into the graph to guarantee the execution order. A practical implementation is similar to the one of the specialized QR decomposition in Listing 4.4.

Since the updates in Step 7 and 8 of Algorithm 22 need to be computed block by block to create a proper graph, each update task will either compute

$$Y_{i,l:l_h} \leftarrow Y_{i,l:l_h} - A_{i,k:k_h} X_{k:k_h,l:l_h} B_{l:l_h,l:l_h} - C_{i,k:k_h} X_{k:k_h,l:l_h} D_{l:l_h,l:l_h}, \quad 1 \leq i < k,$$

or

$$Y_{k:k_h,j} \leftarrow Y_{k:k_h,j} - A_{k:k_h,k:m} X_{k:m,l:l_h} B_{l:l_h,j} - C_{k:k_h,k:m} X_{k:m,l:l_h} D_{l:l_h,j}, \quad l_h < j \leq n,$$

which leads to multiple computations of

$$X_{k:k_h,l:l_h} B_{l:l_h,l:l_h} \quad \text{and} \quad X_{k:k_h,l:l_h} D_{l:l_h,l:l_h}$$

and

$$A_{k:k_h,k:m} X_{k:m,l:l_h} \quad \text{and} \quad C_{k:k_h,k:m} C_{k:m,l:l_h}.$$

Hence, a naive implementation of the right-hand side updates will perform a lot more operations than necessary. We avoid this by introducing four auxiliary matrices $\hat{W}, \tilde{W}, \hat{U}$, and $\tilde{U}$ of dimension $m \times n$ with the same block partitioning as X and Y. After we solved for the block $X_{k:k_h,l:l_h}$ we precompute

$$\tilde{W}_{k:k_h,l:l_h} = A_{k:k_h,k:m} X_{k:m,l:l_h},$$
$$\hat{W}_{k:k_h,l:l_h} = C_{k:k_h,k:m} X_{k:m,l:l_h},$$
$$\tilde{U}_{k:k_h,l:l_h} = X_{k:k_h,l:l_h} B_{l:l_h,l:l_h},$$

and

$$\hat{U}_{k:k_h,l:l_h} = X_{k:k_h,l:l_h} D_{l:l_h,l:l_h}.$$

Additional dependencies to $\tilde{W}, \hat{W}, \tilde{U}$, and $\hat{U}$ are introduced into the dependency list for the updates in Steps 7 and 8. Although this requires $4mn$ memory, we use this scheme for the sake

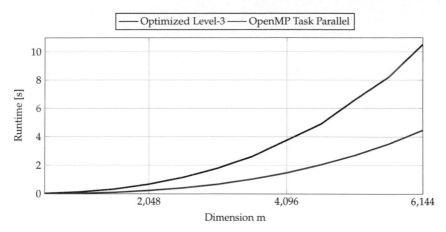

Figure 6.9.: Performance comparison of the optimized level-3 solver and the OpenMP 4 task parallel implementation.

of simplicity. Otherwise, we have to implement a complicated memory management here, which cannot be represented with OpenMP directives, directly.

This idea together with the aligned local copy approach for the small subproblems and the eigenvalue reordering leads to the performance shown in Figure 6.9. We see an additional performance gain factor of 2.36 compared to the standard level-3 version in Figure 6.7, which was only using the parallelization provided by the BLAS library.

Further acceleration techniques like using GPU accelerators are not evaluated here, since we have to fall back to the level-3 algorithm approach. The explicit parallelization using OpenMP tasks is not possible with OpenMP's accelerator interface.

6.4. Triangular Solvers and the Approximated Schur Decomposition

In the previous section we optimized the level-2 and level-3 parts of the block solver for the triangular systems. Now, we switch back to general matrices again, where the generalized Schur form is necessary to rewrite our problem such that it can be solved efficiently. Instead of using the QZ algorithm, we employ our Divide-Shift-and-Conquer and the Divide-Scale-and-Conquer algorithms from Chapters 3 and 4 to obtain the (quasi) upper triangular matrices A, B, C, and D. However, our algorithms are not as accurate as the QZ algorithm in all cases, as shown in the numerical results in Chapter 5. The approximation error of the generalized Schur form propagates into the solution of the generalized Sylvester equations, as well.

One way to improve the accuracy of the solution of the generalized Sylvester equation becomes evident when we recall Theorem 6.2, to connect the matrix equation with the corresponding standard linear system. We know that the generalized Sylvester equation

$$AXB + CXD = Y$$

is equivalent to the solution of the standard linear system

$$\mathcal{A}\,\mathrm{vec}(X) = \left(B^{\mathsf{T}} \otimes A + D^{\mathsf{T}} \otimes C\right)\mathrm{vec}(X) = \mathrm{vec}(Y)\,.$$

For standard linear systems $Ax = b$ it is known that the accuracy can be improved using an iterative refinement approach. Thus, it can also be applied to the generalized Sylvester equation, as well. The iterative refinement procedure for $Ax = b$ was studied many times in the past decades. Several convergence results were obtained by Wilkinson [174], Moler [128], and Stewart [156]. Newer results presented by Demmel [62] show a connection to Newton's method applied to $f(x) = b - Ax$. Higham [88] derived error bounds if all computations are performed in single or in double precision. The mixed precision case was handled by Buttari et al. [50] during the development of fast linear system solvers on hardware platforms that allow fast single precision computations. Carson and Higham [52] showed how three precision levels can be used to recover even sharper error bounds.

The general iterative refinement scheme for a linear system $Ax = b$ of dimension n is given by

1: $x_0 \leftarrow 0$
2: **for** $k = 0, 1, 2, \ldots$ **do**
3:     $r_k = b - Ax_k$
4:     Solve $Aw_k = r_k$ approximately
5:     Update $x_{k+1} = x_k + w_k$
6: **end for**

If Step 3 and Step 5 are evaluated in higher precision, i.e. double precision, and Step 4 in lower precision, i.e. single precision, Buttari et. al. [50] showed:

$$\lim_{k \to \infty} \|x - x_k\| \le \beta \left(1 - \alpha\right)^{-1} \|x\|, \tag{6.51}$$

where $\alpha = \psi(n)\kappa(A)\epsilon_s$ and $\beta = \xi(n)\kappa(A)\epsilon_d$. Thereby, $\psi(n)$ and $\xi(n)$ are slowly growing polynomials in n and ϵ_s and ϵ_d are the machine precision in single and double precision, respectively. Thus, if A is not too ill-conditioned with respect to the single precision machine precision [50], we have

$$\beta \left(1 - \alpha\right)^{-1} < 1$$

and the iterative refinement will converge up to the accuracy which can be achieved in double precision.

For the generalized Sylvester equation $AXB + CXD = Y$, the iterative refinement reads

1: $X_0 = 0$
2: **for** $k = 0, 1, 2, \ldots$ **do**
3:     $R_k = Y - AX_kB - CX_kD$
4:     Solve $AW_kB + CW_kD = R_k$ approximately
5:     Update $X_{k+1} = X_k + W_k$
6: **end for**

If we use the triangular solution techniques we can reuse the generalized Schur decompositions in every iteration step. This reduces the computation time, and we benefit from the optimized triangular level-3 solver from Section 6.3. Since the computation of the generalized Schur decomposition consumes most of the time, solving two or three times with the triangular solver will not increase the runtime dramatically. Integrating the precomputation of the generalized Schur decomposition into the iterative refinement we obtain Algorithm 23. Although the matrices A_S, B_S, C_S, and D_S are an approximation and thus we only obtain an approximate solution W_k, we perform the correction with the help of the original data. Regarding the approximation error for the generalized Schur decomposition from Chapter 5, we see that upper triangular matrices are more accurate than truncating the result of the QZ algorithm single precision. In this way, as long as the matrix $(A_S \otimes B_S + C_S \otimes D_S)$ does not get too ill-conditioned with respect to ϵ_s, Algorithm 23 will

---

Algorithm 23 Iterative refinement for the generalized Sylvester equation

1: Compute $\left(QA_SZ^H, QC_SZ^H\right) = (A, C)$
2: Compute $\left(UD_SV^H, UB_SV^H\right) = (D, B)$
3: $X_0 = 0$
4: **for** $k = 0, 1, \dots$ **do**
5: $\quad R_k = Q^H\left(Y - AX_kB - CX_kD\right)V$
6: $\quad$ Use a triangular solver to solve $A_SW_kB_S + C_SW_kD_S = R_k$
7: $\quad$ Update $X_{k+1} = X_k + ZW_kU^H$
8: **end for**

---

converge as for the classical linear system. The additional operations, namely the multiplication with Q, Z, U, and V, do not influence the accuracy since they are unitary matrices.

Regarding the convergence criteria for the iterative refinement, we adjust the ones proposed in LAPACK [10]. First, they allow a maximum number of 30 iterations to be taken. Second they use the following residual check for $\mathcal{A}x = b$ of dimension n to guarantee convergence:

$$\|r_k\| \leq \sqrt{n}\,\|x\|\,\|\mathcal{A}\|\,\epsilon_d. \tag{6.52}$$

Since we have $\mathcal{A} = \left(B^\mathsf{T} \otimes A + D^\mathsf{T} \otimes C\right)$ as system matrix in our case, we use the 2-norm to handle the Kronecker product properly. This leads to the following condition:

$$\|\mathrm{vec}(R_k)\|_2 \leq \sqrt{mn}\,\|\mathrm{vec}(X_k)\|_2 \left\|B^\mathsf{T} \otimes A + D^\mathsf{T} \otimes C\right\|_2 \epsilon_d.$$

By replacing the vector norms with the matrix valued counterpart, we obtain

$$\|R_k\|_F \leq \sqrt{mn}\,\|X_k\|_F \left\|B^\mathsf{T} \otimes A + D^\mathsf{T} \otimes C\right\|_2 \epsilon_d. \tag{6.53}$$

The quantity $\left\|B^\mathsf{T} \otimes A + D^\mathsf{T} \otimes C\right\|_2$ cannot be computed cheaply, because either we have to set up the matrix or use an iterative eigenvalue solver. Hence, we want to approximate the value by a cheaper norm. Therefore, we use

$$\left\|B^\mathsf{T} \otimes A + D^\mathsf{T} \otimes C\right\|_2 \leq \left\|B^\mathsf{T} \otimes A\right\|_2 + \left\|D^\mathsf{T} \otimes C\right\|_2$$
$$= \|A\|_2\,\|B\|_2 + \|C\|_2\,\|D\|_2.$$

Since $\|A\|_2 \leq \|A\|_F$ we have

$$\left\|B^\mathsf{T} \otimes A + D^\mathsf{T} \otimes C\right\|_2 \leq \|A\|_F\,\|B\|_F + \|C\|_F\,\|D\|_F, \tag{6.54}$$

which can be computed cheaply. In order to avoid overflows during the computation of the Frobenius norm, we can use the Hölder inequality $\|A\|_2 \leq \sqrt{\|A\|_1\,\|A\|_\infty}$ to approximate the involved two-norms. Finally, this leads to the convergence criterion

$$\|R_k\|_F \leq \sqrt{mn}\,\|X_k\|_F \left(\|A\|_F\,\|B\|_F + \|C\|_F\,\|D\|_F\right)\epsilon_d\tau, \tag{6.55}$$

where τ is an additional security factor typically set to one. In a practical implementation we will perform one step more after the criterion (6.55) is fulfilled. Since Demmel [62] showed that the iterative refinement is a Newton scheme, this will not worsen the result. Furthermore, if we have computed R for the residual check, the most expensive part of the next iteration step is already done. Thus, we secure our solution by finalizing the iteration step. Regarding the remaining matrix equations, mentioned at the beginning of this chapter, we obtain the convergence criteria

---

Equation	Criterion
GSYLV (GSYLV)	$\|R_k\|_F < \sqrt{mn}\, \|X_k\|_F \left(\|A\|_F \|B\|_F + \|C\|_F \|D\|_F\right) \epsilon_d \tau$
SYLV (SYLV)	$\|R_k\|_F < \sqrt{mn}\, \|X_k\|_F \left(\|A\|_F + \|D\|_F\right) \epsilon_d \tau$
DSYLV (DSYLV)	$\|R_k\|_F < \sqrt{mn}\, \|X_k\|_F \left(\|A\|_F \|B\|_F + 1\right) \epsilon_d \tau$
LYAP (LYAP)	$\|R_k\|_F < 2m\, \|X_k\|_F \|A\|_F\, \epsilon_d \tau$
GLYAP (GLYAP)	$\|R_k\|_F < 2m\, \|X_k\|_F \|A\|_F \|C\|_F\, \epsilon_d \tau$
STEIN (STEIN)	$\|R_k\|_F < m\, \|X_k\|_F \|A\|_F^2\, \epsilon_d \tau$
GSTEIN (GSTEIN)	$\|R_k\|_F < m\, \|X_k\|_F \left(\|A\|_F^2 + \|C\|_F^2\right) \epsilon_d \tau$
GCSYLV (GCSYLV)	$\left\|\left[R_k^{(1)}\ R_k^{(2)}\right]\right\|_F < \sqrt{mn}\, \left\|\left[X_k^{(1)}\ X_k^{(2)}\right]\right\|_F \left(\|A\|_F + \|B\|_F + \|C\|_F + \|D\|_F\right) \epsilon_d \tau$

Table 6.2.: Convergence criteria for the iterative refinement.

given in Table 6.2. Thereby, the residual of the generalized coupled Sylvester equation splits into two parts as the right-hand side and the solution.

In this way, we use the iterative refinement, well-known for classical linear systems, to reduce the influence of our approximated solution of the generalized eigenvalue problem. Furthermore, regarding the theory of Buttari et al. [50] one can think of a mixed precision solver for the generalized Sylvester equation.

NUMERICAL EXPERIMENTS FOR THE MATRIX EQUATIONS

Contents

7.1. Hardware and Software Setup

The numerical experiments for the solution of the matrix equations are performed using the same hardware and software setup as for the generalized eigenvalue problem, described in Section 5.1. The only difference is that no GPUs are used in the implementation of our triangular solvers since no GPU aware algorithms are presented in Chapter 6.

In order to measure the performance increase, we compare our implementation with existing software. Since only the RECSY [100, 101] library and the SLICOT [29] library provide solvers for nearly the full set of equations mentioned in Chapter 6, we use them for comparison. Existing old implementations by Bartels and Stewart [19] or by Gardiner et al. [77] cover only a few special cases. The implementations in LAPACK cover only a subset of the mentioned equations as well.

Our algorithms are implemented in Fortran 95/2003 to be able to use the modern language features, which allow better optimizations by the compiler. Tuning parameters, like the block sizes, are set at runtime using a configuration script written in LUA [95] similar to the interface describe in Section 5.1.2. In this way, machine specific parameters can be set without recompiling the whole software. Our library implements the concepts presented in Chapter 6 for the equations mentioned there, including all reasonable transposition cases. This allows to compare the performance for all equations with RECSY, which is the fastest library previously known to us.

The determination of the optimal block sizes and the performance evaluation are done using random matrix pencils with uniform $(-1, 1)$ distribution of the entries. The matrices are reduced to generalized Schur decomposition with the help of the QZ algorithm. Later, for testing the iterative refinement, this step is done using the Divide-Shift-and-Conquer and the Divide-Scale-and-Conquer algorithms. Since most of the real world example problems used in Chapter 5 have their origin in system theory, modeling of physical processes, or (semi) discretization of partial differential equations, we use them as coefficient matrices of a generalized Lyapunov equation. The BCSST 13 problem is not used here, since it has infinite eigenvalues and, i.e., one of the coefficient matrices is singular. The corresponding matrix equations require additional analysis before they are useful, e.g. [159, 160].

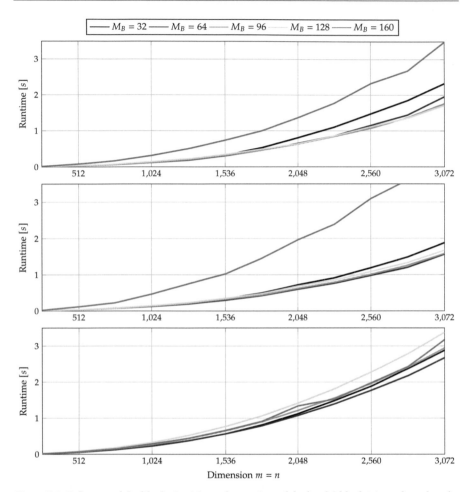

Figure 7.1.: Influence of the block size M_B on the runtime of the level-3 block triangular solver for GSYLV. (From top to bottom: Haswell, Skylake, and POWER 8 system)

7.2. Performance of Triangular Solvers

Since our triangular solvers, the level-3 approach and the explicitly parallelized one, rely on a block partitioning of the involved matrices, it is necessary to determine the optimal partitioning in dependence of the hardware parameters. Figures 7.1 and 7.2 show the influence of the block sizes on the runtime for solving the generalized Sylvester equation measured on our three hardware systems. Regarding the level-3 case, in Figure 7.1, we see that a moderate block size of 64 leads to reasonable runtimes for almost all problem dimensions, independent of the hardware architecture. On both Intel® systems, a too large block size yields an intermediate performance loss, while the

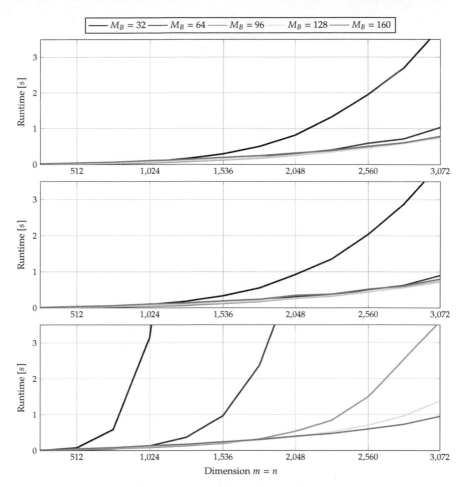

Figure 7.2.: Influence of the block size M_B on the runtime of the OpenMP 4 triangular solver for (GSYLV). (From top to bottom: Haswell, Skylake, and POWER 8 system)

remaining choices still lead to a good performance. The POWER 8 system behaves slightly different in the experiments. The runtime for a fixed problem dimension is not influenced as much by the block size as on the Intel® systems. Comparing the Intel® and the POWER 8 systems, we see that on the Intel® systems the optimal block size lies between 64 and 128 while the POWER 8 system benefits from smaller block sizes like 32 or 64. Using the explicitly parallelized implementation for the solution of the triangular generalized Sylvester equation, we obtain Figure 7.2. Besides the runtime reduction in the optimal cases, we see that a bad selection of the block size yields a worse performance than in the level-3 case before. On both Intel® systems, this only occurs for a too small block size, like 32, while the larger ones still provide a good performance. On the POWER 8

Equation	$M_B = 32$	$M_B = 64$	$M_B = 96$	$M_B = 128$	$M_B = 160$
SYLV	$m < 512$	$m < 2304$	$m \geq 2304$	–	–
DSYLV	–	$m < 1792$	$m < 3072$	$m \geq 3072$	–
LYAP	–	$m < 3072$	$m \geq 3072$	–	–
STEIN	–	$m < 3072$	$m \geq 3072$	–	–
GSYLV	–	$m < 2048$	$m < 2816$	$m \geq 2816$	–
GCSYLV	–	$m < 2048$	$m \geq 2048$	–	–
GLYAP	–	$m < 2048$	$m \geq 2048$	–	–
GSTEIN	–	$m < 2048$	$m \geq 2048$	–	–

Table 7.1.: Optimal block size for the level-3 triangular solvers, Haswell system.

Equation	$M_B = 32$	$M_B = 64$	$M_B = 96$	$M_B = 128$	$M_B = 160$
SYLV	–	$m > 0$	–	–	–
DSYLV	–	$m < 2304$	$m \geq 2304$	–	–
LYAP	–	$m > 0$	–	–	–
STEIN	–	$m > 0$	–	–	–
GSYLV	–	$m > 0$	–	–	–
GCSYLV	–	–	$m > 0$	–	–
GLYAP	–	$m > 0$	–	–	–
GSTEIN	–	$m > 0$	–	–	–

Table 7.2.: Optimal block size for the level-3 triangular solvers, Skylake system.

system, smaller block sizes result in a too large runtime when the problem dimension increases. In this case selecting a large block size, like 128 to 160, will lead to the maximum performance.

Since the eight presented matrix equations in Chapter 6 differ in structure and the amount of required data for the solution, we repeat the optimal block size benchmarks for the remaining equations as well. The results are shown in Tables 7.1, 7.2, and 7.3 for the level-3 approaches and in Tables 7.4, 7.5, and 7.6 for the explicitly parallelized implementation. In case of the level-3 approach on the Intel® systems, we see that choosing a block size of 64 is almost always a good choice. Only on the older Haswell system, we have to increase the block size to 96 for an increasing problem dimension. The POWER 8 system requires only a block size of 32 or 64, as already seen in Figure 7.1. Especially, the standard Sylvester equation (SYLV), the discrete time Sylvester equation (DSYLV), and the standard Lyapunov equation (LYAP) lead to an optimal block size of 32, independent of the problem dimension. As already shown for the generalized Sylvester equation, the situation changes when we move on to the explicitly parallelized implementation. There, a more detailed selection of the optimal block size is required, as shown in Tables 7.4, 7.5, and 7.6. In case of the Intel® systems, slightly larger block sizes, compared to the level-3 approach, are required on average to achieve the optimal performance. On the POWER 8 system, we already need a block size of 160 for moderate problem sizes. From Figure 7.2, we know that a too small block size will lead to a large performance loss. The boundaries given in Tables 7.1 to 7.6 are implemented in the LUA configuration file to select the block size at runtime.

After determining the optimal block sizes, we compare the performance with the existing software. For the comparison, we use the RECSY software package and the generalized Sylvester equation. A benchmark presented by Jonsson and Kågstöm [100, 101] shows that RECSY is competitive against other existing software, especially old implementations and hence the old implementations are not covered here. Figure 7.3 shows the performance of our triangular generalized Sylvester solvers compared to the RECSY implementation with and without explicit parallelization. First, we see that the explicit parallelization done in the RECSY software package only leads to

Equation	$M_B = 32$	$M_B = 64$	$M_B = 96$	$M_B = 128$	$M_B = 160$
SYLV	$m > 0$	–	–	–	–
DSYLV	$m > 0$	–	–	–	–
LYAP	$m > 0$	–	–	–	–
STEIN	$m < 2560$	$m \geq 2560$	–	–	–
GSYLV	$m < 1536$	$m \geq 1536$	–	–	–
GCSYLV	–	$m > 0$	–	–	–
GLYAP	$m < 2816$	$m \geq 2816$	–	–	–
GSTEIN	$m < 2304$	$m \geq 2304$	–	–	–

Table 7.3.: Optimal block size for the level-3 triangular solvers, POWER 8 system.

Equation	$M_B = 32$	$M_B = 64$	$M_B = 96$	$M_B = 128$	$M_B = 160$
SYLV	$m < 768$	$m < 2048$	$m < 2816$	$m \geq 2816$	–
DSYLV	$m < 768$	$m < 1536$	$m < 2816$	$m \geq 2816$	–
LYAP	$m < 1280$	$m < 3072$	$m \geq 3072$	–	–
STEIN	$m < 1024$	$m < 2048$	$m \geq 2048$	–	–
GSYLV	$m < 768$	$m < 1536$	$m < 2560$	$m \geq 2560$	–
GCSYLV	–	$m < 2560$	$m \geq 2560$	–	–
GLYAP	$m < 768$	$m < 2048$	$m < 2816$	$m \geq 2816$	–
GSTEIN	$m < 768$	$m < 2048$	$m < 2816$	$m \geq 2816$	–

Table 7.4.: Optimal block sizes for the explicitly parallelized triangular solvers, Haswell system.

a negligible performance increase. Furthermore, for large-scale problems, the variant with explicit parallelization is even slower than the one relying on the features of the BLAS library. The reason for this behavior is that once a part of a program runs inside an OpenMP parallel region, all BLAS and LAPACK routines are called without parallelization. In this way, the large OpenMP parallel regions, which are used in the RECSY library to set up their explicit parallelization approach, lead to a performance loss. Our explicitly parallelized implementation is not affected by this problem, since we focus on the creation of many tasks which then utilize all available CPU cores. In this way, our explicitly parallelized implementation does not slow down, unlike the RECSY approach, for an increasing problem size.

Tables 7.7 to 7.14 provide the comparison between RECSY and our implementations for the remaining equations on the Skylake system. Since the explicitly parallelized version of RECSY has only a marginal advantage over its non-parallelized one and the fact that for large problems, the explicit parallelized version is even slower, we do not use this for comparison. The tables show that our level-3 approach is constantly faster than the RECSY implementation. Depending on the structure, i.e. whether the updates on the right-hand side inside the algorithm are one or two-sided, the impact of our optimizations increases. Especially, for equations, where each part of the solution is multiplied from left and right with the coefficient matrices, we reach speedups of a factor of nearly two. Only for the LYAP and the CGSYLV case, we only reach a speedup slightly larger than one. Regarding our explicitly parallelized OpenMP 4 implementation we obtain a speedup of at least two up to a factor of 6.38. This shows that a BLAS level-3 formulation with optimized level-2 solvers leads to a high performance. However, the parallel capabilities of the BLAS library are not fine-grained enough to obtain the highest performance. The explicit parallelization, which takes care of the algorithmic structure outside each BLAS call, yields further performance improvements. The numerical results on the Haswell and on the POWER 8 system show the same qualitative behavior and thus are omitted here.

The numerical experiments show that we are able to construct fast triangular solvers for Sylvester-

Equation	$M_B = 32$	$M_B = 64$	$M_B = 96$	$M_B = 128$	$M_B = 160$
SYLV	$m < 1\,024$	$m < 2\,560$	$m \geq 2\,560$	–	–
DSYLV	$m < 768$	$m < 2\,048$	$m \geq 2\,048$	–	–
LYAP	$m < 1\,536$	$m \geq 1\,536$	–	–	–
STEIN	$m < 768$	$m < 2\,048$	$m \geq 2\,048$	–	–
GSYLV	$m < 512$	$m < 2\,048$	$m < 2\,816$	$m \geq 2\,816$	–
GCSYLV	$m < 2\,816$	$m \leq 2\,816$	–	–	–
GLYAP	$m < 768$	$m < 2\,048$	$m < 2\,816$	$m \geq 2\,816$	–
GSTEIN	$m < 768$	$m < 2\,048$	$m < 2\,816$	$m \geq 2\,816$	–

Table 7.5.: Optimal block size for the explicitly parallelized triangular solvers, Skylake system.

Equation	$M_B = 32$	$M_B = 64$	$M_B = 96$	$M_B = 128$	$M_B = 160$
SYLV	$m < 512$	$m < 1\,280$	$m < 2\,304$	$m < 2\,816$	$m \geq 2\,816$
DSYLV	$m < 512$	$m < 1\,280$	$m < 2\,304$	$m < 2\,816$	$m \geq 2\,816$
LYAP	$m < 768$	$m < 1\,792$	$m \geq 1\,792$	–	–
STEIN	$m < 512$	$m < 1\,280$	$m < 2\,816$	$m \geq 2\,816$	–
GSYLV	$m < 512$	$m < 1\,024$	$m < 1\,792$	$m < 2\,304$	$m \geq 2\,304$
CGSYLV	$m < 512$	$m < 1\,280$	$m < 2\,048$	$m < 2\,560$	$m \geq 2\,560$
GLYAP	$m < 512$	$m < 1\,280$	$m < 2\,304$	–	$m \geq 2\,304$
GSTEIN	$m < 512$	$m < 1\,280$	$m < 2\,304$	–	$m \geq 2\,304$

Table 7.6.: Optimal block size for the explicitly parallelized triangular solvers, POWER 8 system.

	RECSY	Level-3		OpenMP 4	
Dimension m	Runtime	Runtime	Speedup	Runtime	Speedup
$1\,024$	0.18	0.13	1.41	0.05	3.95
$2\,048$	0.84	0.55	1.55	0.16	5.21
$3\,072$	2.12	1.29	1.64	0.35	6.13
$4\,096$	4.33	2.45	1.77	0.74	5.84
$5\,120$	6.91	3.80	1.82	1.26	5.49
$6\,144$	11.20	5.78	1.94	2.15	5.20

Table 7.7.: Runtime and speedup of the SYLV solver on the Skylake system.

	RECSY	Level-3		OpenMP 4	
Dimension m	Runtime	Runtime	Speedup	Runtime	Speedup
$1\,024$	0.19	0.10	1.95	0.03	5.69
$2\,048$	0.82	0.46	1.78	0.14	5.68
$3\,072$	2.06	1.16	1.78	0.38	5.35
$4\,096$	3.87	2.46	1.57	0.90	4.32
$5\,120$	6.20	4.08	1.52	1.58	3.94
$6\,144$	9.64	6.61	1.46	2.74	3.52

Table 7.8.: Runtime and speedup of the DSYLV solver on the Skylake system

type matrix equations, as given in Chapter 6, without using the recursive block approach. Furthermore, reformulations in our level-3 BLAS approach enable us to use fine-grained parallelization techniques provided by modern compilers. Additionally, the rewritten level-2 solvers for the small subproblems can be optimized by modern compilers to adjust them to the used hardware properly.

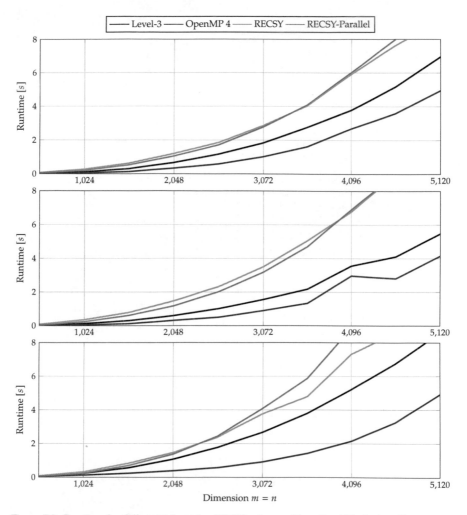

Figure 7.3.: Runtime for different triangular GSYLV solvers with optimal block size. (From top to bottom: Haswell, Skylake, and POWER 8 system.)

Dimension m	RECSY Runtime	Level-3 Runtime	Level-3 Speedup	OpenMP 4 Runtime	OpenMP 4 Speedup
1 024	0.08	0.06	1.22	0.03	2.77
2 048	0.32	0.27	1.20	0.09	3.66
3 072	0.78	0.64	1.23	0.20	3.83
4 096	1.39	1.28	1.09	0.67	2.10
5 120	2.28	1.96	1.16	0.81	2.80
6 144	3.45	3.05	1.13	1.57	2.20

Table 7.9.: Runtime and speedup of the LYAP solver on the Skylake system.

Dimension m	RECSY Runtime	Level-3 Runtime	Level-3 Speedup	OpenMP 4 Runtime	OpenMP 4 Speedup
1 024	0.12	0.06	2.07	0.03	4.40
2 048	0.54	0.25	2.15	0.10	5.58
3 072	1.32	0.59	2.22	0.24	5.46
4 096	2.58	1.31	1.97	0.54	4.80
5 120	3.97	1.98	2.01	0.91	4.35
6 144	6.20	3.31	1.88	1.56	3.97

Table 7.10.: Runtime and speedup of the STEIN solver on the Skylake system.

Dimension m	RECSY Runtime	Level-3 Runtime	Level-3 Speedup	OpenMP 4 Runtime	OpenMP 4 Speedup
1 024	0.33	0.14	2.42	0.05	6.38
2 048	1.42	0.67	2.13	0.26	5.47
3 072	3.49	1.57	2.22	0.72	4.84
4 096	6.80	3.43	1.98	1.72	3.96
5 120	10.64	5.39	1.97	3.06	3.47
6 144	16.60	8.93	1.86	5.42	3.06

Table 7.11.: Runtime and speedup of the GSYLV solver on the Skylake system.

Dimension m	RECSY Runtime	Level-3 Runtime	Level-3 Speedup	OpenMP 4 Runtime	OpenMP 4 Speedup
1 024	0.27	0.24	1.11	0.06	4.29
2 048	1.12	1.05	1.06	0.22	5.08
3 072	2.65	2.45	1.08	0.58	4.58
4 096	5.08	4.79	1.06	1.36	3.74
5 120	7.82	7.37	1.06	2.38	3.28
6 144	11.90	11.43	1.04	4.18	2.85

Table 7.12.: Runtime and speedup of the CGSYLV solver on the Skylake system.

Dimension m	RECSY Runtime	Level-3 Runtime	Level-3 Speedup	OpenMP 4 Runtime	OpenMP 4 Speedup
1 024	0.19	0.08	2.46	0.04	4.69
2 048	0.85	0.38	2.23	0.16	5.36
3 072	2.09	0.86	2.42	0.43	4.92
4 096	4.15	1.92	2.16	0.98	4.24
5 120	6.32	2.93	2.16	1.72	3.68
6 144	9.75	4.84	2.02	3.01	3.24

Table 7.13.: Runtime and speedup of the GLYAP solver on the Skylake system.

Dimension m	RECSY Runtime	Level-3 Runtime	Level-3 Speedup	OpenMP 4 Runtime	OpenMP 4 Speedup
1 024	0.20	0.09	2.10	0.04	4.87
2 048	0.89	0.44	1.99	0.16	5.58
3 072	2.12	1.00	2.11	0.43	4.98
4 096	4.21	2.22	1.89	0.98	4.28
5 120	6.45	3.41	1.89	1.73	3.73
6 144	10.19	5.76	1.77	3.01	3.38

Table 7.14.: Runtime and speedup of the GSTEIN solver on the Skylake system.

Problem	Haswell Runtime	Skylake Runtime	POWER 8 Runtime	Forward Error	Relative Residual
BCSST 07	1.01	1.01	1.23	5.02e-10	1.70e-14
BCSST 19	4.95	5.22	6.98	4.67e-07	7.50e-15
BCSST 09	15.15	17.49	19.79	1.57e-11	7.82e-14
BCSST 12	44.59	45.75	48.20	1.61e-08	2.08e-14
Filter 2D	57.76	51.69	65.24	1.31e-11	1.15e-12
BCSST 26	72.25	65.90	85.87	4.05e-10	7.46e-14
BCSST 23	346.45	302.54	482.67	9.43e-05	2.30e-14
BCSST 21	542.45	486.92	722.65	4.62e-11	1.85e-14
Steel 5	2 346.45	2 394.72	4 557.31	5.47e-12	1.82e-10
Flow Meter	8 638.64	10 266.94	55 493.85	2.51e-08	7.41e-13
BCSST 25	46 402.17	55 677.29	241 070.70	1.50e-04	1.29e-12
Chip Cooling	83 416.58	106 048.60	445 241.40	1.25e-09	1.99e-12
Steel 20	132 470.50	161 448.90	695 405.70	1.08e-10	2.21e-09
Gyro	out of mem.	454 857.42	3 077 919.70	4.51e+02	2.79e-11

Table 7.15.: Runtime and accuracy of the SLICOT Generalized Lyapunov solver
(POWER 8 Runtime for the Gyro example estimated.)

7.3. Solution of Generalized Lyapunov Equations

The solution of generalized Lyapunov equations is an important task in systems and control theory [12]. For this reason and due to the fact that some of our real world example matrices have their origin in systems and control theory, we use this equation for an overall test. This combines the fast approximation of the generalized Schur decomposition from Chapters 3 and 4 with the fast Sylvester-type matrix equation solvers. The reference result is obtained using the SLICOT library [29, 149], which is the foundation of the generalized Lyapunov equation solvers in MATLAB [28, 126], GNU Octave [164], and the SLICOT interface (Slycot) for Python. Beside the runtime we compare the quality of the results. Therefore, we use two measures. First, the normalized forward error defined as

$$\frac{\left\| X - \hat{X} \right\|_F}{\|X\|_F}, \tag{7.1}$$

where X is the true solution and $\hat{X}$ the computed one. In order to know X exactly, we set all entries of X to one and compute the resulting right-hand side Y. Afterwards we solve the equation and compare the computed $\hat{X}$ with the real solution. The second measure is the normalized residual defined by

$$\frac{\left\| Y - A\hat{X}C^H - C\hat{X}A^H \right\|_F}{\|Y\|_F}. \tag{7.2}$$

Table 7.15 shows the qualitative measures of the SLICOT generalized Lyapunov equation solver and the runtime on our three hardware systems. We see that beside the already long runtimes for computing the generalized Schur decomposition, compare Tables 5.17, 5.18, and 5.19 in Chapter 5, the runtime of the triangular solver has a large impact on the overall runtime as well. Especially on the POWER 8 system, Table 7.15 shows a dramatic increase of the runtime for the large-scale problems. In general, for problems larger than 10 000, like BCSST 25, Chip Cooling, Steel 20, or Gyro, the solution takes more than half a day up to several days on the Skylake and the POWER 8 system. In this way solving generalized Lyapunov equations directly is impractical even on current computer systems. Regarding the accuracy, the Gyro example is critical. Here, the SLICOT solver is not able to achieve a relative forward error smaller than one, although it reaches a sufficient relative residual. Both examples, the Gyro and BCSST 25 are ill-conditioned and thus a though task for direct solution techniques anyway, as already shown in the results for the SLICOT reference results in Table 7.15. In the experiments with the iterative refinement, this problem still exists, but is a bit reduced in both cases.

We reduce the runtime of the overall procedure by first approximating the generalized Schur decomposition either by using the Divide-Shift-and-Conquer algorithm or the Divide-Scale-and-Conquer algorithm. Both use the randomization based subspace extraction since Chapter 5 showed that this results in the best runtimes. In order to extend the comparison to systems with and without GPUs, we include both cases in the experiments. Since the generalized Schur decomposition is only approximated, we use the iterative refinement technique, shown in Section 6.4, to improve the accuracy of the result. Thereby, we set the maximum number of iterations to 30 and the security factor τ in the stopping criterion to 0.1. The triangular equations are solved with the explicitly parallelized OpenMP 4 solver, since the previous section showed that this is the fastest solver. The Gyro example is solved only with the level-3 BLAS solver, since it needs less additional memory during the computation and thus can be used without employing the swap-memory for this example.

Tables 7.16, 7.17, and 7.18, show the results for the Divide-Shift-and-Conquer approach. We see that our approach yields a better forward error and a better residual for almost all example

Problem	without GPUs		with GPUs			Forward	Relative
	Runtime	Speedup	Runtime	Speedup	Iter.	error	residual
BCSST 07	0.66	1.54	0.68	1.49	2	7.47e-14	1.63e-15
BCSST 19	3.54	1.40	3.44	1.44	2	4.18e-12	5.18e-16
BCSST 09	2.55	5.94	3.14	4.82	2	2.75e-15	2.40e-15
BCSST 12	6.27	7.12	7.11	6.27	2	1.67e-11	1.88e-15
Filter 2D	7.42	7.79	7.87	7.34	2	2.17e-17	3.33e-14
BCSST 26	38.01	1.90	41.29	1.75	2	1.06e-15	2.06e-15
BCSST 23	138.98	2.49	121.51	2.85	2	7.00e-11	1.17e-15
BCSST 21	90.54	5.99	79.24	6.85	2	2.81e-18	1.29e-15
Steel 5	185.39	12.66	127.64	18.38	2	1.01e-17	1.65e-12
Flow Meter	1 004.35	14.44	617.04	23.50	2	6.19e-12	2.03e-14
BCSST 25	11 015.07	4.21	6 165.59	7.53	2	1.15e-07	8.21e-13
Chip Cooling	7 154.10	11.66	6 743.63	12.37	2	5.37e-15	3.18e-14
Steel 20	7 548.03	17.55	4 307.21	30.76	2	4.53e-18	2.92e-12
Gyro	out of memory.						

Table 7.16.: Iterative refinement for the generalized Lyapunov equation with Divide-Shift-and-Conquer approximation on the Haswell system.

Problem	without GPUs		with GPUs			Forward	Relative
	Runtime	Speedup	Runtime	Speedup	Iter.	error	residual
BCSST 07	0.78	1.29	0.79	1.27	3	7.59e-14	1.80e-15
BCSST 19	4.59	1.14	3.72	1.40	3	1.57e-09	5.15e-16
BCSST 09	3.13	5.59	3.42	5.12	3	2.79e-15	2.17e-15
BCSST 12	7.50	6.10	7.82	5.85	3	1.24e-11	1.82e-15
Filter 2D	8.27	6.25	8.29	6.23	3	4.74e-18	1.86e-14
BCSST 26	43.66	1.51	40.47	1.63	3	9.07e-16	2.02e-15
BCSST 23	137.20	2.21	114.23	2.65	3	5.86e-11	1.31e-15
BCSST 21	89.80	5.42	75.28	6.47	3	2.69e-18	1.22e-15
Steel 5	182.60	13.11	117.27	20.42	3	1.01e-17	2.11e-12
Flow Meter	982.07	16.63	481.08	33.95	4	1.63e-11	4.87e-14
BCSST 25	9 638.51	5.78	4 494.16	12.39	3	1.20e-05	2.08e-11
Chip Cooling	7 161.04	14.81	5 156.76	20.56	3	1.47e-15	2.63e-14
Steel 20	7 597.39	21.25	2 900.26	55.67	3	3.80e-18	4.50e-12
Gyro	37 086.13	12.26	15 698.41	28.97	3	3.44e-07	1.88e-15

Table 7.17.: Iterative refinement for the generalized Lyapunov equation with Divide-Shift-and-Conquer approximation on the Skylake system.

matrices. On the Skylake system only the BCSST 25 example results in a worse residual than the SLICOT approach but with a better forward error. Regarding the POWER 8 system the BCSST 25 has the same issues, too. Although its residual is not as good as in the SLICOT case, the result is still sufficient for most of the applications. The Gyro example shows much better relative residuals and forward errors than the SLICOT solution. Even if we use the iterative refinement approach together with the generalized Schur decomposition obtained from the QZ algorithm, we end up with a relative forward error of 2.34e-6. This is still one order of magnitude worse compared to solving the Lyapunov equation with the approximation of the Divide-Shift-and-Conquer algorithm. A reason for this behavior are perturbed eigenvalues computed by the QZ algorithm, compare the results for the Gyro example in Sections 5.3.1 and 5.3.2.

Problem	without GPUs		with GPUs		Iter.	Forward error	Relative residual
	Runtime	Speedup	Runtime	Speedup			
BCSST 07	0.61	2.03	0.61	2.01	2	3.49e-14	3.14e-15
BCSST 19	3.76	1.86	3.98	1.75	2	2.02e-11	1.06e-15
BCSST 09	2.70	7.33	3.02	6.56	2	5.37e-15	5.16e-15
BCSST 12	6.36	7.58	6.40	7.53	2	3.76e-11	4.71e-15
Filter 2D	7.80	8.36	7.57	8.62	2	3.90e-18	3.33e-14
BCSST 26	36.29	2.37	33.76	2.54	2	5.38e-15	4.16e-15
BCSST 23	125.52	3.85	87.51	5.52	2	2.43e-10	4.14e-15
BCSST 21	83.50	8.65	58.07	12.44	2	3.88e-18	6.06e-15
Steel 5	175.18	26.02	89.26	51.06	2	2.15e-18	2.10e-12
Flow Meter	881.40	62.96	373.22	148.69	2	5.86e-12	2.12e-14
BCSST 25	8 906.50	27.07	3 853.34	62.56	2	1.21e-05	1.22e-10
Chip Cooling	7 169.73	62.10	2 301.23	193.48	2	4.59e-15	3.28e-14
Steel 20	7 462.54	93.19	2 352.17	295.64	2	4.16e-18	6.18e-12
Gyro	38 795.52	79.33	12 176.88	252.79	3	1.66e-05	1.58e-14

Table 7.18.: Iterative refinement for the generalized Lyapunov equation with Divide-Shift-and-Conquer approximation on the POWER 8 system.

Regarding the number of required iterations in the iterative refinement we see that in most cases it is enough to call the triangular solver twice. A look at the runtime and the speedup shows that our approximation and iterative refinement approach reduces the runtime drastically. For large-scale problems, we achieve a speedup of 17 to 21 on the Intel® system without using GPUs. The usage of GPUs gives speedup of 30 to 56 there, as well. On the POWER 8 system, the speedups are even higher, up to 93.19 on the CPU, and 295.64 with the support of GPUs. The main reason for this is the unacceptable performance of the triangular generalized Lyapunov equation solver from SLICOT and the slow QZ procedure on the POWER 8 system, too.

Using the Divide-Scale-and-Conquer algorithm for the approximation of the generalized Schur decomposition Tables 7.19, 7.20, and 7.21 are obtained. The forward error as well as the relative residual are comparable to the values achieved by the Divide-Scale-and-Conquer algorithm in most cases. Furthermore, the comparison to the reference solution computed with SLICOT shows that our results are more accurate, except of a single case. The BCSST 19 example gives only a comparable forward error on the POWER 8 system, see Table 7.21. Regarding the Gyro example again, the relative forward error decreases even more. Compared to the initial solution with the help of SLICOT we gain eleven orders of magnitude. Even the iterative refinement with the generalized Schur decomposition, computed by the QZ algorithm of LAPACK, is three orders of magnitude worse than using the Divide-Scale-and-Conquer approximation.

The iterative refinement procedure needs only two iteration steps in all cases to reach the accuracy. Since the number of iteration steps is almost the same as in the Divide-Shift-and-Conquer case, the runtime and the speedup gets dominated by the runtime of the generalized Schur approximation. Nevertheless, we achieve large speedups for medium and large-scale problems on all systems. On the Skylake system, the long runtime for the Gyro example, caused by the insufficient GPU memory, influences the runtime for the Lyapunov equation as well. Again, the POWER 8 system achieves a much higher speedup due to the unsatisfying performance of the QZ algorithm, its higher performance during the Divide-Scale-and-Conquer algorithm, and the slow triangular solver in SLICOT, as explained in the Divide-Shift-and-Conquer case.

The results show that a good approximation of the generalized Schur decomposition can be used in combination with iterative refinement to accelerate the solution of matrix equations. Thereby,

Problem	without GPUs		with GPUs		Iter.	Forward error	Relative residual
	Runtime	Speedup	Runtime	Speedup			
BCSST 07	1.34	0.76	1.30	0.78	2	1.38e-12	2.66e-15
BCSST 19	7.69	0.64	7.70	0.64	2	8.74e-08	5.49e-16
BCSST 09	6.12	2.48	6.03	2.51	2	1.69e-17	1.41e-15
BCSST 12	11.19	3.99	10.77	4.14	2	2.59e-12	1.74e-15
Filter 2D	16.10	3.59	15.51	3.72	2	4.96e-18	2.93e-14
BCSST 26	56.62	1.28	57.19	1.26	2	1.17e-15	1.41e-15
BCSST 23	190.22	1.82	189.90	1.82	2	1.14e-07	9.40e-16
BCSST 21	132.15	4.10	130.26	4.16	2	1.93e-18	9.46e-16
Steel 5	189.85	12.36	156.28	15.01	2	6.54e-18	1.50e-12
Flow Meter	2 356.77	6.15	1 316.78	11.01	2	5.85e-12	2.07e-14
BCSST 25	17 815.44	2.60	8 147.44	5.70	2	1.45e-11	3.00e-15
Chip Cooling	21 258.05	3.92	9 751.57	8.55	2	2.56e-15	2.24e-14
Steel 20	15 158.92	8.74	7 318.30	18.10	2	5.99e-18	3.97e-12
Gyro	out of memory.						

Table 7.19.: Iterative refinement for the generalized Lyapunov equation with Divide-Scale-and-Conquer approximation on the Haswell system.

Problem	without GPUs		with GPUs		Iter.	Forward error	Relative residual
	Runtime	Speedup	Runtime	Speedup			
BCSST 07	1.46	0.69	1.37	0.73	2	1.37e-12	2.66e-15
BCSST 19	8.16	0.64	8.47	0.62	2	8.75e-08	5.49e-16
BCSST 09	6.37	2.74	6.30	2.78	2	1.97e-17	1.40e-15
BCSST 12	12.04	3.80	12.98	3.52	2	3.15e-12	1.73e-15
Filter 2D	18.33	2.82	19.84	2.61	2	4.47e-18	2.88e-14
BCSST 26	63.03	1.05	61.33	1.07	2	1.29e-15	1.44e-15
BCSST 23	202.10	1.50	151.82	1.99	2	6.54e-08	7.71e-16
BCSST 21	138.70	3.51	104.66	4.65	2	2.67e-18	1.09e-15
Steel 5	197.09	12.15	140.41	17.06	2	2.83e-18	1.44e-12
Flow Meter	982.53	16.62	945.42	17.27	2	6.35e-12	2.13e-14
BCSST 25	18 822.72	2.96	5 466.70	10.18	2	1.82e-11	3.18e-15
Chip Cooling	23 263.03	4.56	6 217.97	17.05	2	5.54e-15	2.35e-14
Steel 20	16 360.09	9.87	4 984.20	32.39	2	4.22e-18	4.13e-12
Gyro	298 568.18	1.52	209 302.24	2.17	2	7.92e-09	1.86e-15

Table 7.20.: Iterative refinement for the generalized Lyapunov equation with Divide-Scale-and-Conquer approximation on the Skylake system.

we preserve the accuracy of the direct solution technique. Depending on the capabilities of the hardware and the size of the problem, speedups of more than 17 up to 295 can be achieved.

Problem	without GPUs		with GPUs		Iter.	Forward error	Relative residual
	Runtime	Speedup	Runtime	Speedup			
BCSST 07	1.40	0.88	1.44	0.86	2	2.09e-12	4.13e-15
BCSST 19	8.03	0.87	8.09	0.86	2	7.24e-07	1.13e-15
BCSST 09	6.18	3.20	5.45	3.63	2	3.91e-17	2.11e-15
BCSST 12	13.36	3.61	9.47	5.09	2	6.17e-12	5.57e-15
Filter 2D	19.30	3.38	13.03	5.01	2	6.58e-18	2.89e-14
BCSST 26	66.28	1.30	41.64	2.06	2	4.65e-15	5.29e-15
BCSST 23	199.55	2.42	103.60	4.66	2	4.99e-07	2.93e-15
BCSST 21	138.42	5.22	74.80	9.66	2	3.04e-18	2.52e-15
Steel 5	192.46	23.68	100.51	45.34	2	5.46e-18	1.43e-12
Flow Meter	2 309.71	24.03	620.77	89.40	2	5.81e-12	1.74e-14
BCSST 25	17 729.15	13.60	3 536.39	68.17	2	5.87e-11	3.48e-14
Chip Cooling	21 764.33	20.46	4 080.66	109.11	2	3.26e-15	3.11e-14
Steel 20	15 391.56	45.18	3 440.00	202.15	2	7.79e-18	4.07e-12
Gyro	278153.92	11.06	68 211.45	45.12	2	8.89e-10	1.57e-14

Table 7.21.: Iterative refinement for the generalized Lyapunov equation with Divide-Scale-and-Conquer approximation on the POWER 8 system.

CONCLUSIONS

In this thesis, divide-and-conquer algorithms for the approximation of the generalized Schur decomposition of arbitrary but regular matrix pairs was developed. Thereby, the main idea was to employ deflating subspaces of the matrix pair to split its spectrum recursively until only trivial-to-solve problems are left. The initial challenge was to develop a generic approach, which heavily relies on level-3 BLAS and LAPACK algorithms, such as the matrix-matrix multiply, the LU decomposition, or the QR decomposition. This restriction originates, on the one hand, on the fact that with new computer architecture and accelerator devices, these operations have mostly become available in optimized versions, first. On the other hand, these operations are known to achieve a very good performance on almost all current computer architecture. This includes distributed systems as well, although they are not covered in this thesis. The restriction to mainly use these operations as building blocks in a generalized eigenvalue solver results in two different ways to compute deflating subspaces of a matrix pair.

First, we used the generalized matrix sign function to compute the deflating subspace corresponding to the eigenvalue either in the left or right open complex half-plane. The introduction of a shift parameter selection strategy allowed the application of the generalized matrix sign function recursively, which results in the Divide-Scale-and-Conquer algorithm in Section 3.2.3. Beside a naive implementation we presented optimized versions with one- and two-sided initial transformations. Different strategies to accelerate and optimize the CPU-only and the GPU accelerated computation of the generalized matrix sign function have been regarded, too.

The second way to obtain a deflating subspace for the divide-and-conquer framework is the matrix disc function, especially since its computation is possible with the inverse free iteration. In contrast to the generalized matrix sign function it provides a deflating subspace, which divides the eigenvalues into the ones lying inside and the ones outside of the unit circle. With the help of a scaling parameter selection strategy this leads to the Divide-Scale-and-Conquer algorithm in Section 3.3.2. The naive implementation of the inverse free iteration was extended by a directed acyclic graph scheduling approach to properly support current multi-core CPUs. Further optimizations focusing on CPU-only and GPU-accelerated systems were investigated as well.

Since we developed different implementations to compute the deflating subspaces, we performed an a priori selection of the fastest algorithm depending on the problem size and the capabilities of the hardware. The experiments required for this tuning showed that not only the number of necessary operations needs to be minimized to obtain the fastest algorithm but the operations need to be parallelizable or well suited for accelerator devices. In case of the generalized matrix sign function we show that all efficiency improving techniques require a minimum problem size to be faster than the naive algorithm. Working with CPUs only the band reduction, developed in Section 4.2.3, have been the fastest variant for large-scale problems. Having GPUs available the computation of the generalized matrix sign function benefits from the fast inversion of a matrix or the solution of a linear system with many right-hand sides using the Gauss-Jordan elimination scheme, depending on the size of matrix and the memory on the GPU.

For the inverse free iteration we had fewer options to reduce the operation count or accelerate the computation. In case of the naive implementation we evaluated different algorithms to compute the QR decomposition. Multi-core CPUs allow an extension of an already existing structure preserving scheme [124] to reduce the number of required operations. In the GPU case, we developed a multi-GPU aware QR decomposition, which outperforms standard software packages like MAGMA [3] even in the single GPU case. All implementations allow to easily adjust the algorithms to new CPU and GPU architectures using a few tuning parameters. These need to be determined once for each new architecture.

The divide-and-conquer idea together with the optimization and tuning of the main opera-tion inside the Divide-Shift-and-Conquer and the Divide-Scale-and-Conquer algorithm proved to drastically reduce the time to compute a generalized Schur decomposition. The generalized matrix sign function based Divide-Shift-and-Conquer algorithm accelerated the computation for an order 20 209 matrix pair from 27 hours to 59 minutes, from 33 hours to 37 minutes, and from 72 hours to 28 minutes on the Haswell, Skylake, and POWER 8 architecture, respectively. Using the inverse free iteration based Divide-Scale-and-Conquer algorithm these times reduced to 109 minutes on the Haswell architecture, to 72 minutes on the Skylake architecture, and to 46 minutes on the POWER 8 architecture but with improved accuracy competitive to the QZ algorithm. Even improved variants of the QZ algorithm, like the KKQZ used as well in the numerical examples, could hardly compete with the divide-and-conquer approach, especially when GPUs came into play. Even large scale matrix pairs, like the Gyro example of order 34 722, can be handled within a few hours instead of several days, now.

As an application of large-scale generalized Schur decompositions we came up with the direct solution of Sylvester-type matrix equations in Chapter 6. Thereby, we extend the classical level-3 block concept used in many LAPACK algorithms to Bartels-Stewart like algorithms [19, 78]. In contrast to the recursive blocking approach [100, 101], we showed how the standard blocking approach needs to be optimized to enable recent compilers to adjust the code to the capabilities of the hardware. First ideas of this were already used in the implementation of the recursive blocking approach, but with further implementation specific improvements and a classical regular block partitioning we obtained a faster algorithm. With a rearrangement of the developed level-3 BLAS enabled solver we derived a novel parallel solver, which uses the internal data dependencies of the Sylvester-type matrix equations to increase the overall parallel capabilities. This solver achieved a huge speedup compared to the fastest public available software, the recursive blocking approach. The algorithms only have the block size as tuning parameters and, thus, can be easily adjusted to new computer architectures. Theorem 6.8 showed that special block size selections lead to already known algorithms, like the Bartels-Stewart algorithm [19], the Gardiner et al. approach [78], or even the recursive blocking approach.

In order to reduce the errors in the solution, which were introduced by having only an approx-imation of the generalized Schur decomposition, we extended the iterative refinement approach to Sylvester-type matrix equations. Practical examples with the generalized Lyapunov equation showed that in most cases two or three steps of the iterative refinement were enough to recover or improve the accuracy obtained with the generalized Schur decomposition from the QZ algorithm. Regarding the runtime for the 20 209 × 20 209 example, we reduce the time to solution compared to the SLICOT library [149] from 37 hours to 72 minutes on the Haswell architecture, from 45 hours to 48 minutes on the Skylake architecture, and from 8 days to 39 minutes on the POWER 8 architecture in the best case.

Future Research Perspectives With the derivation of the divide-and-conquer idea in Chapter 3 this thesis showed that even algorithms which require significantly more operations than the classical approach can compete with them, if they are better suited for current hardware. The thesis only provided two ways to turn the basic divide-and-conquer approach from Section 3.1 into a concrete algorithm. The search for further ways to compute the deflating subspaces and to guarantee a working recursion need be done in the future. Although many operations, like the computation of the generalized matrix sign function and the inverse free iteration as well as the large matrix-matrix multiplies, are aware of the GPU acceleration, the rank revealing QR decompositions are still CPU only. Although MAGMA contains a GPU port of the column pivoted QR decomposition from LAPACK, its parallel speedup is theoretically limited to two. Newer approaches, like the column pivoted QR decomposition with random sampling, used in our implementation, does not provide a GPU accelerated variant yet. Here, developing a hybrid GPU-CPU code will lead to further performance enhancements. The panel factorization in the host part of the GPU accelerated decomposition needs further tuning and other approaches, like the CholeskyQR [74], have to be taken into account.

Regarding the implementation details of the Divide-Shift-and-Conquer and the Divide-Scale-and-Conquer algorithms there is room for improvements. Especially reducing the memory footprint, which is much higher than for the QZ algorithms, would be beneficial. Furthermore, combining the generalized matrix sign function and matrix disc function based approaches, such that in case of stagnation, the algorithm switches to a different splitting approach, needs to be regarded as well. Since both algorithms have advantages and disadvantages, one has to decide which one to used. Furthermore, to achieve a good trade-off between accuracy and runtime one can use the Divide-Shift-and-Conquer algorithm for good conditioned problems and the Divide-Scale-and-Conquer for ill-conditioned ones. The eigenvalue distribution has a large influence on the runtime as well. Since the deflating subspaces are invariant under certain eigenvalue transformations, see Section 2.4, one can introduced initial spectral transformations such that the spectral splitting emerges more equally sized problems in each step. Especially for large scale problems, this can lead to large runtime reductions.

Another application, where the computation of specific deflating subspaces is a key ingredient for the solution, is the Uniliteral Quadratic Matrix Equation $AX^2 + BX + C = 0$ [90]. For this matrix equation one can develop a modified divide-and-conquer scheme, which stops when the desired deflating subspace is known.

Appendices

EXPLANATORY EXPERIMENTS

Beside the numerical experiments in Chapters 5 and 7 many explanatory and motivating experiments have been performed to support the work in Chapters 4 and 6. These experiments have been perfomed on the same hardware and software setup described in Section 5.1.1. In this appendix we describe the problem setups for each experiment briefly. All experiments use IEEE double precision arithmetic. Random matrices are generated using the LAPACK random number generator LARNV with and an initial seed $(1, 1, 1, 1)$ and a uniform $(-1, 1)$ distribution. The updated seed after a call to LARNV is used in the subsequent call.

A.1. Examples in Chapter 4

Figure 4.4 The experiment utilizes the Haswell architecture. We vary the number of used threads for the BLAS and LAPACK operations in the MKL library from 1 to 16. For each number of threads, we ran the QZ algorithm implemented as GGES in LAPACK for ten different subsequent generated matrix pairs of order 1 024. The experiment is repeated ten times for each matrix pair and the average for each number of threads is computed.

Example 4.9 The example uses matrix pairs of order 64, where 1 024 of them are randomly generated. Each matrix pair is passed to a single thread executing GGES. We use 16 threads to parallelize this procedure with the of an OpenMP parallel loop. The experiment is repeated ten times and the average runtime is taken. It is performed on the Haswell architecture.

Figure 4.5 The examples uses the Haswell architecture again. The problem dimension is increased from 10 to 300 in steps of 10. For each problem size, 100 eigenvalue problems are solved in parallel with 16 threads and the GGES implementation from LAPACK. The experiment is repeated 25 times and the average runtime is taken for each problem size.

Figure 4.6 The benchmark is run on the Intel® Skylake system. The problem dimensions vary from 500 to 10 000 in steps of 500. The block size in the algorithm is set to 192. The reference result is obtained by using DTRMM from BLAS and DLACPY from LAPACK. The experiment is repeated 25 times and the average runtime is taken for each problem size. All 16 CPU cores are used.

Figure 4.7 The benchmark is run on the Intel® Skylake system. The problem dimensions vary from 500 to 10 000 in steps of 500. The block size in the algorithm is set to 32 for problems smaller than 5 000 and to 128 for problems from size 5 000 on. The block size in the called TRMM3 routine is set to 192. The reference results are obtained by using DTRSV and DTRSM from BLAS as described in Section 4.2.2. The experiment is repeated 25 times and the average runtime is taken for each problem size. All 16 CPU cores are used.

Tables 4.2 and 4.3 The experiment is done on all three architectures since there are big differences between them. For each problem size from 2 000 to 10 000 with a step size of 2 000 we perform each computation ten times and take the average runtime. On each system all CPU cores are utilized.

Figure 4.8 The experiment uses the Haswell architecture. For a varying matrix dimension from 100 to 3700 in steps of 100 the experiments is repeated 200 times and the average runtime is taken. In case of the parallel execution all 16 CPU cores are used.

Figure 4.11 The experiment uses the Skylake architecture. For a varying matrix dimension from 2 000 to 20 000 in steps of 2 000 the experiments is repeated ten times and the average runtime is taken. The LU and the QR decomposition are used from the MKL and the Tile-QR comes from a Fortran reimplementation of the PLASMA approach [5, 50]. The MKL uses 16 threads.

Table 4.5 The experiment uses the Haswell architecture. For a varying matrix dimension from 2 000 to 20 000 in steps of 2 000 the experiments is repeated ten times and the average runtime is taken. The OpenMP 4 enabled algorithm uses all 16 CPU cores.

Figure 4.13 The experiment is performed on the Haswell architecture. For a varying matrix dimension from 2 000 to 20 000 in steps of 2 000 the experiments is repeated ten times and the average runtime is taken. The MKL uses all 16 available CPU cores for parallelization.

Figure 4.11 The experiment uses the POWER 8 architecture. For a varying matrix dimension from 2 048 to 20 480 in steps of 2 048 the experiments is repeated ten times and the average runtime is taken. The MAGMA library was used in version 2.5.0. The GPU code utilized either a single or both GPUs of the system and the host part uses 20 threads for parallelization.

Figure 4.17 The experiment uses the POWER 8 architecture. For a varying number of rows from 2 048 to 43 008 in steps of 2 048 and a fixed number of 128 columns the experiments is repeated ten times and the average runtime is taken. The $WY - QR$ is used there comes directly from the reference LAPACK implementation and uses only BLAS calls of the ESSL. This is necessary since IBM's ESSL does not provide a full featured LAPACK interface and some routines cannot be used from outside. The code used 20 threads.

Figure 4.18 The experiment uses the POWER 8 architecture again. For a varying number of columns m from 2 048 to 40 960 in steps of 2 048 two different benchmarks are done. First, we use a square $m \times m$ matrix and, second, we use a $2m \times m$ matrix. For each matrix the experiment is performed ten times and the average runtime is taken. The MAGMA library is used in version 2.5.0. The host part of the algorithm used up to 20 threads.

A.2. Examples in Chapter 6

All experiments in Chapter 6 are performed on the Skylake architecture for the generalized Sylvester equation. Thereby all coefficient matrices are of the same size. Figures 6.1 to 6.6 use problems of order m from ten to 2 100 with a step size of one. The experiment in Figure 6.7 uses problems of dimension 1 500 to 2 100. For Figure 6.8 the problem dimension varies from 512 to 6 144 in step of 512. All experiments are repeated ten times and the average runtime is taken.

BIBLIOGRAPHY

[1] *Data Alignment to Assist Vectorization.* `https://software.intel.com/en-us/articles/data-alignment-to-assist-vectorization`. Accessed, January 7th, 2019. 172

[2] *HIP : C++ Heterogeneous-Compute Interface for Portability.* `https://github.com/ROCm-Developer-Tools/HIP`. 53

[3] *MAGMA – Matrix Algebra on GPU and Multicore Architectures.* `https://icl.utk.edu/magma`. 104, 196

[4] *OpenCL BLAS (clBLAS).* `http://github.com/clmathlibraries/clblas`. 21

[5] *PLASMA – Parallel Linear Algebra Software for Multicore Architectures.* `https://icl.utk.edu/plasma`. 127, 202

[6] *rocBLAS – Basic Linear Algebra Subprograms for the Radeon Open Compute platform.* `https://github.com/ROCmSoftwarePlatform/rocBLAS`. 21

[7] J. C. ADAMS, B. T. SMITH, J. T. MARTIN, W. S. BRAINERD, AND J. L. WAGENER, *FORTRAN 95 Handbook*, MIT Press, Cambridge, MA, USA, 1997. 170

[8] B. ADLERBORN, B. KÅGSTRÖM, AND D. KRESSNER, *A parallel QZ algorithm for distributed memory HPC systems*, SIAM J. Sci. Comput., 36 (2014), pp. C480–C503. 13, 118

[9] G. M. AMDAHL, *Validity of the single processor approach to achieving large scale computing capabilities*, in Proceedings of the April 18-20, 1967, Spring Joint Computer Conference, AFIPS '67 (Spring), New York, NY, USA, 1967, Association for Computing Machinery, p. 483–485. 65

[10] E. ANDERSON, Z. BAI, C. BISCHOF, J. DEMMEL, J. DONGARRA, J. DU CROZ, A. GREENBAUM, S. HAMMARLING, A. MCKENNEY, AND D. SORENSEN, *LAPACK Users' Guide*, SIAM, Philadelphia, PA, third ed., 1999. 2, 10, 13, 14, 15, 26, 79, 92, 179

[11] E. ANDERSON, Z. BAI, AND J. DONGARRA, *Generalized QR factorization and its applications*, Linear Algebra Appl., 162–164 (1992), pp. 243–271. 15

[12] A. C. ANTOULAS, *Approximation of Large-Scale Dynamical Systems*, vol. 6 of Adv. Des. Control, SIAM Publications, Philadelphia, PA, 2005. 1, 2, 151, 152, 154, 190

[13] Z. BAI AND J. DEMMEL, *Design of a parallel nonsymmetric eigenroutine toolbox, Part I*, in Proceedings of the Sixth SIAM Conference on Parallel Processing for Scientific Computing, R. F. S. et al., ed., Philadelphia, 1993, SIAM, pp. 391–398. *See also:* Tech. Report CSD-92-718, Computer Science Division, University of California, Berkeley, CA 94720. 31

[14] Z. BAI, J. DEMMEL, J. DONGARRA, A. RUHE, AND H. VAN DER VORST, EDITORS, *Templates for the Solution of Algebraic Eigenvalue Problems: A Practical Guide*, no. 11 in Software Environ. Tools, SIAM, Philadelphia, 2000. 10

[15] Z. Bai, J. Demmel, and M. Gu, *An inverse free parallel spectral divide and conquer algorithm for nonsymmetric eigenproblems*, Numer. Math., 76 (1997), pp. 279–308. 4, 5, 24, 44

[16] O. M. Baksalary, S. Bernstein, D, and G. Trenkler, *On the equality between rank and trace of an idempotent matrix*, Appl. Math. Comput., 217 (2010), pp. 4076–4080. 31

[17] G. Ballard, J. Demmel, L. Grigori, M. Jacquelin, N. Knight, and H. Nguyen, *Reconstructing householder vectors from tall-skinny QR*, J. Parallel Distr. Com., 85 (2015), pp. 3–31. 5, 99, 100, 107, 109, 110, 129

[18] S. Barrachina, P. Benner, and E. S. Quintana-Ortí, *Efficient algorithms for generalized algebraic Bernoulli equations based on the matrix sign function*, Numer. Algorithms, 46 (2007), pp. 351–368. 67, 124

[19] R. H. Bartels and G. W. Stewart, *Solution of the matrix equation $AX + XB = C$: Algorithm 432*, Comm. ACM, 15 (1972), pp. 820–826. 2, 3, 5, 6, 153, 154, 155, 156, 166, 181, 196

[20] A. N. Beavers and E. D. Denman, *A computational method for eigenvalues and eigenvectors of a matrix with real eigenvalues*, Numer. Math., 21 (1973), pp. 389–396. 30

[21] P. Benner, *Contributions to the Numerical Solution of Algebraic Riccati Equations and Related Eigenvalue Problems*, Dissertation, Fakultät für Mathematik, TU Chemnitz–Zwickau, 09107 Chemnitz (Germany), Feb. 1997. 4, 5, 40, 41, 42, 44, 67, 94

[22] P. Benner and R. Byers, *Disk functions and their relationship to the matrix sign function*, in Proc. European Control Conf. ECC 97, Paper 936, BELWARE Information Technology, Waterloo, Belgium, 1997. CD-ROM. 42, 43

[23] P. Benner, P. Ezzatti, E. S. Quintana-Ortí, and A. Remón, *Matrix inversion on CPU-GPU platforms with applications in control theory*, Concurrency and Comput.: Pract. Exper., 25 (2013), pp. 1170–1182. 80, 81

[24] P. Benner, M. Köhler, and C. Penke, *GPU-accelerated and storage-efficient implementation of the QR decomposition*, in Parallel Computing is Everywhere, Proceedings of the International Conference on Parallel Computing, ParCo 2017, 12-15 September 2017, Bologna, Italy, S. Bassini, M. Danelutto, P. Dazzi, G. R. Joubert, and F. Peters, eds., vol. 32 of Advances in Parallel Computing, IOS Press, 2017, pp. 329–338. 92

[25] ——, *GPU-accelerated implementation of the storage-efficient QR decomposition*, in 2nd Workshop on Power-Aware Computing 2017 (PACO2017), Ringberg Castle, Germany, 5-8 July 2017, June 2017, p. 6. 92

[26] P. Benner, M. Köhler, and J. Saak, *A cache-aware implementation of the spectral divide-and-conquer approach for the non-symmetric generalized eigenvalue problem*, Proc. Appl. Math. Mech., 14 (2014), pp. 819–820. 4, 7, 61

[27] ——, *Fast approximate solution of the non-symmetric generalized eigenvalue problem on multicore architectures*, in Parallel Computing: Accelerating Computational Science and Engineering (CSE), M. Bader, A. Bodeand, H.-J. Bungartz, M. Gerndt, G. R. Joubert, and F. Peters, eds., vol. 25 of Advances in Parallel Computing, IOS Press, 2014, pp. 143–152. 4, 7, 24, 35, 61

[28] P. Benner, D. Kressner, V. Sima, and A. Varga, *Die SLICOT-Toolboxen für MATLAB*, at-Automatisierungstechnik, 58 (2010), pp. 15–25. 190

[29] P. BENNER, V. MEHRMANN, V. SIMA, S. V. HUFFEL, AND A. VARGA, *SLICOT - a subroutine library in systems and control theory*, in Applied and Computational Control, Signals, and Circuits, B. N. Datta, ed., vol. 1, Birkhäuser, Boston, MA, 1999, ch. 10, pp. 499–539. 181, 190

[30] P. BENNER, E. QUINTANA-ORTÍ, AND G. QUINTANA-ORTÍ, *State-space truncation methods for parallel model reduction of large-scale systems*, Parallel Comput., 29 (2003), pp. 1701–1722. 5, 163

[31] P. BENNER AND E. S. QUINTANA-ORTÍ, *Solving stable generalized Lyapunov equations with the matrix sign function*, Numer. Algorithms, 20 (1999), pp. 75–100. 4, 5, 67, 70, 75, 124, 163

[32] P. BENNER, E. S. QUINTANA-ORTÍ, AND G. QUINTANA-ORTÍ, *Balanced truncation model reduction of large-scale dense systems on parallel computers*, Math. Comput. Model. Dyn. Syst., 6 (2000), pp. 383–405. 5

[33] P. BENNER, E. S. QUINTANA-ORTÍ, AND G. QUINTANA-ORTÍ, *PSLICOT routines for model reduction of stable large-scale systems*, in Proc. 3rd NICONET Workshop on Numerical Software in Control Engineering, Louvain-la-Neuve, Belgium, January 19, 2001, 2001, pp. 39–44. 5, 163

[34] P. BENNER AND J. SAAK, *Linear-quadratic regulator design for optimal cooling of steel profiles*, Tech. Rep. SFB393/05-05, Sonderforschungsbereich 393 *Parallele Numerische Simulation für Physik und Kontinuumsmechanik*, TU Chemnitz, D-09107 Chemnitz (Germany), 2005. 2, 6, 118

[35] P. BENNER AND S. W. R. WERNER, *MORLAB-3.0 – model order reduction laboratory*, Sept. 2017. see also: http://www.mpi-magdeburg.mpg.de/projects/morlab. 29

[36] J. BEWERSDORFF, *Algebra für Einsteiger*, Springer Fachmedien Wiesbaden, Wiesbaden, Jan. 2013. 10, 22, 23

[37] J. BEZANSON, A. EDELMAN, S. KARPINSKI, AND V. B. SHAH, *Julia: A fresh approach to numerical computing*, SIAM Review, 59 (2017), pp. 65–98. 2

[38] D. BILLGER, *The butterfly gyro*, in Dimension Reduction of Large-Scale Systems, vol. 45 of Lecture Notes in Computational Science and Engineering, Springer-Verlag, Berlin/Heidelberg, Germany, 2005, pp. 349–352. 118, 119

[39] D. A. BINI, B. IANNAZZO, AND B. MEINI, *Numerical solution of algebraic Riccati equations*, vol. 9 of Fundamentals of Algorithms, SIAM Publications, Philadelphia, 2012. 16, 18, 19

[40] C. BISCHOF AND C. VAN LOAN, *The WY representation for products of Householder matrices*, SIAM J. Sci. Statist. Comput., 8 (1987), pp. S2–S13. Parallel processing for scientific computing (Norfolk, Va., 1985). 2, 5, 22, 64, 65, 75, 92, 96, 104, 112, 127

[41] C. H. BISCHOF, *Incremental condition estimation*, SIAM J. Matrix Anal. Appl., 11 (1990), pp. 312–322. 26, 66, 135

[42] C. H. BISCHOF AND G. QUINTANA-ORTÍ, *Algorithm 782: Codes for rank-revealing QR factorizations of dense matrices.*, ACM Trans. Math. Software, 24 (1998), pp. 254–257. 4, 25, 26, 33, 65

[43] ——, *Computing rank-revealing QR factorizations of dense matrices*, ACM Trans. Math. Software, 24 (1998), pp. 226–253. 4, 25, 26, 33, 65

[44] L. S. BLACKFORD, J. CHOI, A. CLEARY, E. D'AZEVEDO, J. DEMMEL, I. DHILLON, J. J. DONGARRA, S. HAMMARLING, G. HENRY, A. PETITET, K. STANLEY, D. WALKER, AND R. C. WHALEY, *ScaLAPACK User's Guide*, vol. 4 of Software, Environments and Tools, SIAM Publications, Philadelphia, PA, USA, 1997. 2, 21, 85

[45] F. BORNEMANN, *Funktionentheorie*, Mathematik Kompakt, Birkhäuser, Basel, 2016. 18

[46] K. BRAMAN, R. BYERS, AND R. MATHIAS, *The multishift QR algorithm. I. Maintaining well-focused shifts and level 3 performance*, SIAM J. Matrix Anal. Appl., 23 (2002), pp. 929–947. 2

[47] ——, *The multishift QR algorithm. II. Aggressive early deflation*, SIAM J. Matrix Anal. Appl., 23 (2002), pp. 948–973. 2

[48] F. BROQUEDIS, J. CLET-ORTEGA, S. MOREAUD, N. FURMENTO, B. GOGLIN, G. MERCIER, S. THIBAULT, AND R. NAMYST, *hwloc: a Generic Framework for Managing Hardware Affinities in HPC Applications*, in PDP 2010 - The 18th Euromicro International Conference on Parallel, Distributed and Network-Based Computing, IEEE, ed., Pisa, Italy, Feb. 2010. 116

[49] P. BUSINGER AND G. H. GOLUB, *Handbook series linear algebra. Linear least squares solutions by Householder transformations*, Numer. Math., 7 (1965), pp. 269–276. 4, 25, 26, 65

[50] A. BUTTARI, J. DONGARRA, J. KURZAK, P. LUSZCZEK, AND S. TOMOV, *Using mixed precision for sparse matrix computations to enhance the performance while achieving 64-bit accuracy*, ACM Trans. Math. Software, 34 (2008), pp. 1–22. 22, 178, 180, 202

[51] A. BUTTARI, J. LANGOU, J. KURZAK, AND J. DONGARRA, *Parallel tiled QR factorization for multicore architectures*, Concurrency and Comput.: Pract. Exper., 20 (2008), pp. 1573–1590. 2, 5, 71, 92, 93, 99, 127

[52] E. CARSON AND N. J. HIGHAM, *Accelerating the solution of linear systems by iterative refinement in three precisions*, SIAM J. Sci. Comput., 40 (2018), pp. A817–A847. 178

[53] S. CATALÁN, J. R. HERRERO, E. S. QUINTANA-ORTÍ, R. RODRÍGUEZ-SÁNCHEZ, AND R. VAN DE GEIJN, *A case for malleable thread-level linear algebra libraries: The LU factorization with partial pivoting*, e-print arXiv:1611.06365, arXiv, 2016. cs.DS. 86, 111

[54] T. CHAN, *Rank revealing QR factorizations*, Linear Algebra Appl., 88/89 (1987), pp. 67–82. 25

[55] J. CHOI, J. J. DONGARRA, L. S. OSTROUCHOV, A. P. PETITET, D. W. WALKER, AND R. C. WHALEY, *Design and implementation of the scaLAPACK LU, QR, and cholesky factorization routines*, Scientific Programming, 5 (1996), pp. 173–184. 2

[56] E. K. CHU, *The solution of the matrix equations $AXB - CXD = E$ and $(YA - DZ, YC - BZ) = (E, F)$*, Linear Algebra Appl., 93 (1987), pp. 93–105. 153, 154

[57] K. DACKLAND AND B. KÅGSTRÖM, *Blocked algorithms and software for reduction of a regular matrix pair to generalized Schur form*, ACM Trans. Math. Software, 25 (1999), pp. 425–454. 2, 22

[58] T. A. DAVIS AND Y. HU, *The University of Florida sparse matrix collection*, ACM Trans. Math. Software, 38 (2011), pp. Art. 1, 25. 118

[59] J. DEMMEL, L. GRIGORI, M. HOEMMEN, AND J. LANGOU, *Communication-optimal parallel and sequential QR and LU factorizations*, SIAM J. Sci. Comput., 34 (2012), pp. A206–A239. 5, 99, 100, 106, 107, 108, 129

[60] J. DEMMEL AND B. KÅGSTRÖM, *The generalized Schur decomposition of an arbitrary pencil $a - \lambda b$ – robust software with error bounds and applications. Part II: software and applications*, ACM Trans. Math. Software, 19 (1993), pp. 175–201. 10

[61] ——, *The generalized Schur decomposition of an arbitrary pencil A − λB − robust software with error bounds and applications. Part I: theory and algorithms*, ACM Trans. Math. Software, 19 (1993), pp. 160–174. 10

[62] J. W. DEMMEL, *Applied numerical linear algebra*, Society for Industrial and Applied Mathematics (SIAM), Philadelphia, 1997. 178, 179

[63] E. D. DENMAN AND A. N. BEAVERS, *The matrix sign function and computations in systems*, Appl. Math. Comput., 2 (1976), pp. 63–94. 30

[64] G. DIETRICH, *A new formulation of the hypermatrix Householder-QR decomposition*, Comp. Meth. Appl. Mech. Eng., 9 (1976), pp. 273–280. 102

[65] J. DONGARRA, M. GATES, A. HAIDAR, J. KURZAK, P. LUSZCZEK, S. TOMOV, AND I. YAMAZAKI, *Accelerating numerical dense linear algebra calculations with gpus*, Numerical Computations with GPUs, (2014), pp. 1–26. 2, 80, 92

[66] J. J. DONGARRA, J. D. CROZ, I. S. DUFF, AND S. HAMMARLING, *A set of Level 3 Basic Linear Algebra Subprograms*, ACM Trans. Math. Software, 16 (1990), pp. 1–17. 38, 64, 71

[67] J. J. DONGARRA, F. G. GUSTAVSON, AND A. KARP, *Implementing linear algebra algorithms for dense matrices on a vector pipeline machine*, SIAM Rev., 26 (1984), pp. 91–112. 2

[68] Z. DRMAČ AND Z. BUJANOVIĆ, *On the failure of rank-revealing QR factorization software − a case study*, ACM Trans. Math. Software, 35 (2009), pp. Art. 12, 28. 27, 65

[69] J. A. DUERSCH AND M. GU, *Randomized QR with column pivoting*, SIAM J. Sci. Comput., 39 (2017), pp. C263–C291. 65

[70] I. S. DUFF, R. G. GRIMES, AND J. G. LEWIS, *Sparse matrix test problems*, ACM Trans. Math. Software, 15 (1989), pp. 1–14. 118

[71] J. W. EATON, D. BATEMAN, S. HAUBERG, AND R. WEHBRING, *GNU Octave version 4.4.0 manual: a high-level interactive language for numerical computations*, 2018. 2

[72] E. ELMROTH AND F. G. GUSTAVSON, *Applying recursion to serial and parallel QR factorization leads to better performance*, IBM Journal of Research and Development, 44 (2000), pp. 605–624. 75, 94, 112

[73] J. G. F. FRANCIS, *The QR–transformation, part I, II*, Comput. J., 4 (1961), pp. 265–271, 332–345. 1, 13, 22

[74] T. FUKAYA, R. KANNAN, Y. NAKATSUKASA, Y. YAMAMOTO, AND Y. YANAGISAWA, *Shifted CholeskyQR for computing the QR factorization of ill-conditioned matrices*, arXiv e-prints, (2018), p. arXiv:1809.11085. 197

[75] F. R. GANTMACHER, *Theory of Matrices*, vol. 1, Chelsea, New York, 1959. 42

[76] ——, *Theory of Matrices*, vol. 2, Chelsea, New York, 1959. 42

[77] J. D. GARDINER AND A. J. LAUB, *A generalization of the matrix-sign-function solution for algebraic Riccati equations*, Internat. J. Control, 44 (1986), pp. 823–832. 4, 32, 33, 181

[78] J. D. GARDINER, A. J. LAUB, J. J. AMATO, AND C. B. MOLER, *Solution of the Sylvester matrix equation AXB + CXD = E*, ACM Trans. Math. Software, 18 (1992), pp. 223–231. 2, 5, 153, 158, 166, 196

[79] G. H. Golub, S. Nash, and C. F. Van Loan, *A Hessenberg–Schur method for the problem $AX+XB = C$*, IEEE Trans. Autom. Control, 24 (1979), pp. 909–913. 2, 158

[80] G. H. Golub and C. F. Van Loan, *Matrix Computations*, Johns Hopkins Studies in the Mathematical Sciences, Johns Hopkins University Press, Baltimore, fourth ed., 2013. 1, 3, 9, 12, 13, 15, 21, 22, 26, 29, 39, 52, 60, 62, 64, 66, 73, 76, 82, 92, 103, 110, 136, 151, 153, 161

[81] K. Goto and R. A. van de Geijn, *Anatomy of high-performance matrix multiplication*, 34 (2008), pp. Art. 12, 25. 21, 72, 173

[82] R. Granat and B. Kågström, *Algorithm 904: The SCASY library—parallel solvers for Sylvester-type matrix equations with applications in condition estimation, part II*, ACM Trans. Math. Software, 37 (2010), pp. 33:1–33:4. 160

[83] ——, *Parallel solvers for Sylvester-type matrix equations with applications in condition estimation, part I: Theory and algorithms*, ACM Trans. Math. Software, 37 (2010), pp. 32:1–32:32. 160

[84] G. Hager and G. Wellein, *Introduction to High Performance Computing for Scientists and Engineers*, CRC Press, Inc., Boca Raton, FL, USA, 1st ed., 2010. 7, 62, 84, 85

[85] S. J. Hammarling, *Numerical solution of the stable, non-negative definite Lyapunov equation*, IMA J. Numer. Anal., 2 (1982), pp. 303–323. 3, 156

[86] H. V. Henderson and S. R. Searle, *The vec-permutation matrix, the vec operator and Kronecker products: a review*, Linear Multilinear Algebra, 9 (1981), pp. 271–288. 153

[87] M. A. Heroux, R. A. Bartlett, V. E. Howle, R. J. Hoekstra, J. J. Hu, T. G. Kolda, R. B. Lehoucq, K. R. Long, R. P. Pawlowski, E. T. Phipps, A. G. Salinger, H. K. Thornquist, R. S. Tuminaro, J. M. Willenbring, A. Williams, and K. S. Stanley, *An overview of the trilinos project*, ACM Trans. Math. Software, 31 (2005), pp. 397–423. 2

[88] N. J. Higham, *Accuracy and Stability of Numerical Algorithms*, SIAM Publications, Philadelphia, 1996. 9, 15, 31, 103, 178

[89] N. J. Higham, *Functions of Matrices: Theory and Computation*, Applied Mathematics, SIAM Publications, Philadelphia, 2008. 16, 30, 31

[90] N. J. Higham and H.-M. Kim, *Numerical analysis of a quadratic matrix equation*, IMA J. Numer. Anal., 20 (2000), pp. 499–519. 197

[91] C. A. R. Hoare, *Algorithm 64: Quicksort*, Comm. ACM, 4 (1961), pp. 321–. 51

[92] ——, *Algorithm 65: Find*, Comm. ACM, 4 (1961), pp. 321–322. 51, 53

[93] D. Hohlfeld and H. Zappe, *An all-dielectric tunable optical filter based on the thermo-optic effect*, Journal of Optics A: Pure and Applied Optics, 6 (2004), pp. 504–511. 118

[94] IEEE, *IEEE Std 1003.1-2001 Standard for Information Technology — Portable Operating System Interface (POSIX) System Interfaces, Issue 6*, IEEE, New York, NY, USA, 2001. Revision of IEEE Std 1003.1-1996 and IEEE Std 1003.2-1992, Open Group Technical Standard Base Specifications, Issue 6. 60

[95] R. Ierusalimschy, L. H. de Figueiredo, and W. C. Filho, *Lua—an extensible extension language*, Software: Practice and Experience, 26 (1996), pp. 635–652. 5, 118, 181

[96] INTEL CORPORATION, *Intel Math Kernel Library (MKL)*. https://software.intel.com/en-us/mkl. 21, 60, 64, 77, 83

[97] INTERNATIONAL BUSINESS MACHINES CORPORATION, *IBM Engineering and Scientific Subroutine Library (ESSL)*. https://www-03.ibm.com/systems/power/software/essl/. 21, 64, 77, 83

[98] A. JAZLAN, V. SREERAM, AND R. TOGNERI, *Cross Gramian based time interval model reduction*, in Proceedings of the 5th Australien Control Conference (AUCC), 2015, pp. 274–276. 151

[99] E. JONES, T. OLIPHANT, P. PETERSON, ET AL., *SciPy: Open source scientific tools for Python.* 2

[100] I. JONSSON AND B. KÅGSTRÖM, *Recursive blocked algorithms for solving triangular systems–part I: One-sided and coupled Sylvester-type matrix equations*, ACM Trans. Math. Software, 28 (2002), pp. 392–415. 2, 3, 5, 6, 159, 160, 181, 184, 196

[101] ——, *Recursive blocked algorithms for solving triangular systems–part II: Two-sided and generalized Sylvester and Lyapunov matrix equations*, ACM Trans. Math. Software, 28 (2002), pp. 416–435. 2, 3, 5, 6, 159, 160, 167, 181, 184, 196

[102] B. KÅGSTRÖM, *A direct method for reordering eigenvalues in the generalized real Schur form of a regular matrix pair (A, B)*, in Linear algebra for large scale and real-time applications (Leuven, 1992), vol. 232 of NATO Adv. Sci. Inst. Ser. E Appl. Sci., Kluwer Acad. Publ., Dordrecht, 1993, pp. 195–218. 2, 3, 13, 14, 15, 173

[103] B. KÅGSTRÖM AND D. KRESSNER, *Multishift variants of the QZ algorithm with aggressive early deflation*, SIAM J. Matrix Anal. Appl., 29 (2007), pp. 199–227. 2, 13, 21, 22, 66, 118

[104] B. KÅGSTRÖM, D. KRESSNER, E. S. QUINTANA-ORTÍ, AND G. QUINTANA-ORTÍ, *Blocked algorithms for the reduction to Hessenberg-triangular form revisited*, BIT, 48 (2008), pp. 563–584. 2, 13, 21, 22, 66, 118

[105] B. KÅGSTRÖM AND P. POROMAA, *LAPACK-style algorithms and software for solving the generalized Sylvester equation and estimating the separation between regular matrix pairs*, ACM Trans. Math. Software, 22 (1996), pp. 78–103. 2, 14, 156, 159, 163

[106] B. KÅGSTRÖM AND P. VAN DOOREN, *A generalized state-space approach for the additive decomposition of a transfer matrix*, Numer. Lin. Alg. Appl., 1 (1992), pp. 165–181. 29

[107] B. KÅGSTRÖM AND L. WESTIN, *Generalized Schur methods with condition estimators for solving the generalized Sylvester equation*, IEEE Trans. Autom. Control, 34 (1989), pp. 745–751. 1, 2, 14, 154, 156, 163

[108] C. KENNEY AND A. J. LAUB, *On scaling Newton's method for polar decomposition and the matrix sign function*, in Proc. 1990 Amer. Control Conf., San Diego, CA, May 1990, pp. 2560–2564. 31

[109] ——, *The matrix sign function*, IEEE Trans. Autom. Control, 40 (1995), pp. 1330–1348. 30, 31

[110] D. E. KNUTH, *The art of computer programming. Vol. 2: Seminumerical algorithms*, Addison-Wesley Publishing Co., Reading, Mass.-London-Don Mills, Ont, 1969. 51, 54, 57

[111] M. KÖHLER, C. PENKE, J. SAAK, AND P. EZZATTI, *Energy-aware solution of linear systems with many right hand sides*, Comput. Sci. Res. Dev., 31 (2016), pp. 215–223. 7, 80, 81, 82, 85, 87, 89, 90

[112] M. KÖHLER AND J. SAAK, *On BLAS level-3 implementations of common solvers for (quasi-) triangular generalized Lyapunov equations*, ACM Trans. Math. Software, 43 (2016), pp. 3:1–3:23. 8, 160, 175

[113] ——, *On GPU acceleration of common solvers for (quasi-) triangular generalized Lyapunov equations*, Parallel Comput., 57 (2016), pp. 212–221. 8

[114] ——, *A GPU accelerated Gauss-Jordan elimination on the OpenPOWER platform – a case study*, Proc. Appl. Math. Mech., 17 (2017), pp. 845–846. 7

[115] ——, *Frequency scaling and energy efficiency regarding the Gauss-Jordan elimination scheme with application to the matrix-sign-function on OpenPOWER 8*, Concurrency and Comput.: Pract. Exper., 31 (2018). 7

[116] J. G. KORVINK AND E. B. RUDNYI, *Oberwolfach benchmark collection*, in Dimension Reduction of Large-Scale Systems, P. Benner, D. C. Sorensen, and V. Mehrmann, eds., vol. 45 of Lecture Notes in Computational Science and Engineering, Springer Berlin Heidelberg, 2005, pp. 311–315. 118

[117] D. KRESSNER, *Block variants of Hammarling's method for solving Lyapunov equations*, ACM Trans. Math. Software, 34 (2008), pp. Art. 1, 15. 156

[118] J. KURZAK AND J. DONGARRA, *Implementing linear algebra routines on multi-core processors with pipelining and a look ahead*, in PARA 2006: Applied Parallel Computing. State of the Art in Scientific Computing, Springer, Berlin, Heidelberg, 2006, pp. 147–156. 2

[119] C. LAWSON, R. HANSON, D. KINCAID, AND F. KROGH, *Basic linear algebra subprograms for FOR-TRAN usage*, ACM Trans. Math. Software, 5 (1979), pp. 303–323. 170

[120] R. B. LEHOUCQ, D. C. SORENSEN, AND C. YANG, *ARPACK users' guide*, Software, Environments and Tools, (1998). 35

[121] A. N. MALYSHEV, *Computing invariant subspaces of a regular linear pencil of matrices*, Siberian Math. J., 30 (1989), pp. 559–567. 4, 5, 44

[122] ——, *Parallel algorithm for solving some spectral problems of linear algebra*, Linear Algebra Appl., 188/189 (1993), pp. 489–520. 44

[123] M. MARQUÉS AND V. HERNÁNDEZ, *Parallel algorithm for the quasi-triangular generalized sylvester equation on a shared memory multiprocessor*, in Algorithms and Architectures for Real-Time Control 1992, P. J. Fleming and W. H. Kwon, eds., IFAC Postprint Volume, Pergamon, 1992, pp. 23 – 28. 163

[124] M. MARQUÉS, E. S. QUINTANA-ORTÍ, AND G. QUINTANA-ORTÍ, *Specialized spectral division algorithms for generalized eigenproblems via the inverse-free iteration*, in Applied Parallel Computing. State of the Art in Scientific Computing, B. Kågström, E. Elmroth, J. Dongarra, and J. Waśniewski, eds., vol. 4699 of Lecture Notes in Comput. Sci., Springer-Verlag, 1 2007, pp. 157–166. 5, 95, 96, 97, 196

[125] P. MARTINSSON, G. QUINTANA ORTÍ, N. HEAVNER, AND R. VAN DE GEIJN, *Householder QR factorization with randomization for column pivoting (HQRRP)*, SIAM J. Sci. Comput., 39 (2017), pp. C96–C115. 4, 26, 27, 65, 133

[126] THE MATHWORKS, INC., *The MATLAB Control Toolbox, Version 5*, Cochituate Place, 24 Prime Park Way, Natick, Mass, 01760, 2000. 190

[127] ——, *MATLAB Version 6.5*, Cochituate Place, 24 Prime Park Way, Natick, Mass, 01760, 2002. 2

[128] C. B. Moler, *Iterative Refinement in Floating Point*, J. Assoc. Comput. Mach., 14 (1967), pp. 316–321. 6, 178

[129] C. B. Moler and G. W. Stewart, *An algorithm for generalized matrix eigenvalue problems*, SIAM J. Numer. Anal., 10 (1973), pp. 241–256. 1, 9, 11, 13, 21, 22, 24, 29, 36, 39, 55, 56, 59, 62, 63, 66, 118

[130] C. Moosmann, E. B. Rudnyi, A. Greiner, and J. G. Korvink, *Model order reduction for linear convective thermal flow*, in THERMINIC 2004, Sophia Antipolis, France, 2004, pp. 317–321. 118

[131] NVIDIA Corporation, *CUDA Toolkit Documentation*. https://docs.nvidia.com/cuda/. 86

[132] ——, *nVidia CUBLAS Library*. https://developer.nvidia.com/cublas/. 21, 64, 87

[133] ——, *nVidia CUSOLVER Library*. https://docs.nvidia.com/cuda/cusolver/. 104

[134] Oberwolfach Benchmark Collection, *Butterfly gyroscope*. hosted at MORwiki – Model Order Reduction Wiki, 2004. 118, 119

[135] ——, *Steel profile*. hosted at MORwiki – Model Order Reduction Wiki, 2005. 118

[136] ——, *Tunable optical filter*. hosted at MORwiki – Model Order Reduction Wiki, 2019. 118

[137] *OpenBLAS*. http://www.openblas.net, 2016. 21, 60, 64, 83, 120, 170

[138] *OpenMP*. http://openmp.org. 53, 60, 99, 176

[139] C. Penke, *GPU-accelerated computation of the matrix disc function and its applications*, Master's thesis, Otto-von-Guericke-Universität, Magdeburg, Germany, Jan. 2017. 92

[140] T. Penzl, *Numerical solution of generalized Lyapunov equations*, Adv. Comp. Math., 8 (1997), pp. 33–48. 3, 6, 156

[141] J. Poulson, B. Marker, R. A. van de Geijn, J. R. Hammond, and N. A. Romero, *Elemental*, ACM Trans. Math. Software, 39 (2013), pp. 1–24. 21, 73

[142] E. S. Quintana-Ortí, G. Quintana-Ortí, X. Sun, and R. van de Geijn, *A note on parallel matrix inversion*, SIAM J. Sci. Comput., 22 (2001), pp. 1762–1771. 5, 80, 81, 82

[143] G. Quintana-Ortí, X. Sun, and C. H. Bischof, *A BLAS-3 version of the QR factorization with column pivoting.*, SIAM J. Sci. Comput., 19 (1998), pp. 1486–1494. 4, 25, 26, 27, 33, 65

[144] J. D. Roberts, *Linear model reduction and solution of the algebraic Riccati equation by use of the sign function*, Internat. J. Control, 32 (1980), pp. 677–687. (Reprint of Technical Report No. TR-13, CUED/B-Control, Cambridge University, Engineering Department, 1971). 4, 30, 31, 40, 161, 163

[145] Y. Saad, *Numerical Methods for Large Eigenvalue Problems*, Manchester University Press, Manchester, UK, 1992. 35

[146] R. Schreiber and B. Parlett, *Block Reflectors: Theory and Computation*, SIAM J. Numer. Anal., 25 (1988), pp. 189–205. 102, 103

[147] R. Sedgewick, *Implementing Quicksort programs*, Comm. ACM, 21 (1978), pp. 847–857. 53, 54, 57

[148] R. Sedgewick and K. Wayne, *Algorithms*, Addison-Wesley Professional, 4th ed., 2011. 54, 55, 57

[149] *SLICOT*, https://www.slicot.org. 2, 3, 6, 190, 196

[150] B. T. Smith, J. M. Boyle, J. J. Dongarra, B. S. Garbow, Y. Ikebe, V. C. Klema, and C. B. Moler, *Matrix Eigensystem Routines—EISPACK Guide*, vol. 6 of Lecture Notes in Comput. Sci., Springer-Verlag, New York, second ed., 1976. 2

[151] F. Song, H. Ltaief, B. Hadri, and J. Dongarra, *Scalable tile communication-avoiding QR factorization on multicore cluster systems*, in 2010 ACM/IEEE International Conference for High Performance Computing, Networking, Storage and Analysis, IEEE, 2010, pp. 1–11. 99, 127

[152] D. C. Sorensen and Y. Zhou, *Bounds on eigenvalue decay rates and sensitivity of solutions to Lyapunov equations*, Tech. Rep. TR02-07, Dept. of Comp. Appl. Math., Rice University, Houston, TX, June 2002. 5

[153] ———, *Direct methods for matrix Sylvester and Lyapunov equations*, J. Appl. Math, 2003 (2002), pp. 277–303. 156, 157, 158

[154] W.-H. Steeb and Y. Hardy, *Matrix calculus and Kronecker product*, World Scientific Publishing Co. Pte. Ltd., Hackensack, NJ, second ed., 2011. A practical approach to linear and multilinear algebra. 154

[155] G. W. Stewart, *On the sensitivity of the eigenvalue problem $Ax = \lambda Bx$*, SIAM J. Numer. Anal., 9 (1972), pp. 669–686. 1, 3, 9, 11, 12, 13, 24, 66

[156] ———, *Introduction to Matrix Computations*, Academic Press, New York, 1973. 6, 178

[157] J. E. Stone, D. Gohara, and G. Shi, *Opencl: A parallel programming standard for heterogeneous computing systems*, Computing in Science Engineering, 12 (2010), pp. 66–73. 53

[158] V. Strassen, *Gaussian elimination is not optimal*, Numer. Math., 13 (1969), pp. 354–356. 69

[159] T. Stykel, *Analysis and Numerical Solution of Generalized Lyapunov Equations*, Dissertation, TU Berlin, 2002. 181

[160] ———, *Stability and inertia theorems for generalized Lyapunov equations*, Linear Algebra Appl., 355 (2002), pp. 297–314. 181

[161] X. Sun and E. S. Quintana-Ortí, *The generalized Newton iteration for the matrix sign function*, SIAM J. Sci. Comput., 24 (2002), pp. 669–683. 4, 33, 37, 67, 76, 77, 92

[162] X. Sun and E. S. Quintana-Ortí, *Spectral division methods for block generalized Schur decompositions*, Mathematics of Computation, 73 (2004), pp. 1827–1847. 4, 11, 12, 24, 25, 28, 32, 33, 37, 44, 46, 65

[163] The MORwiki Community, *Convection*. MORwiki – Model Order Reduction Wiki, 2019. 118

[164] The Octave Developers, *GNU Octave*. http://octave.org. 190

[165] S. Toledo, *Locality of reference in LU decomposition with partial pivoting*, SIAM Journal on Matrix Analysis and Applications, 18 (1997), pp. 1065–1081. 2

[166] S. Tomov, J. Dongarra, and M. Baboulin, *Towards dense linear algebra for hybrid GPU accelerated manycore systems*, Parallel Comput., 36 (2010), pp. 232–240. 2, 80

[167] S. Tomov, R. Nath, H. Ltaief, and J. Dongarra, *Dense linear algebra solvers for multicore with GPU accelerators*, in Proc. of the IEEE IPDPS'10, Atlanta, GA, Apr. 2010, IEEE Computer Society, pp. 1–8. 2, 80

[168] *The top500 list*. Available at http://www.top500.org. 6

[169] P. Van Dooren, *The computation of Kronecker's canonical form of a singular pencil*, Linear Algebra Appl., 27 (1979), pp. 103–121. 10

[170] C. F. Van Loan, *The ubiquitous Kronecker product*, J. Comput. Appl. Math., 123 (2000), pp. 85–100. Special Issue: Numerical analysis 2000, Vol. III. Linear algebra. 153

[171] F. G. Van Zee and R. A. van de Geijn, *BLIS: a framework for rapidly instantiating BLAS functionality*, 41 (2015), pp. Art. 14, 33. 21, 64, 73

[172] R. C. Whaley and A. Petitet, *Automatically Tuned Linear Algebra Software (ATLAS)*. http://math-atlas.sourceforge.net/. 21, 118

[173] R. C. Whaley, A. Petitet, and J. J. Dongarra, *Automated empirical optimization of software and the ATLAS project*, Parallel Comput., 27 (2001), pp. 3–35. 7, 21

[174] J. H. Wilkinson, *Rounding Errors in Algebraic Processes*, Prentice-Hall, Englewood Cliffs, NJ, 1963. 6, 178

[175] ——, *Kronecker's canonical form and the QZ algorithm*, Linear Algebra Appl., 28 (1979), pp. 285–303. 10

[176] P. Yalamanchili, U. Arshad, Z. Mohammed, P. Garigipati, P. Entschev, B. Kloppenborg, J. Malcolm, and J. Melonakos, *ArrayFire - A high performance software library for parallel computing with an easy-to-use API*, 2015. 104

[177] I. Yamazaki, S. Tomov, and J. Dongarra, *One-sided dense matrix factorizations on a multicore with multiple GPU accelerators*, Procedia Computer Science, 9 (2012), pp. 37–46. 22, 85

[178] A. Yarkhan, J. Kurzak, and J. Dongarra, *QUARK users' guide – QUeueing And Runtime for Kernels*, tech. rep., Electrical Engineering and Computer Science, Innovative Computing Laboratory, University of Tennessee, 2011. 93